Star Bird
CALLIOPE'S MAZE

July 2022 Edition

The characters and events portrayed in this book are fictitious. Any similarity to real persons, living or dead is coincidental and not intended by the author.

Schultz, Robert, 1961-
Starbird III; Calliope's Maze: A ScFi Novel/ by Robert Schultz –
1st edition
RJS Publishing

ISBN-13: 978-0-9960448-7-5
Titling & Cover by Anina Laird Swallow
Summary:

Audra Atlanta has been dead for two years. So it comes as a bit of a shock when she walks into the *Athena's* conference room during a Command briefing. Now she must grapple with her body's response to being brought back from the dead by the miraculous skill of Doctor Caidin Mantose and his powers of the Thane.

Caught trying to make good his escape from the Albion Battlecruiser, *Tarzana*, Gunnar Conrad finds himself a 'guest' of the Queen Captain, Drax Blair and Albion Thane, Blinda Koss. Prize in hand, Drax orders CJ Barker to take the *Tarzana* to Calliope, the Castellian home world where the Hadrian Rite of Pintar is being held.

Discovering there's a saboteur onboard the *Constellation*, Dakota and the crew continue to struggle with trying to restore power and control to their ship before their Albion captors can reach their final destination, Calliope.

With the revelation of a possible way home, Rick must now mount a search and rescue operation for the *Constellation* and her crew. Not only to keep it from falling into the hands of those who would subvert its vast capabilities, but to bring his friend Gunnar back to Audra.

In a galaxy half explored beyond the boundaries of space and time.

In a galaxy pulled apart by impending civil war.

In a galaxy where the knowledge to command the elements can make or take life.

Two friends battle overwhelming odds to allow others the choice, to do the right thing.

"You want us to believe an orphan girl who's been imprisoned in a subterranean biosphere by herself for most of her life is supposed to rule a galaxy that's on the brink of civil war?"
General Richard Alexander Niker, Commander, Athena SAC09, Kalamarion Flight Ministry

"We're either going to implode and break into all-out war that will destroy most of Hadrian as we know it, or... a few of us will do the right thing and sanity will rebalance itself."
Colonel Casey Janae "CJ" Barker, Tarzana, Albion Flagship Battlecruiser

"If life isn't an adventure, then we're doing it all wrong."
Colonel Gunnar Lee Conrad, Constellation, SAC10, Kalamarion Flight Ministry

A Note from the Author

The feeling of coming to the end of a trilogy you've worked on for so long is bittersweet, to say the least. As I look back to 45 years ago when I first started writing Starbird, I remember penning the first character to paper. An 8 year old Rick Niker. He embodied everything I wanted to be, someone I aspired to be like. A year later, I created Gunnar Conrad. He was more the person I was and was to become. Together, these two characters went on many adventures together. As they grew, Princess Jayda and Audra Atlanta came into being. While there are other prominent characters, these four have always been the staples of the plot that have weaved such an intricate adventure.

A couple of readers have commented that they wished the characters could just sit for a while, sip a cold drink and enjoy watching life pass them by. While it's true that most people never experience so much commotion or personal upheaval in such a short time, that's exactly the reason characters such as these are created, so we can experience adventure vicariously.

It is my sincerest hope that you've enjoyed all these characters in one form or another and the story that you've traveled with them, through their tragedies and triumphs. I have literally laughed and cried with them and now when I think about life without them, I am truly sad, but happy; Bittersweet.

To my good friend
Sonja Porter
Because she's so smart and funny

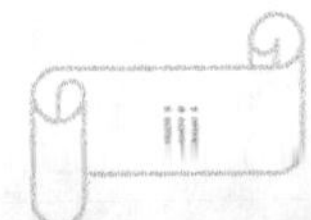

Starbird Assault Corsair Crew assignments

Athena SAC9

Commander/Thane- General Richard Alexander Niker
First Officer- Lieutenant Commander Jayda Niker
Interceptor Pilot 1- Captain Zek Korack
Interceptor Pilot 2- Captain Zak Korack
Chief Engineer- Frank Cooper (Head of Starbird training)
Engineer's Mate- Cacao Walker (Colonian replacement in training)
Turret Gunner- Lance Banco (Colonian replacement in training)
Chief Medical Officer- Kelly Gibraltar (Cross replacement in training)
Medical Assistant- Mayfield Hathorne (Cross replacement in training)
Helm Pilot- Captain Lisa Dayton
Navigator- Robin Mandrel (Cross replacement in training)
Weapons Officer- Karlyn Kinoy (Colonian replacement in training)
Com Officer- First Lieutenant Laura Habba
Science Officer- Mister Toby Mavis
Integrated Artificial Intelligence- CORA 500

Constellation SAC10

Commander- Colonel Gunnar Lee Conrad
First officer- Captain Dakota Abrams
Interceptor Pilot 1- Captain Dakota Abrams
Interceptor Pilot 2- Captain Logan Dalley
Chief Engineer- William "Billy" Moon
Engineering Independent Contractor- Tiana Mantose
Engineer's Mate- Scott Brandon
Turret Gunner- First Lieutenant Hayden Hunter
Chief Medical Officer- Fuji Yamoto MD
Medical Technician- Alder Gantrie
Helm Pilot- First Lieutenant Lynette Starman
Navigator- First Lieutenant Nigel Kramer
Weapons Officer- Second Lieutenant Doran Cartwright
Com Officer- First Lieutenant Lana Nevall
Science Officer- Mister Pippin Habba
Artificial Intelligence Droid- Alex 7001

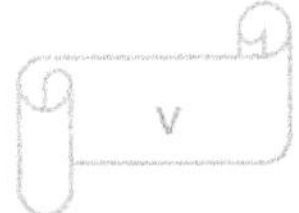

Albion and Colonian Command Staff

Commander of Albion Battlecruiser- Colonel Casey Janae Barker
Commander of Albion Medium Destroyer- Commander Dalton SoKnack
Commander of Albion Medium Fleet- Queen Captain Drax Blair
Albion Supreme Commander- King Commander Thoene Dismon
Albion Thane- Blinda Koss
Stephanie Benetar- Duchess of Teleknee

Additional Characters

Kalamarion Medical Thane- Caidin Mantose
Calypso Pilot- Dãsha Mantose
Hadrian Princess- Jana Tilee Oxlind
Royal Au Pair- Tonnie Cap
Administrator of Cross- Diord Vandmire
Tomplie Pirate Commander- Lou Aura
Castellian Aide- BachTL

Foreword

Colonel Gunnar Lee Conrad personal log; Entry 128

I've tasked Captain Abrams and Tiana Mantose with vetting potential candidates to replace lost crew members. It's been a difficult process as there seems to be a constant stream of Albion and Colonian spies trying to infiltrate Aster Command's security. Thus far, we feel confident that we've caught everyone that's tried to gain access, but it's getting more difficult as qualified applicants are getting fewer and farther between. The hope is to have a full complement of people trained and ready by the time General Niker gets back with the *Athena*.

Colonel Gunnar Lee Conrad personal log; Entry 132

I am concerned about Janox. She spends most of her time alone and seems determined to go back to Reako. I keep telling her that there is nothing to go back to, but she keeps insisting that she must return for some kind of present her nanny promised to give her when she was old enough.

Colonel Gunnar Lee Conrad personal log; Entry 135

I'm getting worried that Rick hasn't returned yet. The urgency to get these Starbirds back to full power has grown to the point that I've decided an Ops mission to Carolon to acquire more Asium is in order. The plan is to load the *Constellation* into the landing bay of the *Realistic* and drop out of light speed operations on the far side of Boris. The *Realistic* will wait out of scanner range of Carolon while we draw off any sentries, then we'll send a team down to retrieve the Asium we need. We then extract the team and run for the *Realistic*. Easy-peasy. CJ isn't happy with the idea but is unable to come up with a suitable alternative.

Colonel Gunnar Lee Conrad personal log; Entry 138

Getting to the Nulark system was uneventful, but as we started inbound to Carolon, we found Janox stowed away in one of the Interceptors. Alex and I went in to get her, but she was so determined to get to Reako that she stunned me a couple of times. So, I assigned Captain Abrams to continue with the Carolon mission while Alex and I took her down in the Interceptor.

Colonel Gunnar Lee Conrad personal log; Entry 140

It turns out Janox wanted us to help her retrieve her nanny, Tonnie. An android. We found her at the bottom of a well. Apparently, she had jumped into to protect Janox. Something about a prime directive… I don't know… The moment we pulled the android out, the whole underground environment started to flood. Alex and I got everyone to the surface of Reako before the whole place went under. It took some doing, but Alex got the android operating again and she recounted the story of how they got down there. Turns out Janox's real name is Jana Tilee Oxlind; she's a Colonian Princess. After a lot of discussion, we decided to go to a place called Dither to retrieve Jana's birth documents and parent's royalty rings. This could repatriate her as the true queen of this galaxy and stop all the fighting.

Colonel Gunnar Lee Conrad personal log; Entry 148

Dither… Things went pretty good with getting Jana's personal records until Blinda Koss showed up. I ran interference for Tonnie and Janox, I mean, Jana, while Alex brought the Interceptor in close for a hot pickup. Tonnie and I took several hits from small arms fire trying to get onboard. I finally got Tonnie inside, but I knew I couldn't make it in. So, I ordered Alex to take them to Cross, knowing he couldn't refuse a direct order. I let go, trusting that somehow I'd survive the fall. I hit the forest canopy pretty hard, but as I'm made of some pretty tough stuff, I didn't break anything. It took me a long time to get back to the Repos Station. That was some tough jungle. I think I slept under a bush for a couple of hours before I found a departing freight transport and stowed away. Of course, my luck was spot on and the ship had engine trouble while descending to the surface of Albia. I guess the pilots did the only thing they could think of; bail out. I was able to get the ship on the ground, but not without cracking it up pretty good.

Colonel Gunnar Lee Conrad personal log; Entry 154

Apparently I've lost some of my memory in the crash. There are a lot of mixed up shadows knocking around in my head that I can't seem to lock down. I don't know, maybe I have a concussion. A really nice couple found me and have taken me in. Eldon and Rayna Grant have been nursing me back to health. Eldon is retired Albion Intelligence, and agreed to help me assume the identity of a Captain Mace Bridger. I can't remember how, but somehow I ended up with his ID band. My memory is slowly coming back. Things are starting to separate into

something a little more recognizable. They drove me to Zepplin; Albia's capital city. While we were enroute, I came to the realization that I needed to let go of Audra. She was gone and there was no way to bring her back. I tossed the Hallavertor program chip out the window as we were traveling. It was a painful moment, but a good turning point for me, I think.

Colonel Gunnar Lee Conrad personal log; Entry 155

Made it to Zeppelin and was able to board the *Tarzana*. The fun part was going after the ring belonging to Jana's father we lost during our firefight on Dither. As fate would have it, I met up with Captain Bridger's widow, Malina Cass. There were some awkward discussions, but she finally agreed to help me find Jana's ring if I would take her with me. Once we retrieved the ring, she volunteered us for a transport assignment so we could escape the ship. While loading for departure, Blinda Koss found me again. That woman really has a sixth sense about her. I created a diversion, keeping Blinda busy long enough for Malina to get away with the ring. Unfortunately, my luck ran out and I was taken prisoner by the great Drax Blair. Not sure what happens now.

*   *   *   *

Captain Dakota Abrams personal log; Entry 1

Colonel Conrad assigned me temporary command of the *Constellation* while he took Alex 7001 and Janox down to the planet Recko. The Nulark operation went pretty good... considering. The landing party teleported to the surface near the old Carolon base. While the away team was retrieving the needed Asium, we drew off most of the Albion forces in orbit over Carolon. As we had only one Interceptor, Captain Dalley was assigned to draw away all their fighters so we could keep the larger ships occupied. We lost track of him sometime after we returned to the other side of the planet to retrieve the away team. Because our orbit had become a hot zone, mission protocols dictated a low pass over the base to pick up our people. With the mission objective complete, we attempted to break orbit, but became boxed in by several Albion destroyers. I ordered the weapons officer to put a Mark V torpedo into one of the Albion ships trying to cut off our escape vector. To my knowledge, this is the first time a Mark V had ever been fired in combat. The resulting carnage was unprecedented. The destroyer was completely obliterated and a

second, severely damaged. The *Constellation* was thrown clear, but not without damage to all our systems.

Captain Dakota Abrams personal log; Entry 2

The Albions sent what they call a Torag after us. It's a single weapons platform designed to incapacitate a ship using powerful pulse energy to scramble internal systems. It worked to a point. For a time, we only had maneuvering thrusters to run with, and as most of our systems were offline, we couldn't fire back. Through the heroic deeds of our chief engineer, Billy Moon, a Pin missile was hand delivered to the Torag, disabling it. Engineer Moon was lost in the attempt.

Captain Dakota Abrams personal log; Entry 4

The ship has been captured and taken aboard one of the Albion destroyers. They've been trying to gain access to the ship, but we've been able to get up just enough shield energy to keep them out. Tiana Mantose and the entire crew have been working non-stop to get our systems operational. Our hope is to work fast enough to stay ahead of the Albion's efforts.

*   *   *   *

Captain Logan Dalley, Pilot in Command of Interceptor 2, flight log; Entry 1

I've just finished intense diversionary operations over Carolon. I launched as the *Constellation* was starting an orbit to the far side of the planet. My orders were to draw off any fighters from the Albion task force in orbit. There were at least two squadrons of Black Tigers and Star Hoppers, probably more. I lost count... After an hour and a half, as I was finishing up with the last of them, I picked up a Kalamarion transponder free floating in the debris field outside Carolon's orbit. It was Engineer Billy Moon's ident signal. He was out there in the middle of nowhere, half dead; just about to run out of air when I got to him. He was in pretty bad shape, but still in one piece. Powering down in the debris field gave me a little time to do some first aid on him. From where we were hiding, we were able to watch the Albions take the *Constellation* aboard one of their destroyers. Now, we're out here alone. The Interceptor has taken a real beating going up against so many fighters all at once, but it's still space worthy for now. Mission protocol calls for a three tier light speed jump back to Aster. As soon as things cool down a little more, we'll make the jump.

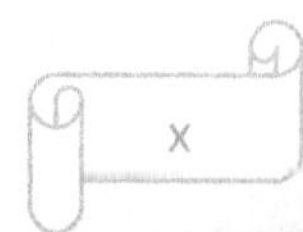

$$* \qquad * \qquad * \qquad *$$

Richard Alexander Niker personal log; Entry 95

I've been tracking the *Athena* for months now. Going in and out of stasis is getting taxing. The Interceptor's sensors keep picking up possible targets and waking me up. Unfortunately, none of them have panned out. I've made short landings on habitual planets to restock on food and water where I can, but I don't linger long as I don't want to risk drawing attention to myself.

Richard Alexander Niker personal log; Entry 132

Six months now and I'm getting a little weary. I've lost the Criterion trail from the *Athena* several times, but using my Thane training I've been able to regain the trail.   The traces of spent fuel have become less perceptible as I go along. I assume CORA 500 is having to power back to conserve energy. The *Athena* can remain active under sub light propulsion almost indefinitely. But I worry about its ability to follow the trail of Jayda's turret pod. Many questions constantly run through my thoughts. Is she all right? Did she even survive the attack? Did the stasis pod functions activate? As long as I can follow the trail, there's still hope.

Richard Alexander Niker personal log; Entry 139

I've come to the surface of a water planet. The trail stops here. It's a unique phenomenon to come across a planet whose surface is entirely water. Deep scans have shown a constantly shifting core surrounded by thick ice. I can only guess that Jayda's pod splashed down here and is somewhere beneath the surface. There are nearby planets farther beyond Acuity. One is actually close enough that its gravity draws water and material from this world.

Richard Alexander Niker personal log; Entry 140

My sensors have picked up a plethora of objects in the depths of this water planet. I'm currently mapping everything that's even remotely close to the pod's size and makeup. So far, nothing but space junk, ship parts and debris.

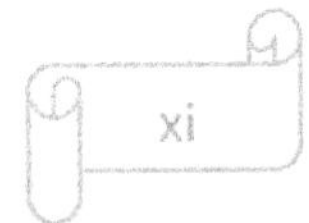

Richard Alexander Niker personal log; Entry 141

I've come upon Audra Atlanta's burial pod; It's still functioning. I've marked its location and am moving on. After I find the *Athena*, I'll come back for it and send it off in another direction. Somehow having it end up here just doesn't sit well with me.

Richard Alexander Niker personal log; Entry 142

Another target, enormous! This one is a ship... intact. There are life readings inside. I'm trying to find a way to dock with it.

Richard Alexander Niker personal log; Entry 143

The most unlikely pair were inside the ship. I was met at the docking hatch by a tall, slender woman. She looks like she's barely thirty years old. Her choice of attire is interesting. Tight fitting mini-skirt and top with knee high, spiked heel boots. Not someone you would expect to find inside a derelict ship. Something about her mannerisms though; she seems almost mechanical. She is the wife of one Caidin Mantose. Caidin looks and sounds so familiar. They are the only occupants of this vast ghost ship. Not sure what to make of them. They say they've been marooned here for a long time, but their stories don't match up. They're hiding something.

Richard Alexander Niker personal log; Entry 146

I think I'm starting to get to the bottom of what's going on here. Caidin is a Thane. Commissioned by the same Ona Tusk that commissioned me. While conscripted to a Drake medical research lab, he built Dãsha from another woman he was tasked to experiment on. He used Nanomech to help reconstruct her. She hardly ages, a byproduct of the Nanomech inside her. Explains why she seems so mechanical. I suspect a conflict between the human and synthetic within her. She is an amazing feat of medical engineering. The Drake were afraid of what Dãsha represented in Bioengineering, so Caidin smuggled her off their planet and fled here to *Calypso.* They've been marooned here ever since. The Drake monitor the old ship to make sure it doesn't move. If it varies from its position more than a couple of degrees off any of its current axis, the Drake come and remind them of their place. Not that this ship could move much anyway. Its systems are mostly dead.

Richard Alexander Niker personal log; Entry 149

Looks like I'm stuck here as well. The Drake discovered my Interceptor docked to the boarding hatch and blew it up. Dãsha showed me Calypso's sensors picking up objects in the nearby ice. It took a moment to figure out it was the *Athena* trying to reach other frozen targets. Not sure what to make of it. Dãsha has indicated that she has no spare parts to make repairs to Calypso, so I've taken it upon myself to figure out how to get this enormous hulk moving. I believe I have repaired most of the damage, but now we have to get it's control systems back online. This requires a restart of all the central processing engines located in another part of the ship. Problem is getting to it. The room is intact and operational, but someone will have to go through the flooded parts of the ship in order to reach it. We've devised a plan to maneuver the ship to displace the water to other areas so someone can pass through to the right room. After much discussion, it's been decided that Caidin and I will pilot the ship through several maneuvers to displace the water, while Dãsha goes to restart the processing engines.

Richard Alexander Niker personal log; Entry 158

We were successful, but not without great cost. The Drake noticed the movement and sent sentry ships in to remind us of our incarceration. Caidin and I were unable to keep the ship from shifting back prematurely and the water reflooded the areas that Dãsha was in, drowning and crushing her. That was tough. I offered to stay and help bury her, but Caidin refused. Strange... While he seemed upset, he wasn't distraught. He told me about a small scout ship in one of the landing bays that was still operational. I showed him how to navigate out of the Acuity and told him I'd catch up to him once I had found my ship. I offered to take him with me, but he wouldn't leave Dãsha. I took off in the scout ship to find the *Athena* after I removed all the monitoring beacons the Drake had placed on his ship.

Richard Alexander Niker personal log; Entry 159

I had to dodge several Drake ships before I could duck into the ice. I found another ship the *Stark*, one of the Kendalon freighters Gunnar and I were originally sent to rescue from the Oneida Caldron. Most of the crew were still onboard. The Drake had taken their Captains prisoner, but the crew overpowered their guards and retreated into the pack ice. I asked them what their cargo was, but they didn't know...

Richard Alexander Niker, *Athena* ship log; Entry 163

The *Athena* met up with the *Stark* a short time ago and I have downloaded all my personal logs from my wristband into the logging files. It's really good to be aboard her again. CORA found Jayda's pod, but it was empty. Her sensors detected ships stuck in the ice inside Acuity, so this is where she came. Right now we have the immediate problem of the Drake drilling their way to get at us. The *Athena* isn't going to be too much help as her power levels are so low and her systems so damaged as to render her weapons nearly useless.

Richard Alexander Niker, *Athena* ship log; Entry 166

The fight was pretty intense, in favor of the Drake until one of their own turned on them. It turned out to be my Interceptor pilots, Captains Zek and Zak Korack. They survived the attack on the *Athena* in the Nulark system and tracked the escape pod here. They took over one of the Drake ships and followed them into the ice. Before our happy reunion could finish, *Calypso* came crashing through the ice to provide us a way out. I went over to visit with Caidin about why he didn't do as I instructed, but when I arrived, I found Dãsha there; very much alive! Using his Thane powers and his vast medical knowledge, he healed her and brought her back from the dead. Amazing!

Richard Alexander Niker, *Athena* ship log; Entry 179

Getting under way, we discovered the contents of the freighters. Complete Starbird prototypes and extra parts: engine assemblies, Interceptors, conventional ordinance. Everything but Asium crystals. All the inventory was unregistered and I couldn't find any documentation indicating their purpose, only that they be delivered at a specified time to a Filon ship, the *Zebra Phar*, over Genis Two. This has brought up a myriad of questions there are no answers to. Right now, the priority is to find Jayda.

Richard Alexander Niker, *Athena* ship log; Entry 180

It's been determined that Jayda was taken to the Conti Nine Station, a free moving facility a short distance from here. She is to be sold into the sex trades! Dãsha insisted that she come to help find her. She was once a sex slave in her former life and could help find Jayda if she is still alive. Long story short. We did find her, or rather Dãsha did and mounted a rescue. There is some disagreement about who rescued who as we sort of got into a fire-fight aboard a buyer's ship. Dãsha and Jayda ended up running cover for us in order to get everybody back aboard the assault craft we were using. After that, we

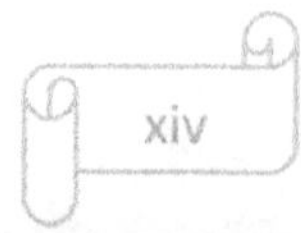

headed to Skadil, a Drake penal colony and mining operation to rescue the Captains of the freighters.

Richard Alexander Niker, *Athena* ship log; Entry 181

Only two of the Kendalon Captains survived their incarceration. We got them out of the Skadil facility without incident, but the sentries discovered we weren't who we were pretending to be. We ended up in a running fire-fight with a pack of Drake assault ships. Our assault craft was damaged and I received plasma burns, nearly losing my hands to a fuel manifold breach. *Calypso* arrived in time to save us. I spent a lot of time in sick bay, but thanks to Caidin's abilities as a medical Thane, my hands were quickly healed.

Richard Alexander Niker, *Athena* ship log; Entry 185

We're enroute back to Aster now, but I sense the need to have these four Starbirds ready to fight or act as a decoy. The biggest problem is being able to properly man them. In combat, there can be as few as five people onboard to run things, but there has to be a central figure, a Commander. We only have three. Captains Fowler, Quayle and Lieutenant Commander Jayda Niker. We need a fourth. As we were discussing the options, Caidin and Dãsha walked into the conference room and indicated they had found a solution. I thought they were going to suggest Dãsha be the fourth, but she quickly put that idea to rest. I nearly had a heart attack when Audra Atlanta walked in; very much alive! Caidin and Dãsha had found her burial pod, brought her aboard *Calypso* and healed her. It was unbelievable to see her. She looks remarkable, considering she's been dead for the last two years. Now, the urgency to find Gunnar and the *Constellation* are of paramount importance!

Star Bird

CALLIOPE'S MAZE

Starbird III: Calliope's Maze

The Real Thing

Once a gallant juggernaut of a ship, all of *Calypso's* military hardware had been removed long ago. She could hardly be considered much of a threat to anyone or anything. General Richard Niker and Dãsha Mantose had run a deep scan of the ship's logs trying to discover her origins and what role she may have played in some epic struggle. But alas, her service records were corrupted after she had been pulled into a salvage yard for breakdown and disposal. Now, *Calypso* was all but a ghost ship; operational, but with no crew to man her and no real military purpose.

Focusing on the ships beyond *Calypso's* main hangar entrance under a blanket of burning lights, Rick Niker paused to survey the scene before him. Several Kendalon freighters were parked against the massive outer bay doors. A small squadron of five Starbird Attack Corsairs took up the rest of the hangar floor space, their white hulls gleaming in the hangar bay lights.

"How are we doing in here?" Rick asked, walking toward the closest ship as Jayda Niker and Dãsha joined him.

Jayda let out an exasperated sigh and looked at Dãsha.

"Are you asking about the *Athena* or the other four?"

"Yes," Rick said, examining the first Starbird in the cluster of five.

"You have difficulty disconnecting," Dãsha commented.

"I just finished disconnecting," Rick responded. "Now it's time to reconnect."

"Two of the four are operational," Jayda reported. "This one and the next one."

"And the *Athena*?"

"Captain Zek is doing the final calibrations on the Asium chamber now."

"What's left on those other two birds?" Rick asked, turning toward the *Athena* sitting in the middle of the cluster.

"Asium chamber installation and calibration," Jayda said. "Richard..."

"That's impressive," Rick mumbled. "Too bad we don't have any Asium to fit them with."

"Richard," Jayda repeated, a little more insistent.

"At least they can operate on a limited basis."

"Richard!"

Rick turned impatiently back to Jayda. She did her best to give him the eye, but he was far too focused on other matters to pick up on it. Dãsha finally took a couple of steps closer.

"Your friend is in need of your attention, Richard. She is asking questions that no one but you can answer."

Rick looked back at Jayda who was giving him a stern expression.

"Audra has a right to know what's going on with Gunnar," Dãsha persisted, raising her hands to her hips.

"Yeah, I've been meaning to sit down with her," Rick said looking up at Dãsha. "Just been a little preoccupied."

"Right now, Audra's mental and emotional health are of paramount importance. The effects of her long stasis, coupled with having her structures repaired and brought back from the brink of death, weigh considerably on her. While I am certain she is a strong woman, there would be frailties in anyone who has endured what she has experienced. We need to do our utmost to see to her needs."

"Shouldn't Caidin be giving me this lecture?" Rick asked, looking into her black eyes.

"Were he here, he would. He is attending to Audra as we speak."

Rick glanced back at Jayda, then stepped around Dãsha toward the *Athena*.

"Have Captains Fowler and Quayle meet me in my quarters in ten minutes."

As Rick and Jayda entered the ship, they stepped directly into sickbay. Caidin looked up from a display device he was studying.

"This is an amazing ship," Caidin gleamed. "The tech you have in just this room alone is nothing short of astonishing."

"How's she doing?" Rick asked, stopping next to Caidin.

The older gentleman tucked the viewing device under his arm.

"You should ask her yourself. She's in your quarters resting. Audra's a strong woman, but even the strong require assistance in healing."

"What about her mental state, her mind?"

"She's experiencing disorientation and severe emotional stress, but given time, that should settle. Right now, answers to questions will go a long way in helping her return to a sense of normalcy. She keeps asking about Gunnar."

"Yeah…" Rick sighed.

The hall door suddenly opened behind them.

"Richard, Captains Fowler and Quayle are waiting in your quarters," Dãsha announced.

Rick looked back at Caidin.

"I need her if we're to accomplish what's to come."

"If you're asking me if she can perform command duties, I can't say for sure. It will be entirely up to her."

Caidin motioned to Dãsha as the group headed forward to the Commander's quarters. There was plenty of room inside the outer chamber that doubled as a living room and office. Not as stark as a normal office, Jayda and Rick had decorated it as they would have on their home world. After everyone made themselves comfortable, Rick sat down at his desk and started poring through the information scrolling up the screens in front of him. He stopped the data several times to study it closer, then continued. He was as focused as Jayda had ever seen him. He had been so fixated on the data in front of him, he had nearly forgotten he wasn't alone. Jayda finally nudged him.

"Hey, I think you're trying to overthink this."

"It's no longer just the *Athena* and the crew we're dealing with here," Rick said slowly, still riveted to the information in front of him. "It's all our ships and their crews. We have to be positive *Calypso* can make this jump safely."

"I am positive," Dãsha stated frostily. "I have run all the calculations myself and inspected all of the components that will be required for this operation. I have even performed a comprehensive inspection of the ship's hull integrity."

"Believe her, I got to help," Jayda mumbled.

Rick gave his wife a double-take.

"*Calypso* will jump and drive at full light speed without fear of catastrophic failure."

"But now there's more than just the three of us," Rick replied.

"I understand your concerns, Richard," Dãsha countered. "Tell me, when have you known me to be in error... About anything technical?"

Rick finally stopped the data and turned to the group in his quarters. Dãsha sat next to her husband, Caidin, while Captains Fowler and Quayle leaned against the wall. Rick's eyes swung to Lieutenant Commander Audra Atlanta, nestled comfortably in a chair near the door. A smile instantly developed as he looked at her. The longer he looked, the harder it was to keep from becoming teary-eyed.

"I still can't believe this," he said, shaking his head.

"Will you stop," Jayda choked up. "You're not even the emotional one!"

"Come on you two," Audra chuckled, bewildered. "It's really me, I swear!"

"I know it's you," Rick sniffled, wiping his eyes. "It's just you've been dead..."

"Technically, not dead," Caidin corrected.

"Not sure what else you would call it," Jayda quibbled.

"It's still a bit of a shock..." Rick said, clearing his throat.

"A bit?" Jayda repeated. "We had accepted that you were dead, and not coming back, and now, here you are, sitting here, and... and... will you stop smiling? ... for heaven's sake!"

"I'm sorry," Audra gushed. "I feel about as turned around as I can right now. I don't feel like I've been dead."

"You've experienced being dead before?" Dãsha asked stoic.

"I know I died... I remember it... sort of... It's like I just woke up from a long nap, except I don't feel rested at all. More sick and achy all over."

"You did just wake up, this morning," Jayda snickered, wiping her eyes.

"What you are feeling is normal," Dãsha said.

"I know you and Caidin have tried to explain some of this to me," Audra said. "But I can't even begin to believe anything about what's happened to me as being normal. Right now, I can't seem to focus or remember anything in sequence."

"It's really quite simple," Rick said, clearing his throat again. "Doctor Yamoto put you in the stasis pod directly after you..." Rick glanced over at Caidin, then shrugged slightly. "Well, maybe it isn't that simple."

"You died, Audra," Jayda finished. She looked down at Rick, then back over at Caidin. "It's as simple as that."

"Yes, I think I remember that part," Audra agreed.

"We buried you in space," Rick said after a brief hesitation. "Since then, we've been chased halfway across this galaxy by the Indigenous populations trying to get their hands on our ships and technology."

"So how long have I been... dead?" Audra asked.

"Two years," Rick said with a sigh. "It's been a lot of fun."

"Two years," Audra repeated, trying to digest. "If you buried me in space... How did you find me, and how did you...?"

"We sent Gunnar on a scouting mission to find Asium in a system called, the Nulark. He got caught up in the middle of some local military action and was separated from his ship helping one of the locals. While Jayda and I were looking for him, the *Athena* came under attack. We lost several good people that day." Rick fell silent, thinking.

"I barely got away in the escape pod," Jayda cut in.

"The turret," Audra clarified, trying to concentrate on the chain of events.

"The pod automatically put me into stasis," Jayda continued. "Next thing I know, I'm waking up in an alien craft being carted off to some brothel to be sold to the highest bidder."

"Sounds like an adventure," Audra said.

"Not one I'd care to repeat," Jayda reassured her.

"I located your pod while searching for Jayda and the *Athena*," Rick said.

"Richard and I mounted a rescue for Jayda," Dãsha jumped in. "Although I'm not sure who ended up rescuing who."

"You're never going to let that one go, are you?" Rick tipped his head at Dãsha.

"So where is Gunnar?" Audra asked. "You said something about him being in a lot of trouble."

All eyes turned back to Rick who was deep in thought.

"Is he all right?" Audra persisted.

"Huh?" Rick broke from his trance. "I'm sure he's fine. You know how he is," Rick smiled.

Everyone chuckled softly, the room falling silent as they held their thoughts to themselves. Rick finally sat forward, looking over at Dãsha and Caidin.

"If you're confident in *Calypso*, then I am. How soon can we get to light speed?"

"We can be ready within the hour," Dãsha said, helping Caidin to his feet.

"Captains Fowler and Quayle will accompany me to see that all the ships are secure here in the holds, then we'll join you on the bridge of *Calypso*."

"Where would you like me?" Audra asked, working her way out of her chair. She looked exhausted. Dãsha quickly stepped forward to help her.

"You just relax here until I'm finished, then we'll all go back to *Calypso's* bridge together. It's kind of a long trek across a ship that size," Rick said. "I imagine you could use something to eat as well."

Everyone left the room, except for Jayda and Audra.

"Was I really gone that long?" Audra asked after a long moment of silence. "Two years?"

"Give or take a couple of months," Jayda nodded. "Remember, I've been in stasis for six of those months. Talk about coming out of that not feeling very chipper."

"I have zero concept of what's happening to me," Audra complained quietly. She wiped her eyes and closed them as she leaned against the wall.

"I'm having a hard time with it too."

"Why would Rick say Gunnar was probably in a lot of trouble?"

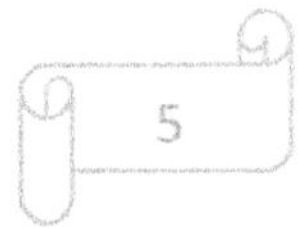

Jayda drew in a deep breath and held it for a moment, looking at nothing. Finally exhaling, she looked back at Audra.

"Gunnar didn't do very well after you... you know..."

"Died..."

"Yeah, Fuji did everything she could think of to help him, but nothing seemed to work. I've never seen anyone sink into a depression that low before. It got so bad, Rick and Fuji created a Hallavertor program to help him deal with his grief."

"A program? Of what?"

"You..."

Audra developed an odd look, trying to process what Jayda meant.

"Of me...?"

"Yes, Rick and CORA 500 created all the technical parameters."

Audra slowly made her way to the Lexan doors of the bed chambers, then looked back at Jayda.

"Do you have it here?"

Jayda considered a moment.

"I think so."

"Can I see it?"

"We have a copy of it here, but the actual operating version is on a chip key Gunnar keeps on the *Constellation*." Jayda slid onto the desk seat and started working the terminal. "All the learning and personality trait files are stored there. We only have the basic startup copy."

The door to Rick and Jayda's room slid open and both women stepped inside. As the Hallavertor came on, their surroundings instantly changed to a green landscape of rolling hills with a single tree standing near the top of one of the hills. As they moved nearer to the tree, a figure in a blue dress stepped out.

"Hello, I'm Audra Atlanta." She stepped closer, holding her hand out.

Audra froze, staring straight at an exact image of herself wearing her favorite shimmering blue dress. She glanced back at Jayda who stood motionless. While Rick had shown Jayda all the technical parameters associated with the program, this was the first time she had witnessed it firsthand. Both Audras came in close, circling each other, studying every detail.

"I'm confused," the hologram spoke softly, trying to process the encounter. "May I touch you?"

"What are the control parameters set to?" Audra looked back at Jayda again.

"Rick and Fuji decided it needed to be unrestricted. I wasn't fully onboard with the idea, but I don't pretend to know anything about psychology, least of all how Gunnar functions."

"My sensors will allow me to positively identify you," the hologram said quietly.

"Remember," Jayda whispered. "It's governed by a class four learning AI. Right now, it can only process what you'll allow it to."

Audra looked back at herself.

"Yes."

Audra took the hologram's hand. It felt warm and real, just as anything real should feel.

"You are Lieutenant Commander Audra Atlanta." The program developed a confused look. "But... you are deceased. How is it you are here, like this?"

"Freeze program," Audra said quietly as she let go of the hand. The image instantly froze in place. Audra took a step back, bringing her fingers to her quivering lips.

"Audra?" Jayda whispered from behind. "Are you all right?"

"No," she said, shaking her head and bolting from the room. Jayda was right behind her. "No, I'm not all right." Audra stopped as she reached the outer door and slumped against the wall, wiping tears from her eyes. "You mean to tell me, Gunnar has been clinging to that *thing* in there?"

"Audra," Jayda began, putting her hand on Audra's shoulder. "I don't know what Gunnar's been doing, if he's even looked at it at all. Rick hasn't had time to fill me in and I haven't asked. I've been playing catch up myself."

"This is all happening too fast," Audra tried to smile, but the tears continued spilling from her brown eyes.

"We had to do something; Gunnar was coming apart."

Audra looked at Jayda, confusion firmly planted in her expression.

"I-I need time to think," she said rolling to the door and out into the hall.

Jayda held back, understanding what Audra might be feeling right now. After sitting quietly thinking for several minutes, she returned back to her room to shut off the Hallavertor. It felt a little eerie seeing the image of her best friend standing in her room. But it had been even more of a shock to see Audra walk out of Caidin's medical room onboard *Calypso*. Still, Jayda at least had a small sense of what Audra might be going through. Having blacked out right before the *Athena's* escape pod had activated its stasis systems, Jayda had been kept alive until she was found floating on the surface of Acuity six months later. Audra had slept for two years before waking up in a strange place. Jayda finally deactivated the Hallavertor and turned to go find Rick.

*   *   *   *

Audra sat unmoved in the *Athena's* ball turret, tears working their way down her cheeks. Her mind was a swirl of confusion. *This is all just a bad dream… I remember the accident and Gunnar holding me in sickbay. He looked so lost; so broken. Then everything went dark. I really died! And I've been dead for two years! I can only imagine what Gunnar has gone through during that time… alone.* She recalled an odd sensation of floating endlessly through a dreamscape. Memories of her childhood replayed in vivid detail, along with a feeling of wellbeing coupled with an intense warmth from within. Then she opened her eyes and looked up at a ceiling. *There's an older gentleman looking down at me, smiling. Caidin Mantose? Why does that name sound so familiar? Two years! Had Gunnar sunk so low that he had found love in the arms of a computer-generated image of me? What else has he done? Who else has he turn to, to be with?* Her emotions continued to swirl; it was all she could do to keep from breaking out into sobs. *Rick said he thought Gunnar was probably in a lot of trouble. What does that mean? What kind of trouble is he in this time?*

Trying to divert her attention to something else, she looked around at the interior of the turret pod trying to imagine what it must have been like for Jayda to be in stasis for six months. *To remain motionless, asleep for that long… Was it the same as what I experienced? I've never actually been in stasis before. My insides felt like they've been beaten up… I'm sore all over; every joint aches… I'm exhausted.*

She carefully reached over and turned on the turret's systems and working the controls, moved the gun ball silently around its available axis. Maneuvering the turret, she dipped the controls down so she could see the other ships in *Calypso's* hanger bay. The *Athena* was surrounded by four other Starbirds. *Rick indicated they had been sealed in the holds of the freighters which were now parked against the docking bay doors. Freighters? Didn't we try to rescue some at the Oneida?* As she continued to move the turret around, she noticed Rick and Captains Fowler and Quayle moving from one Starbird to another. The realization that Jayda could tell her only so much about Gunnar suddenly rang true in her mind. She abruptly parked the turret and stepped into the turbo elevator. As the door to engineering slid open, she nearly ran into the tall slender form of Dãsha Mantose. Surprised, Audra stepped backwards on wobbly legs. Looking up into the black eyes of the tall beauty, Audra grasped the elevator threshold.

"I am sorry," Dãsha said, stepping back. "It was not my intention to startle you, Commander."

"Audra, please call me Audra." She wasn't sure she was fit for duty just yet. Her legs felt like they were going to crumble beneath her. She leaned against the open door frame of the elevator and closed her eyes.

"I observed your entry into the turret and thought it might be a good idea to check on your condition. You appear to require assistance."

"I'm just so tired and ache all over; even my insides."

"It is a form of hibernation sickness. In your case, your symptoms are amplified four fold." Dãsha held out a container of liquid. "Your body is currently in great need. You need to drink the mixture in this bottle. It is called Ganaush."

As Audra guzzled the fluid, she was somewhat astonished she wasn't even stopping to take a breath. She emptied the bottle and gasped.

"I don't suppose you have any more of this?" Audra asked, licking her lips.

"I thought you might be in a bad way," Dãsha said, producing another container.

Audra went at it again, in the same manner as the first, breathing heavily once the contents were gone. As she finished, Dãsha pulled up a couple of boxes and sat down.

"Now, you need to sit for a couple of minutes and let your body absorb what you've consumed. You'll start to feel better soon."

"Probably going to have to pee for a week before all this is over."

"You can rest assured of that fact," Dãsha replied.

"How do you know so much about what's happening to me?" Audra asked, glad to take a seat.

Dãsha squirmed a bit, trying to get comfortable in the long pants and top she was wearing. She fussed with her long, jet black hair for a moment, then turned back to Audra.

"Because I've been where you are now."

Audra looked straight ahead for a long moment, then turned her eyes to the slender woman.

"You realize we're talking about being dead?"

"Certainly," Dãsha responded casually.

"You've been dead..."

"Yes," Dãsha agreed.

"How?"

"It would seem to me that you're having a hard enough time dealing with what's happening to you without trying to understand what's happened to me."

"Maybe, but I think it might go a long way for me if I tried."

"Very well," Dãsha said, turning to the short brunette. "I had hoped your Jayda would have been able to explain at least a little bit to you, but I understand her emotional response to your return. Besides, she was unaware of everything Richard has experienced."

"Understandable," Audra responded, "as I don't exactly understand everything that's happened."

"In your absence, your friend Richard has become what's called a Thane. As I understand it, a Thane is a protector of persons or things, given the ability to use the upper functions of their brain to manipulate matter at the subatomic level. My husband has been given this ability as well."

"Who gave it to them and why?"

"Ona Tusk. Her origin and true purpose are unknown to me. I only know that she shows up in the most unlikely places when circumstances require it. As Richard has a scientific mind and is a gifted warrior, his talents and abilities follow along those lines. My Caidin's talents are in the medical realm. We met Richard in Acuity while he was searching for Jayda. During his search, he came upon your burial pod and marked its location. We were marooned on *Calypso*. Richard made it possible for us to make repairs which have allowed us to escape our exile in Acuity. In return, Caidin and I retrieved your pod and Caidin used his Thane powers to revive you."

"But how do you revive someone who's dead? Especially after two years?"

"Your doctor was unable to repair your injuries and you passed away in your medical bay."

"I remember most of that," Audra confirmed. "Some of it is a little foggy."

"Naturally," Dãsha agreed. "When a human ceases to function, the organic tissues continue on for some time until all of the energy driving those tissues has dissipated."

"Your spirit leaves your body," Audra said.

Dãsha cocked her head slightly, looking at the opposite wall.

"Depending on your beliefs of mortality and the universe, yes. Your doctor placed you into stasis, not knowing that your life energy had not fully dissipated. For every individual, the duration of that process is unique. Technically, you were buried alive."

"Ok, I understand all that," Audra agreed. "But... that doesn't explain..." Audra closed her eyes and leaned back. It was difficult to concentrate on what Dãsha was trying to say as her eyes stung and her extremities felt like she hadn't had any sleep for a couple of days. *No sleep my eye!* she thought.

"Take a couple of deep breaths and try to relax," Dãsha instructed. "I am unaware of any training that can prepare someone for what you've been through."

"You've got that right," Audra grumbled, trying to breathe in deeply.

"As I indicated previously, a Thane can manipulate matter at the subatomic level. I do not pretend to understand exactly how it is done as I have no such ability, but as Caidin has explained it to me, he can zoom into a body or object in his mind and look right in at the molecules that make up the cell structures. With the correct knowledge, he can move molecules around, and repair cell structures back to their original configuration."

"So you've said that you... died?" Audra asked, starting to find relaxation in Dãsha's explanation. The question was more in jest than serious.

"Yes."

The answer didn't register in Audra's mind at first. Her thoughts wandered about for a moment, then her eyes suddenly popped open.

"So how did...?"

"To understand the answer to that, first you must know who and what I am," Dãsha responded patiently.

"You're not some kind of an android are you?"

Dãsha cocked her head again, thinking on the question.

"Probably not as you would define one."

"I don't understand, either you are, or you aren't."

"Well put. I am a product of Caidin's intellect."

"His intellect? I don't understand?"

"I am human, the same as you. My original identity was Danis Knox. I used to be a slave in the sex trades. As I came to the end of my usefulness, I was transferred to a Drake research and development facility where I was to be used for medical testing and eventual termination."

"Usefulness as a sex slave?" Audra repeated, looking at Dãsha.

"I know," Dãsha smiled. "Please be patient, you will understand my appearance shortly. I was assigned to Caidin's department, where I came to know him. He took great interest in me, and I him. As we had fallen in love, he kept my testing somewhat benign. It wasn't long before Caidin discovered that all the years of abuse I had been subjected to as a slave had taken its toll on my body and mind; I was dying. It was during this discovery that Caidin was visited by Ona Tusk and given the abilities of the Thane. After much reasoning, I requested he remake me in the image of his late wife, Dãsha, and replace my tortured memories with hers. Certainly, he could remove and replace only so much using his memories of her. There were many gaps, but the alternatives were far worse. The life of Danis Knox is not one any human should ever have to endure or remember. It took him many days to make the changes to me, both physical and mental. During the process, there were certain things he had to do in order to solve unanticipated problems; one of which was to introduce Nanomech into my system. They have provided me with great benefits as well as challenges that even now, I struggle to work through."

"I actually understand everything you've told me so far," Audra said. "I just wish I didn't still feel like I was going to curl up and bawl my eyes out."

"Comprehension is a good sign that your mind is recovering. Your body will take longer to recover however. Please be patient with yourself." Dãsha curled her eyes up, trying to recall what she had been explaining. "Where was I?"

"You have Nanomech..."

"Yes, well done. In helping make repairs to *Calypso*, it required making several ship maneuvers in order to displace great amounts of water that had collected in these landing bays, so I could reactivate the main systems of the ship. While doing so, we were attacked by the Drake, which disrupted those maneuvers and displaced the water while I was trying to get back to a safe zone. I became caught in the water's hydraulic force and crushed. Richard retrieved my body and brought me back to Caidin, where he repaired my damaged structures and revived me. It took about a day to do so. It took several days to revive you."

By the time Dãsha had finished her explanation, Audra just wanted to find a bed and sleep for a week.

"Do you know anything about Gunnar?"

"No, how could I?"

"I'm sorry," Audra muttered, somewhat lost in her relaxation. "You're right, there's no way you could have."

"I do know that Richard is now your Gunnar's protector, both have an important role to play in repairing the rift that has caused the inhabitants of this galaxy to fragment."

Audra looked off into nothing.

"Gunnar... my Gunnar... chosen to help mend a broken galaxy?"

"I am certain those were the words Richard used," Dãsha confirmed. "Looks like everyone is ready to head back to *Calypso*," Dãsha said standing up as several others entered the engine room. "Feeling well enough to travel?"

Audra remained in a trance, her mind wandering aimlessly until Rick dropped to a knee in front of her.

"Hey, you with us?" he asked.

"Yeah," Audra blinked a couple of times, then looked at Rick and nodded. "I'm all here, I think. Just a little distracted." She looked up at Dãsha, who was holding out her hand. "Dãsha gave me something that has really helped."

"Ready to head over to *Calypso's* bridge for light speed operations?" Rick asked.

"As long as it doesn't involve a worm hole," Audra said, taking Dãsha's hand.

"I promise," Rick assured her as they left the *Athena*. It turned out that it was impossible for Audra to keep up, having to stop several times to rest.

"I'm not so much tired as I ache," she explained, leaning against a wall.

"That is normal as well," Dãsha said.

"You said, hibernation sickness..." Audra affirmed.

"Dãsha, why don't you and Caidin go ahead of us. We'll catch up to you on the bridge?" Rick suggested.

The tall beauty nodded, turned without a word and moved quickly toward a far corridor.

"You and Gunnar have been in stasis before," Audra said, trying to stretch. "Is it really like this?"

"No, not as bad for us," Rick admitted. "But we weren't all busted up and dead when we were in stasis, and definitely not for as long as you. We just got a slight headache and maybe a little nauseated."

"I definitely have that in spades."

"Just give me another moment," Audra insisted, rubbing her legs. "I think I'll be fine."

"Here," Jayda offered, pulling Audra's arm over her shoulder.

"Take as much time as you need," Rick said. "With what you've been through, we've no right to hurry you... Even though we really are in a hurry," he added, smiling.

"Speaking of going through things," Audra said, looking at Rick. "What happened to Gunnar? You know, after I was gone."

Rick looked over at Jayda. Her expression was similar to Audra's.

"Jayda showed me the program..." Audra stated simply.

Rick took a deep breath and let it out.

"He didn't do very well at all after you were gone. He tried to force himself into normalcy by helping with repairs, but it wasn't working. The harder he tried, the further he sank. It got to the point where he would barely come out of his quarters. With the help of Fuji and CORA, we created a Hallavertor program designed to help him find closure and come to terms with his grief. It was an ambitious project that required unlimited program parameters. The AI protocols we established in the program were a little beyond anything that had ever been attempted, but we felt that he had to believe he was interacting with you in order to come to terms with his loss." Rick glanced at Jayda. "Jayda was against such high settings, but Fuji and I felt like it was the only way to get him to respond."

"Did he use the program?"

Rick hesitated, looking back at Jayda.

"Yes, he did... but I don't know to what extent. He got separated from the *Constellation* shortly after I recommended he use it. When we were setting up the Aster Command base, we had several opportunities to talk, but he was so busy helping with the installation of the base and repairs on his ship, he didn't talk much about it. I will say it seemed like he was doing better on Aster, but it's hard to say what to attribute that to. Fuji still seemed concerned about his state of mind, but I suspected for other reasons." Rick paused again. "He did seem to get along with one of the locals he was helping out; CJ Barker. She appeared to keep him grounded and focused most of the time."

"She?"

"Nothing romantic or intimate; nothing like that, just good friends. Almost like they were helping each other through a tough time."

"You told me he's probably in a lot of trouble." Audra said, letting Jayda help get her walking again. "Where is he now that he's in trouble?"

"The *Constellation* and the *Athena* can't operate at full power without all six Asium channeling crystals. As you're aware, number five becomes unstable and cracks under heavy loads."

"Yes, I think I'm qualified to publish a technical article on that," Audra said.

"Gunnar and I found Asium on a planet called Carolon in the Nulark system. This is sort of where all our troubles started. I sent Gunnar in with the *Constellation* to see about getting some Asium to refine and get us back up to full power. Well, I've already outlined what happened there. That took a while to untangle. Anyway, once we were able to establish the command base on Aster, he wanted to go back to Carolon to retrieve the Asium ore. I told him I didn't think it was a good idea for him to go in without the *Athena* for backup, considering the tensions between the warring factions of this galaxy. Apparently he went in anyway and got into some trouble. Captain Abrams sent out an automated distress signal, not Gunnar, meaning he was either incapacitated or not even onboard. We don't know for sure what happened or where he is."

"Fighter pilot mentality," Audra muttered, shaking her head slightly.

"I'm sure there's more to it than, 'he just drove in there by himself with just half a ship'," Rick reassured her.

"One can only hope," Audra replied. "How long ago was the distress signal received?"

"Not very long," Jayda responded. "But it's unclear how old it is."

"Hence the urgency to get to light speed as quickly as possible," Rick said, ushering them into the elevator. Moments later they were stepping out into the cathedraled bridge of *Calypso*.

The control room felt vastly different. Rick was somewhat taken aback at how alive the ship now felt. Every control surface was dancing with activity, indeed when Caidin had said he thought the great ship would be glad to be of use again, Rick thought he was overdramatizing things, but now it seemed true. The ship did feel more alive, almost happy to be functioning again. Seeing Dãsha first, Rick stepped ahead of Audra and Jayda, heading straight for the light drive panels. Caidin came to his feet.

"As promised," Caidin said with a smile. "She's ready to jump."

"Thanks for all your help," Rick said, putting a hand to the older gentleman's shoulder.

"She's all yours. Let's go find your friend. Dãsha?" Caidin said, turning to his wife.

"If you'll assist me at the controls, Richard," Dãsha said, stepping onto the pilot's platform. Rick joined her at the controls and started strapping himself in.

"If you ladies will have a seat next to me," Caidin instructed Jayda and Audra. "We've never had *Calypso* at light speed before and have no idea what this ride will be like."

They joined the other crewmen that had come to the bridge to watch the operation. As Rick finished strapping himself in he looked out at the starfield that lay before them.

"Nice to look through these windows for a change instead of staring at black," he said admiring the majesty of the cosmos. He glanced over at Dãsha who was preoccupied with her pre-maneuvers checklist.

"From here," Dãsha began, "the light speed operations are surprisingly simple. I have already input all the required data into *Calypso's* navigational controls and programmed the lightspeed controls to respond to those parameters. I have only to make a few adjustments here, activate the Light speed engines and engage them, then monitor everything as we go."

"So what am I doing here?" Rick asked, looking at the controls in front of him.

"Eye candy for the rest of the crew," Dãsha responded without looking up.

Rick looked over at Dãsha, her black eyes slowly shifting to him.

"Really? Eye candy?"

Dãsha held her expression, the corners of her mouth finally turning up slightly. She dropped her gaze back to the screen in front of her, made several adjustments, then looked up. The great bow of the ship was only visible as a black shadow against the infinite starfield ahead of them. Moments later, there was a flurry of lights and displays flashing information all around the bridge, then the starfield flashed in front of them and the great ship bolted into light speed toward the Nulark system.

Guess who's coming to dinner?

Gunnar Conrad looked around carefully as his escort moved him toward a set of elevators. Reaching the doors, the guards nudged him forward.

"We gonna walk around all day?" he asked, continuing down the corridor toward another smaller elevator.

"If I had to wait on ship's personnel, I'd never get anywhere," Drax Blair said from behind. "I have my own personal elevators."

"Just yours," Gunnar replied snidely.

"Well, command personnel use them too."

Once everyone had boarded both elevators, it was a quick ride to the executive levels of the *Tarzana*. Taking the lead, Drax walked them through several turns in larger, better lit hallways until the group stopped at the entrance to a small dining room. Peering inside, Gunnar saw a transparent wall, revealing a starfield in motion as the backdrop to an intimate dining setting. As the guards secured their weapons and departed, Gunnar counted the table service. There were four place settings. He passed Drax and Blinda a glance; neither returned the look as they removed their cloaks and sat down.

"You said you couldn't refuse food," Drax said looking up at him.

"Are we expecting anyone else?"

"Ship's commander. Please..." she said, motioning to the chair in front of him.

"I'd like to be able to see out the window," he said, taking the chair facing away from the only door.

"Wherever you would like," Drax agreed.

Gunnar had to wonder if she was always this cheery. CJ's description of her painted an entirely different picture. He could only surmise that she had gained her prize; his capture.

"Shall we start with drinks while we wait?" Drax asked, tapping several buttons on a control next to her place setting. A platter on mechanical arms extended from around the chandeliered ceiling. On each were a number of drinks in varying colors. "May I suggest the Ditairian wine. It's a great vintage."

Gunnar looked at the platter of drinks in front of him, indecisive.

"It's the red one," Drax instructed as she and Blinda took their glasses. After watching the two women take several sips, Gunnar tried his. He wasn't much for liquor of any kind, but the wine had a sweet taste and only a slight bite.

"There, that's better, isn't it?"

"It'll make you feel warm inside," Gunnar agreed. He felt something touch his shin and froze, trying to figure out what it was.

"So, tell us, where have you been hiding out since you left the safety of the *Tarzana*?"

"The safety?" Gunnar repeated. "I think the correct term should be, escaped to safety."

"Come, come," Drax grinned. "As you can see, we have all the finest amenities here."

"Your goons were going to use me as a lab rat." Something rubbed his shin again, giving him cause to glance under the table.

"But you would have been comfortable," Drax chuckled.

"So, I suppose I'm here for the same reason?" Gunnar suddenly realized what was happening under the table and gave Blinda a glance. She kept her eyes on Drax and her glass to her lips.

"I really don't know why you're here," Drax straightened. "Yes, I have purposes for you, but I didn't bring you here, you did. So, naturally, I'm curious. Why are you here?"

Gunnar took another drink, but became aware of someone entering the room.

"Ah, good. The commander has arrived," Drax exclaimed, standing up. "Now we can begin dinner."

"Apologies for my tardiness," a familiar voice said from behind Gunnar's chair. "The work of a ship's commander is sometimes endless."

"I can think of no one more capable," Drax grinned, motioning toward the chair across the table. Gunnar recognized the curly red hair flowing over the black uniform collar as Colonel CJ Barker stepped around Blinda's seat. CJ stopped short when she looked up to see Gunnar sitting across from her. Drax gushed with glee at their astonished expressions.

"What are you...?" CJ and Gunnar asked in unison.

"How did you get here?" CJ asked quickly.

"What are you doing here?" Gunnar asked immediately.

"Oh, this is tender," Blinda gloated, taking another sip.

CJ turned to Drax, angry.

"This was not part of our deal!"

"No," Drax burst into laughter. "It's not. It's so much better. Had I known things would turn out this way, I wouldn't have tried so hard to begin with."

"Deal? What deal?" Gunnar quickly ramped up from bewilderment to anger.

"Maybe you'd better sit down," Drax suggested, still laughing.

"Not until I get some answers," CJ demanded.

"I'd like some too," Gunnar fumed.

Drax held her hand up, still trying to regain her composure.

"I think I can answer at least a couple of questions. Your boyfriend here will have to provide the rest."

CJ sat down, looking at Gunnar with a mixed expression of confusion and anger. She then turned to Drax, who took a quick drink.

"If you could have only seen your faces," Drax teased.

Gunnar and CJ glared at her, waiting.

"Ok, Ok," she said, trying to wipe the smirk from her lips. "Colonel Conrad, when last you were here, you were given an experimental drug called, Kodiac Blu. Its design is to help subjects to be more cooperative with handling and providing information. No doubt you've been experiencing some undesirable side-effects. If not properly counteracted with the antidote, it's going to kill you. In fact, you should already be dead, but I suspect you're a little different than most humans."

"You might say that," Gunnar agreed, looking across at CJ.

"My baby sister, Casey Janae, is here because I offered her the antidote with certain provisos."

"Baby sister?" Gunnar exclaimed.

CJ bit into her lower lip.

"Oh, you mean she never told you?" Drax glared, half insulted, swinging her gaze at CJ.

"No," Gunnar flared. "Seems she missed that little gem."

"CJ and I share the same father."

"Just good friends growing up?" Gunnar said, feeling a little betrayed.

"I wanted to say something, but couldn't," CJ said, having no defense. "The timing was never right."

"Not too excited about this timing either," Gunnar growled.

"I love front row seats." Blinda snickered softly, even after Drax gave her a look.

"So you've made a deal with the devil, who just happens to be your big sister."

"I did what I had to do," CJ admitted, contrite.

Drax leaned forward a bit.

"I would give her the drug and promised not to pursue you or your friends. In exchange, she would return to Albion service and take her rightful place at my side." Drax turned to CJ. "Casey, Gunnar is here because, well, I guess this is the part where he tells us why."

"He's here because you went back on your word," CJ retorted.

"You know me better, Sister," Drax flared.

CJ toned down a bit at the starkness of Drax's demeanor. A snicker drooled from across the table.

"Shut it, Blinda," Drax lashed, turning to Gunnar, who was still looking at CJ. "Now, Colonel. We're all aquiver."

"Did you get the antidote?" CJ asked quietly.

"What antidote? When I left, you were still in command. I haven't been back yet. I have no idea where anyone is." The room fell silent for a moment, everyone waiting.

"How are you holding up?" CJ asked.

"Not too good. I don't suppose there's any chance you have more of that antidote?" Gunnar asked Drax.

"I gave it all to my sister," the Queen Captain responded looking back at CJ. "Who promptly took this ship and ran it to your good friend on Cross."

"Diord was supposed to get it to you when you returned," CJ finished.

"You left your command?" Gunnar asked.

"I had to do something," CJ responded defensively. "You were starting to come apart and Drax was very convincing."

"Fuji would have figured it out. We're leaps and bounds ahead of anything here in Hadrian."

"She had no idea what to do for you. She could barely identify the problem, let alone know how to treat it. This was the only thing I could do."

Visibly upset, Gunnar clenched his fists. He looked around the table, then put his head in his hands.

"Gunnar, why did you come?" CJ asked.

"Similar story to how you and I got chased all over the place by your sister and her hench ...woman." Blinda smiled and nodded. "By the way, where's the other one? The one that shot me up?"

"Seelix is not in the military. I have little to no control over what she does," Drax said.

"I got separated from my ship in the Nulark system and ended up coming to Albia."

"He was caught with a couple of terrorists on Dither," Blinda said, leaning forward. "Killed several troopers and assumed the identity of a Captain Mace Bridger, one of the officers he made short work of. He then found his way onboard this ship during an Albion resupply where he took up residence with Captain Bridger's wife. A Lieutenant Malina Cass."

CJ's expression shifted slightly, looking at Gunnar.

"Have you found her yet?" Drax asked, looking across at Blinda.

"No," Blinda responded, unmoved. "She registered for supply duty on Drummond and left before we could question her. When she returns, we'll have her brought in."

"So that's how you got here," Drax continued. "But that doesn't explain why you're here. What are you after, Colonel? Hoping to break your girlfriend out?"

"Really?" CJ objected, turning to her sister. "You're gonna go there, again?"

"She's not my girlfriend," Gunnar corrected.

"I will have answers!" Drax demanded, slapping her hand on the table, bringing Gunnar to his feet.

"Don't," Blinda suggested calmly.

Gunnar felt his ears pop, not really understanding what it was all about, but the distraction was enough to quell his anger and he melted back into his chair.

"I got separated from my friends," Gunnar fired back. "This ship is bound to end up someplace close to where they are."

"And where are these friends of yours?"

"Drax," CJ warned. "You promised."

"All right," Drax snapped, holding a hand up. "You're right, I promised. Now please, let's all just relax and have a nice dinner."

As they were served, the room remained silent until the serving arms had retracted.

"With the Queen Captain's permission," Blinda spoke up. "I have a couple of questions I'd like to ask Mr. Conrad." Blinda shifted her eyes to Drax, who slowly pulled her fork from her mouth.

"You've made no promises to anyone," Drax replied coolly.

"Colonel, will you tell us where you're from?" Blinda asked, a strange, playful look developing in her expression.

"A very long way from here."

"So we've been told," Blinda said, glancing at CJ. "But exactly what planet or star system do you hail from?"

"You don't have to answer that," CJ interjected.

"It's ok," Gunnar said. "It's not like there's an Albion or Colonian ship that could ever make it there. I'm from a place called the Mila system. It's a tiny cluster of planets around a blue giant called Orb. Two of the planets have solar capture orbits and are so close to Orb's influence, they have no atmospheres. Commenor is my home world. It sits just outside Orb's solar storm influence. There are several species of humans that live there, including my race, the Dialabrons."

"What's Commenor's military like?"

"We don't have a military."

"Clearly you do, or you wouldn't be here."

"Like the Colonians and Albions, the Mila system is part of a much larger coalition of life forms banding together in a common cause. All alliances require a control point for leadership. The planet Kalamar is centrally located in our galaxy and is where the greatest political and military minds reside, so it makes sense that command and control originates from there. I'm part of the Kalamarion Flight Ministry, a branch of the entire Kalamarion military."

"It sounds like your Kalamar military is quite sizable," Blinda remarked.

Gunnar thought a moment. He felt another brush up the side of his shin.

"Let me put it in terms you might understand." He looked at CJ, then Drax. "If you put all the factions of Hadrian together, you might make up a third of the size of Kalamar Command."

CJ put her glass to her lips, trying not to smile at the drooped looks coming from Drax and Blinda. Gunnar went back to eating and the room was quiet for a few moments.

"Well," Drax said, trying to smile. "It's a good thing we're so far away."

"I don't suppose you'll tell us about your protector?" Blinda asked.

"Thought you might be getting around to that," Gunnar said confidently. "As a Thane, you understand that being a protector doesn't mean that you're glued to someone all the time."

"You're correct; it depends on the need and personal choice of the Thane. Most of the time those two factors run hand in hand."

"I'll just say, he's around, and leave it at that."

Blinda frowned. She would feel a whole lot better knowing exactly where Gunnar's Thane was hiding. Thus far, she hadn't felt his presence, but their last encounter had taught her that he had an uncanny ability to remain undetectable until it was almost too late. As the four neared the end of the meal, Gunnar took a refill on his drink and pushed back a little. He gave Blinda another look and glanced under the table, giving CJ cause to wonder what was going on.

"Now, let me ask a question or two," he said, settling back. "I'm not one who cares a whole lot about the political goings on of my own or anyone else's government. My thinking is politicians have become a bunch of bureaucratic, self-serving blowhards doing their best to make it look like they're serving their constituents while lining their pockets with credits and power. I have little stomach for who they are or what they do."

"That's a pretty bleak outlook," Blinda said.

"I'm right there with you," Drax agreed.

"So I've found it a little weird that I've taken a bit of a fascination with the history of Hadrian's government."

"Any parts in particular?" CJ probed carefully, hoping he didn't get too close to certain subjects.

"How is it, that the Albion military has allowed lawlessness to prevail within its political machine?"

"How's that?" Drax asked.

"Is the military just pawns of a bunch of political thugs, or is it the other way around?"

"Gunnar," CJ chastised.

"What? You've given me the low down on how the Albion machine operates. It's mostly gangster warfare; the Ratronians are worse. How you're able to keep from imploding is far beyond my understanding."

"I don't think you're giving us enough credit," Drax smiled, ignoring the obvious insult.

"You don't participate in thuggery?"

"I'm not going to deny that we do things a little differently here, but as you can see," Drax said looking around, "it works very well. So well, in fact, that we are nearly equal in size to what has taken the Colonians a millennia to achieve."

"I'm not too privy to the laws of Hadrian, but from what CJ tells me, you're bound to obey those laws."

"How dare you call into question my loyalty to the throne of Hadrian and its laws."

"Then you feel perfectly justified in trying to unseat the current ruler, uhm, what's her name, Benetar?"

"Certainly," Drax agreed. "Hadrian law is quite clear. The royal line has followed along the Oxlind bloodline for eons. There has never been a dispute until the Duchess began meddling in royal affairs. An Oxlind must rule or the throne must be passed on to the strongest faction in Hadrian. Since the Colonians can't produce an Oxlind blood heir, by law they should be relinquishing control to the Albion Empire."

"You mean to you," Gunnar suggested firmly.

"I would not refuse such an appointment if asked," Drax agreed, hesitating slightly. "But I am more of a warrior than a politician and would rather ensure the rightful power is positioned on the throne."

"Doesn't Hadrian law require a royal from the Albions, just like it does from the Colonians?"

"No," CJ answered quickly. "The Oxlind line is so old, there was never any development of a royal line in any of the other factions. A new royal line would be created should the Colonians default."

"So as I understand you, if an Oxlind were to surface, and able to prove their legitimacy to the Throne, you would honor their claim?" Gunnar noticed Blinda shifting uncomfortably in her chair. CJ had a worried look as well. Drax looked around the room for a long moment, then back at Gunnar.

"I would not only honor it, but fight anyone who would dispute it. My loyalties are to Hadrian and the Albion Empire." The room fell silent again as dessert was served and they all finished without another word. Drax finally pushed back from the table, gave CJ a glance then looked back at Gunnar.

"This has been a very enlightening evening. I've enjoyed our conversation and you've given me some new things to think about. Blinda and I have some important matters to see to tomorrow, so we need to turn in. I've had some special accommodations prepared for you. Casey will escort you. As we're currently running at Quadra-light, I have little fear of your great escape repeating itself. You would do well to rest up, you have a big day tomorrow." Drax stood up, looking at CJ who came to her feet. "Colonel? Do I need to have Blinda accompany you?"

CJ looked back at the Thane who was licking her spoon.

"No, I think I can manage."

Gunnar came to his feet as CJ moved around next to him.

"Lead the way," he said, following her from the room.

As the Queen Captain started for the door, Blinda stopped her.

"You're not actually going to leave them alone to hatch another plan, are you?"

"Sure, it's cheap entertainment. Where are they going to go?"

"This is your sister we're talking about," Blinda snapped, coming to her feet. "I'm surprised they didn't figure out how to tie us up, jump ship and take dessert with them, even at light speed."

"The evening is still young," Drax said, turning for the door. "You worry too much."

"You would be less careless if the King Commander knew of this."

"Go ahead, send a priority message; use my clearance. But if you do, you better be ready to explain why I have delivered on every promise I've made."

"Be careful not to taunt the King Commander."

"Taunt the King Commander?" Drax repeated. "My loyalties are to Hadrian and the Empire." With that, the Queen Captain breezed from the room, leaving Blinda to consider.

I'll take that

"I wonder how long it took Captain Reader to turn tail and run back to Aster?" Billy asked. He had just awakened from a sound sleep. His headache was gone, but his joints were stiff. He did his best to stretch, but the suit he was still in made that difficult.

"We'll know soon enough," Logan responded, focused on his instruments. They had been traveling at lightspeed for hours, having made several course changes at predetermined waypoints along the way. As per the mission protocols, if anyone were to get separated they could always jump back to Aster alone, making sure that no one followed.

"Do I have enough time to get out of this suit?" Billy asked. He started working the mechanism to get his gloves off and work his way to the shoulder seals.

"Yeah, plenty of time for that. By the way, I took care of your leg while you were asleep."

Billy pulled the top part of his suit over his head and dropped it against the back wall.

"I think this is the only fighter that you can change your clothes in while someone else is flying it." He looked down at the hole in the leg of his suit, then started squirming his way out. "Hey, that's a pretty good job," he said, examining the clear med-plast gel covering the closed gash in his leg. "You could have waited till I was awake; I could have helped."

"And listen to you whine and moan about how much it hurts and how bad you had it outside? I was glad to pass on that one."

"Whine and moan? I don't think so."

"Do you remember your little incident at the Darthfort-Three Star base?"

"Oh come on," Billy grunted, kicking the pants off; boots and all. "Those were plasma burns from that idiot Major Hutch kicking on his punch thrusters before he was safely away from the launch tube."

"He was certainly an idiot," Logan agreed. "But you sure made a lot of noise while they were smearing that burn paste on you."

"You'd holler too if someone were basting your skin with that stuff. Nowadays, they spray on med-gel, but even that can sting a little."

"Yeah, that's why I did it while you were asleep."

"Whatever," Billy grunted, settling behind the Interceptor pilot. "See anything ahead?"

"Not a whole lot. Been keeping an eye on the Asium inverters."

"Uhm, what's the trouble?" Billy asked leaning further forward to examine the readouts.

"Like I said when I picked you up. This bird took a real beating back there at Carolon. The light drive stabilizers are acting really jittery. I thought we were gonna have to drop out of light speed a couple of times while you were asleep."

"Can you tell what or where the problem is?"

"I don't fix them, I fly them," Logan replied.

"Those are some pretty wild readings on the compensation quadratics," Billy pointed out.

"I've flown with a lot worse."

"Yeah, but that was probably in a Tempest where it didn't matter. The Interceptor uses a completely different set of light speed parameters to get the job done. The engineers were able to nearly double the light speed output of this thing by using synchronized transducing duelonics. But the control parameters are way tighter. You can't let it out of the specified limits for very long or the light speed duelonics will become unstable and fry."

"What you're really saying is we need to drop out quick and find someplace to make repairs," Logan said, touching a control. Outside, the stars momentarily elongated and a bright orb appeared in the distance. Several readouts on the console in front of them began to flash warning information and while Logan worked to figure out where they were, Billy reached forward and worked with the controls to silence the alarms.

"Well, you saved the FTL drive, but before we can use it again, I'm going to have to make some repairs," Billy said, reading the information flashing in front of him. "Know any good mechanics?"

"I met one once that could work on anything, ...even knew how to fix things in space."

"I'd like to meet him, someday," Billy snickered.

"Who said it was a *he*?"

"Where are we?"

"According to these Hadrian star charts, somewhere in the Helios system." Logan touched several spots on the glass in front of him, working his scanning equipment. "I don't trust much of this information, but if I read this right, we're in a binary system with five planets." Logan brought up the HUD and swiped through several displays. "Were I Colonel Conrad, I'd be able to just steer to one of these rocks and land."

"So why can't you?"

"I got my flight training through, *Do it only by the book,* Major Heely."

"I see," Billy recognized. "Step out of the norm and you're paralyzed."

"Thankfully, flying wing with Dakota has pushed some of that aside and I've learned a thing or two about how to fly by the seat of your pants." Logan narrowed the view of the star system in the HUD. "This one here looks like a likely starting point," he said, pointing at one of the planets.

"Moyle." Billy looked closer. "Is it deserted?"

"Sensors aren't picking up very much." Logan made several adjustments. "I'm doing good to get this gear to even see the planet. But, it looks habitable. At the very least we should be able to set down long enough to give this bird a good look over and figure out what to do."

"Anything is better than having to climb around outside this thing in space," Billy said leaning back.

"Come, come, Billy. Think of all the walk time you've logged over your career. Heck your last walk should put you in the record books."

"You know what," Billy said, rolling his eyes. "Right now, if I never see another space suit again, I'll count my retirement complete."

Logan grinned broadly and moved the Interceptor's sub-light throttle ahead, steering in the direction of the gold colored speck in the endless starfield.

"We probably should be deploying life sign dampeners," Billy said. "At least people will leave us alone while we're here."

"Unfortunately, the mech dampeners for the ship are all shot up. Maybe anyone passing by will just think the Interceptor is space junk."

"I suppose I could dig around in this thing while you're getting us there," Billy said, moving to one of the many access panels behind the pilot's seat. "I can get to some of the light drive stuff in here, but the main access is on the outside under each wing root."

"You're the *fix it* guy," Logan said, preoccupied with piloting toward the ever growing spec of Moyle. "By the way, what is it with you and Miss Mantose?"

"What do you mean, me and Miss Mantose?" Billy asked, sticking his face into an access hole.

"You two got something going on, or what?"

"Or what…" Billy responded muffled.

"Come on, you two are like an old married couple."

"Where did you come up with that nonsense?"

"Do you really not see how you two interact?"

"I've never touched the girl."

"I would hope not, you're old enough to be her father."

"What are you even talking about? How old do you think I am?"

"Old enough to be her father."

"Not," Billy responded, pulling his head out. "We are well within range of each other."

"Within range," Logan chuckled. "Way to describe your relationship; like a moving target."

Billy plunged his entire arm into the access hole, looking at the back of Logan's head.

"Tiana is great with FTL propulsion and asked that I help her learn more. She's been indispensable in helping me and Cooper get the ship back together. She's a darn sight more help than that goofus Scott Brandon."

"Well, all I know is what the rest of the crew has observed."

"Not sure what you guys think you've observed, but it ain't what you're thinking."

"Uhuh..." Logan continued to monitor the Interceptor's systems as the bright gold sphere of Moyle grew larger. "Weird looking planet," he commented in a low voice.

"We see all kinds of planets," Billy replied. "What makes this one look so weird?"

"I've seen a lot of stuff in space, but this one is a little off the list. Wished my scanning equipment was working better."

"Why?"

"Be nice to know if we can even get down onto the surface."

"Why wouldn't you be able to get down?"

"Have you really been stuck in the back of the ship for that long that you don't remember approach procedures for any unknown planet or space body?"

Billy finally poked his head back up front next to Logan.

"It's gold colored, so what. Milta Nine is gold colored. The algae in its oceans is gold colored and so the water appears gold and the reflection of the atmosphere is gold colored."

"Thank you for the biology lesson, professor. Do you see any oceans on this rock?"

Billy gazed harder at the golden orb filling their vision as Logan brought the Interceptor to a halt. They both looked for a long moment, searching for any clues as to what they might have to deal with to get to the surface.

"I can't see anything, not even mountains," Billy finally said. "Can't these sensors do better?"

"Not much," Logan replied. "Too much dust."

"That's some thick dust."

"Well, to be fair, it's a lot bigger than just dust."

"Can we get under it?"

Logan worked with his scanning gear for a moment longer then turned it off and sat back.

"I guess there's only one way to find out," he said, pulling his harness tighter. "Better hang on to something; this could get a little bumpy."

Billy immediately grasped the back of the seat as the pilot pushed forward on the throttles and drove the Interceptor toward the gold color shrouding the planet. Logan glanced at the shield readouts the moment he detected sparkling apparitions forming on the hull. In an instant, the ride turned from smooth glass to a heavy vibration. The further they plunged into the bright colored cloud layer, the more violent the tiny explosions against the Interceptor's energy shields manifested themselves. Suddenly, the Kalamarion fighter broke through the layer and Logan leveled out. Here there was smooth air, but it was short lived as they entered another layer just as dense and violent as the first. A warning alarm went off on the instrument panel to Logan's left, but he was too busy with the fighter's controls to give it any attention. Billy carefully leaned over for a look and silenced the alarm.

"Well, this is fun," he grunted.

"What's with the alarm over there?" Logan asked, concentrating on the instruments in front of him. "Shield generators?"

"Ignorance is bliss," Billy responded.

"That goes both ways," Logan said.

"I take it we both have news neither wants to hear?"

"Something like that."

"Will it kill us?"

"Likely."

"Feel better not knowing?"

"You tell me."

The fighter suddenly blew through another layer. Logan tried to level the craft out, but something wasn't right.

"It would appear your instruments aren't working," Billy observed, scanning the panel in front of the pilot.

"What was your first clue?"

"The antigrav inducers are indicating we're upside down and in a flat spin."

"That's what happens when you try to do things by the book," Logan grumbled. "We need to see the ground pretty quick or we're going to be auguring into it."

"Who knows how far down this stuff goes," Billy said as they entered another layer of explosive particles.

"There's only one person I know who can fly fighters through crap like this without instruments."

"The Colonel..." Billy affirmed. "What about your auto stabilizers and the autopilot?"

"Yeah, those systems aren't working right now."

"Funny how you take so many things for granted until you actually
need them."

"You use them all the time, you just don't ever know it."

"Emergency jettison?"

"You wanna squeeze in here with me?"

"Better than getting slung out and falling through this crap to get
my flesh pelted off and splatting on wherever the ground ends up
being."

"That had a lot of detail in it," Logan said, pressing himself
sideways in his seat as Billy wormed in next to him.

"Where's my harness?" Billy asked as the turbulence became more
intense.

"Why isn't the anti-grav doing its thing?" Logan complained. Two
people in a seat built for one wasn't very comfortable.

"It is doing its thing," Billy said. "The fact that we're not being
smashed against the windows because we're spinning should be a
major clue."

"Hate this," Logan grumbled.

"You'd rather see the ground coming at you right before you
become a grease spot? I think I'll vote for the not knowing when I'm
going to hit."

"I hate that too."

"Can't you do something; anything?"

"Just hang on," Logan said, arming the ejector system. "Never had
to abandon an Interceptor before. I have no idea what it's going to be
like."

"Probably twice as bad in an atmosphere as it would have been in
space."

"Probably right." Logan checked a readout, then put his hands on
the emergency release. "Ready for this?"

"Do it now before I change my mind." Billy closed his eyes,
tensing up. After waiting several seconds, he popped an eye open and
glanced over at Logan. "Wow, that was smooth. I didn't feel a thing."

"That's because I haven't done anything yet," Logan said looking
outside.

"Why not?"

"Because I think I can see something out there."

"Probably the ground; hit the release!"

"Yeah, it's the ground, but it's quite a ways down there."

Billy strained to see through the exploding gold colored mist,
noticing defined lines appearing through the blur. Continuing to
descend, they suddenly broke through the lower layer, still tumbling.
Logan was already at work when Billy realized they still had quite a
way to go to reach the ground.

"Why the heck are you trying to abandon ship?" Billy asked squirming to get out of the pilot's seat. "We all can't be like the Colonel. Fly this thing!"

Logan refocused. *At least I've got a point of reference now.* Looking at his instruments, his initial assessment of the ship's systems were correct. Nothing coming back at him corresponded with what he was seeing outside. While flight training taught him to always trust his instruments, considering what he had put the Interceptor through over Carolon, he decided to ignore them and go with his instincts. He worked methodically to correct the fighter's tumble, alternating his thruster controls and utilizing the retros that were normally reserved for docking in space. As the ground reached up at them, he activated the landing gear.

"Can you imagine the suck factor if the antigrav weren't working?" Billy commented from behind.

"Have you ever had any Centriff training?"

"Yeah, I didn't like it then either. You gonna straighten her out in time?"

"Gonna be close," Logan grunted, still working with his controls. "I'm not as good at this as Dakota or the Colonel."

"Wished you were," Billy admitted, becoming alarmed at how close to the ground they were.

"Me too." Logan glanced over at several other controls and pulled on a handle. "Never thought I'd ever have to use those."

"Yeah, who needs air brakes in space?"

"One thing's for sure," Logan said, pulling back on the controls as the ground became quite distinct. "I'm throwing away all my manuals on *Flying by the book.*"

"Get us on the ground in one piece first, then we'll talk about your library card," Billy said, tightening his grasp on the back of the pilot's chair.

Logan carefully eased the big fighter out of the tumble and corrected the oscillating yaw as the Interceptor swung away from the ground and glided gracefully over a large barren bluff. Circling slowly, he located a relatively flat area on top and set the ship down. Looking outside, he finally shut down the softly humming engines and after a moment, everything went quiet.

"I really enjoyed that," Billy said, checking several readouts on one of the panels behind the pilot's seat. "Can you tell if we can breathe anything out there?"

"If I'm to believe anything this gear was trying to tell me as we were coming down, it's breathable out there, although it's a little cool."

"Where do you keep your hand scanner?" Billy looked around.

"Third compartment down." Logan pointed toward the hatchway shaft. "You let me take care of looking around outside while you see if you can do something with the FTL engines."

"Yeah, I don't think so." Billy pulled the scanner out and headed down the shaft. "Cure my curiosity first, then I'll focus."

"I guess there's no point in pulling rank," Logan mumbled as he watched the engineer work the lower hatch. After a moment with the scanner, Billy shoved it in his pocket and looked back up at the pilot.

"We're good. Go ahead and cut the shields and we'll get out and stretch our legs."

"Here," Logan said, reaching into a small compartment and tossing Billy a bundle of clothes. "Let's get out of these uniforms so we can get some work done."

"You always carry civilian clothes?" Billy asked, checking the sizing.

"Standard survival issue. Sometimes you need to look a little less military."

"Makes sense."

After a quick change, Logan reached back to the forward console and touched a control as Billy exited the ship. Following the engineer, he stopped next to him, looking out across the bluffs. Billy held the scanner up and studied the readouts while turning in a circle.

"Well, the good news is, we're not going to die breathing this air." He paused a moment to look closer at the scanner readouts. "The Neutrino count is a little high, but manageable as long as we don't stay outside for extended periods of time. I imagine those layers we came through are what's saving our skin from peeling off right now."

"And the bad news?" Logan shielded his eyes from the glare of the binary suns. The terrain was dry with clouds of dust occasionally whipping up by a steady wind scraping across the bluff.

"There's not much water here," Billy finally responded, still looking at the hand scanner.

"As illustrated by the acute lack of vegetation anywhere in sight. At least it's got a tolerable temperature."

"For now," Billy mumbled, still focused.

"Well, we won't be here long enough to worry about it."

"One would hope," Billy stepped back to the hatchway and grabbed something.

"What's wrong?" Logan asked.

Billy raised a pair of electro-binoculars to his eyes and adjusted the controls. He turned both directions slowly, carefully zeroing in on something in the distance.

"Not sure. Thought I saw something out there," he said pointing at a vast valley beyond several lines of bluffs.

"What did you think you saw?"

"Just something reflective; a flash off metal. The hand scanner can't pick it up; probably too far away."

Logan took the binoculars and looked out at the distant valley, scanning back and forth several times. He looked back at the engineer, then started toward one of the cliff edges.

"Probably just a meteor passing through the same muck we just did. You saw an impact flash. I'll check around the perimeter of our little landing platform."

Billy shrugged and turned back to the ship, opening an exterior access panel. After exposing much of the internal workings of the Interceptor, he went to work while Logan busied himself exploring the surrounding area.

*   *   *   *

Several hours slipped by as Logan explored the perimeter of the bluff. Working his way up and down the short face of the cliffs surrounding the bluff, he was just making his way back to the flat area on the top as the binary suns began to touch the horizon. He hadn't heard or seen any signs of life, animal or human. The sky was already a gold color, but now the angle of the suns intensified the color, casting a golden hue over everything. He couldn't help but wonder if he and Billy were the only two humans on this barren planet.

Logan took one last look around, stopping to gaze out across the shallow line of ridges that spanned off for as far as the eye could see. As he turned, he froze. Something was wrong; out of place. Not daring to turn his head, he shifted his eyes side to side, then straight ahead. *It's gone!* He moved carefully to the flat spot where he had left Billy and the Interceptor. A deep hole suddenly developed in the pit of his stomach as he looked frantically around. Looking down, he examined the silted dirt. The landing gear imprints were there; he was in the right spot. He took a couple of deep breaths and looked straight up. He was in hopes Billy was just testing the ship. As the engines didn't make as much noise as his old T-6 Tempest, it could be hovering right overhead and he might not hear it. Raising his wrist communicator, he touched the control on its side.

"Hey, Billy. Where are you?" He continued scanning the skies in the hued light as dusk continued to develop.

"Billy, come in." A smile developed, followed by a half chuckle of both humor and worry. "Ok, you've had your laugh. I get it. You got me. Ha, ha, ha. Very funny. Now come on. If you've got that thing fixed, then let's get out of here." After turning around several times, he stepped to the edge of the bluff and looked in the direction Billy had thought he had seen something earlier. This time, he did see something. He strained hard, watching an object moving in the

32

distance. He brought the electro-binoculars to his eyes. "Ok, I can see you out there. Bring it back and let's get going." There were marker lights flashing around the object, indicating a ship of some type, but the size didn't match the Interceptor. "Billy, come in." He watched the ship descend toward the ground, hover for a moment, then disappear. "Come on. Let me know everything's all right."

Slowly lowering the electro-binoculars, Logan turned around, wondering what to do. He was completely exposed. Looking toward the binaries as they continued to melt into the horizon, he took a step back and stumbled. Looking down, he picked up the hand scanner Billy had been using. In the fading hue, he noticed several sets of footprints in the dirt. *Hello...* He followed them until they disappeared into a blasted area a short distance from where the Interceptor had been sitting. *Ok, we'll do this the hard way.* Turning the scanner on, he adjusted several controls and pointed it near the blasted area, then around the area where the fighter had been. As the light continued to fade, he turned back to where the mystery ship had disappeared and took several more readings. He had been stranded before, but there was always help on the way and he had provisions.

Nuts... I've got next to nothing here. Let's see, survival kit, a couple of bottles of water and this hand scanner. Wonder how long it's gonna be dark and how cold it gets? It's a sure bet the terrain between here and clear the heck out there where that mystery ship landed isn't going to be a walk in the park. Surmising there was little else he could do, he headed down the short cliffs in the direction of the distant valley beyond the lower ridgelines as a few stars manifested themselves through a thick golden haze.

There were no moons to give off any kind of light. As he hiked, Logan's thoughts turned to the time he had spent on the planet Flints. Its moon was close enough at full phase, it appeared to be at least half as bright as the sun during the daytime. *That would have been nice in this circumstance.* Even with his wrist light, he was having difficulty seeing the ground in front of him and it seemed that he either tripped or stumbled every three or four steps. At least it wasn't cold. He had fully expected the temperatures to drop off when the suns went down, but it remained fairly steady.

After several exhausting hours of stumbling through the darkness over the uneven terrain, he finally stopped. He couldn't even focus on the black horizon where the stars stopped and the planet began. Fumbling in the darkness with the tiny wrist light, he found a nook in a formation of rocks. Tired, he curled up and got as comfortable as his accommodations would allow. He was hopeful the place he had chosen wasn't already occupied by something that was liable to poke, prod or stick him. After some worry about what happened to Billy and the Interceptor, he slowly drifted off to sleep.

*   *   *   *

Logan awoke with a start to the sound of a herd of Cantorian racing charkas thundering right past him. Looking around in the dim light, he couldn't see a thing. Blinking several times, he realized it had been a dream. *Hate it when those are so real.* He pulled himself from his sleeping place and stretched stiffly. He wasn't as young as he used to be and somehow he had managed to fall asleep with a rock sticking in his back. Either he had been too tired to care or things had shifted while he slept. Didn't matter, he was sore now.

Not much had changed in the sky. The visible stars were still a deep gold color, but he did notice a lighter glow coming from the direction he had come. Figuring the suns must be on their way, he continued his journey. Hoping to make better time, he didn't stop to consume any of his emergency rations or the hydration capsules he had, but took them as he went.

The ridgelines seemed endless. He would ascend one, only to find several more beyond. As it got lighter, the hike became easier, but the ridgelines still continued on and on. It wasn't until he stopped at the crest of a smaller ridge that he realized he was at the foot of the mountains they had landed in and was now staring out across a vast flat valley. Now that he was finally here, he realized it wasn't even a valley, but a very flat desert; perhaps an ancient lakebed or ocean. Logan pulled out the scanner again and held it up. It's range was somewhat limited, but it was reading something far off in the distance, something he couldn't see. A look through the electro-binoculars only revealed a formless horizon.

After consuming more rations and hydration, he started across the flat dusty landscape. The binaries were behind him now, casting his shadow in front of him. He plodded on for hours as the twin suns slowly made their way overhead. Stopping to rest, he looked around. There were distant mountain ranges all around him.

Moving again, he noticed something odd about the horizon before him. There was a distortion directly ahead. He pulled his scanner out and held it up. It was reading some kind of structure directly in front of him, but there was only a heat distortion. Starting forward again, he looked harder, trying to make sense of what he was seeing. As he got a little closer, he became aware of a buzzing sound. Energy spikes registered on the scanner, giving him cause to step a little more cautiously. With every step, the sound became louder. The readings on the scanner became so erratic, he finally turned it off. The buzzing sounded like it was all around him now. He stopped and looked left, then right. Not sure what to do, he sank to his knees and examined the dirt. Looking back up at the distortion, he listened closely to the

sound. After a moment, he scooped up a handful of dirt and let it slowly sift through his fingers. As he watched it fall, he suddenly pitched it in a wide arc in front of him. The dirt particles came in contact with an unseen energy source, snapping and popping through several horizontal beams.

The highest beam is only a couple of feet high, I think I can jump it. Logan got to his feet and walked along the buzzing fence line, tossing fistfuls of dirt as he went. At a certain point, he found the same odd distortion similar to what was beyond the invisible fence. Tossing the last handful of dirt at the fence, he took a couple of steps back, then ran and jumped. Rolling once, he came right back up on his feet. He brushed himself off and stepped closer to the distortion. Pulling the scanner out again, he looked at it briefly and seeing it wasn't showing him anything he could use, he picked up another handful of dirt and tossed it at the distortion. This time, it dropped straight down after striking an unseen object. Logan stepped closer and brought his hand up to the distortion. There was a slight tingle on his palm and fingertips as he made contact with something solid. The distortion rippled gently to the touch. He brought his other hand up as well, feeling for deviations in the surface tension. It felt like a wall; something metallic. Moving for a considerable distance, he heard the sound of a mechanism behind him and turned to the half torso of someone levitating from the distortion and pointing something at him.

"Hey, can you help me out? I'm looking for…"

There came a muffled, subdued pop. Logan felt something strike him directly in the chest. He recoiled, but remained on his feet. He expected to be thrown. Perhaps it was a dud charge? He looked down where he had been struck and pulled a curious little silver three pronged dart from his chest.

"Ouch!" he said looking at it. "Hey, what did you do that for? I only wanted…" He suddenly felt cold. His vision blurred momentarily, then everything started to spin and a moment later, nothing.

A room at the Inn

CJ and Gunnar walked slowly towards the executive elevator, their hands clasped behind their backs. They glanced at one another a couple of times and smiled slightly, but there was an ominous cloud that hung over their reunion. CJ knew what was to come. This wasn't at all what either of them had in mind when the *Constellation* left Aster.

"I still can't believe you're here," CJ finally said as they approached the elevator door.

"I still can't believe Drax is your big sister," Gunnar said quickly.

"I'm afraid you can't choose your siblings."

"This is the last place I ever figured you'd go."

"What about you? Did you really do all that stuff Blinda said you did?" They stepped into the elevator and started down.

"In hindsight, I can't understand why she didn't catch up to me sooner."

"You have to know I didn't have anything to do with this. Someone is feeding her information."

"I know," Gunnar said, putting his hand to the back of his neck and stretching uncomfortably.

"So how did you get separated from your ship?"

"Our friend... Janox wanted to see how well an Interceptor worked."

"During a mission?"

"Are you guys running any active fighter operations?"

"At Quadra-light? Why do you think Drax left me alone with you?"

"I'd really love to see some engine operations or even some gunnery practice."

CJ looked oddly at Gunnar's tired eyes as he returned her look with one of his own.

"No active fighter operations because we're at speed, but the mechanics are always performing engine run ups in the Black Tiger maintenance bays."

"They always seemed like a troubled ship, prone to a lot of problems."

"It's a great design, but they complicated the parts that should have been simple and simplified the parts that should have been complex."

"Does it fly backwards?"

"Gunnar, you can't leave the ship at speed. It can't be done."

"Oh, I'm completely aware of that. I promise I'm not going to leave the ship; at least not yet. I just want to hear all that engine noise. You know me, sweet tones to my ears."

CJ touched a control on the elevator control panel and the transport car stopped.

"Maintenance bays," CJ commanded. The elevator car abruptly started moving again and presently they stepped out into a noisy hallway lined with windows. "You wanted to hear engine noise, here it is."

Gunnar leaned against the windows, watching the Albion mechanics working on the Manta-Ray shaped fighters below.

"Can we get closer?"

"As close as you want," CJ said, noticing an ugly discoloration on the back of Gunner's neck.

As they stepped onto an open platform, it immediately moved to the main floor. The noise of the Isom engines was deafening, but Gunnar seemed to soak it up as if he had never seen a thruster engine before.

"Can you explain what these guys are doing as we walk?"

"Not right here," CJ said, covering her ears as one of the engines close by started to rev up. She pointed toward several fighters that looked like they were only half assembled. As they walked, and the noise subsided a bit, Gunnar leaned a little closer.

"You asked why I'm here?"

CJ looked at him, then looked around. She suddenly understood why he wanted to see and hear the engines. She pointed to one of the fighters next to them and gestured like she was explaining how it all worked.

"Janox is the blood heir to the Hadrian throne."

"We already knew that, Diord told us," CJ responded, a little confused.

"But he didn't have any verifiable proof. I got separated from the ship because Janox stowed away onboard. We didn't find her until after we started solo operations in the Nulark. We were just passing Reako when she convinced me and Alex to take her for a little sightseeing tour."

"Convinced you?" CJ gave Gunnar an unbelieving glance.

Gunnar just rolled his eyes, waggling his head a bit.

"All right, she used a stunner on me a couple of times; it hurt bad."

CJ burst out laughing and pointed at Gunnar, who looked away a little embarrassed.

"Are you serious? She actually stunned you?"

"I can still feel it," Gunnar said, gesturing to the back of his neck.

CJ walked him around the fighter and pointed at the leading edge of the wings.

"Are you sure that's not the Kodiac Blu churning around inside you?"

"No, but that's where the knot is. Anyway... we found her old home and her Au Pair, Tonnie."

"She actually exists?"

"We reactivated her, which in turn set off some kind of booby trap. The entire place flooded."

"Wait a second," CJ said, grabbing his shoulder. "You reactivated her? How do you reactivate someone?"

"She wasn't a someone, Tonnie is an android. My guess is she isn't even from Hadrian."

"An android," CJ repeated, toning down as the engine noise ebbed.

"When we got back topside, the *Constellation* was gone. There were no indications of it anywhere and the *Realistic* was gone too. I can only assume something went wrong, and they hightailed it back to Aster."

"That was the abort protocol."

"Tonnie confirmed that Janox is the youngest daughter of Tiev and Mila Oxlind, Jana Oxlind. She also explained who was behind the whole Oxlind purge; your King Commander Dismon and the Duchess Benetar. They conspired to get the Duchess empowered into royalty and she would eliminate the blood heirs. If she wasn't accepted as an heir apparent, she and the King Commander would use the youngest to have themselves proclaimed co-rulers. I'm still not sure what your KC gets out of all this."

"Then Diord was right. So how does that get you here?"

"We had to go to Dither where the actual documents and tokens were stashed."

"Come again? Documents and tokens?"

"Jana's lineage records and her parent's royal rings. Without them she can have no legitimate claim to the throne of Hadrian."

"That's where we caught up to you."

"You knew about what was going on?"

"I had no idea. Drax informed me of a terrorist attack on one of the repositories, and sent Blinda down with a Special Forces team."

"Jana lost her father's ring in the escape, so I sent them on to Cross while I came after the ring. Malina and I found it in the command offices. Those guys had no idea what they had. We were trying to get off the ship when your sister caught up to us. I had the ring in my pack, but switched packs with Malina and she got away without Drax and Blinda knowing it, or at least she has a pretty good head start. Hopefully, she'll get the ring to Diord and he'll arrange to get Jana to the right place at the right time." Gunnar closed his eyes for a moment, then motioned to the open cockpit hatch on the underside of the fighter. He and CJ stuck their heads up inside.

"By the way, if anyone should ask, I called Janox, Kara Americ." CJ pointed to the main control panel inside. "Just to hide her identity."

"Where did you come up with that?"

"It's the name of an obscure Bandolair miner I met a long time ago."

"Sounds like an adventure... Not sure what to do at this point," Gunnar admitted.

"Diord needs to get Jana to Calliope during the Rite of Pintar."

"Rite of Pintar? What's that?" Gunnar asked.

"All Hadrian factions are required to gather at the Castellian home world of Calliope for a testament of galactic law. Every head of state for each faction will be represented there. That includes King Commander Dismon and the Duchess Benetar. Present Jana, her documents and tokens in front of everyone there and Hadrian will be forced to see."

CJ and Gunnar came back out from under the fighter cockpit and walked between the rows of ships, back toward the platform elevator.

"My only chance to leave this ship was with Malina," Gunnar said, stopping to rest against the wing of one of the Black Tigers.

"There has to be a way to get you off."

"It's more important that you communicate with Diord and see that he figures out a way to get Jana to that Rite of Pintar."

"Gunnar, if you stay on this ship, you're a dead man," CJ objected.

"I'm a dead man on or off this ship."

"If you haven't gotten the antidote from Diord, that means he still has it. If I can get it back from him, I can get him the message and get you the antidote at the same time."

"Nice try," Gunnar said, stepping onto the elevator platform with CJ. "You can send a message from anywhere. If you try to go back to Cross, your sister will figure that out in quick order, besides, by the time you get me back there, it'll be too late. Your best hope is to send your message from here, as soon as possible and hope Diord can figure something out."

"Gunnar, I can't just let this happen."

"Oh, yes you can."

CJ was increasingly alarmed about how cavalier Gunnar was being about what lay ahead of him. Either he didn't understand, or just didn't care. Perhaps the Kodiac Blu churning inside him had already spelled it out. As they got back on the elevator car and started back up, Gunnar turned to CJ and took her hands.

"Rick never would tell me what he learned on Boris about us that was so all fired important; just that you and I would help bring your galaxy back together. Maybe this is what he meant."

"I don't think he meant for you to die for it."

"If there is one thing I've learned being best friends with that guy is, either I'm going to save his sorry butt or he's going to save mine. I have to believe that formula still works, no matter where we are or what we're doing."

CJ pulled in close and kissed him on the check.

"I wish there was something more I could do."

"You've done all you can do," Gunnar said, trying to smile. Presently, the elevator stopped, and they stepped out into the hallway between four waiting guards.

"We can take him from here," one of them said as the others surrounded Gunnar.

"I'll walk with him to his room," CJ asserted.

"We were ordered by the Queen Captain to take him..."

"I don't care if the King Commander is standing here with a gun pointed at his face," CJ bristled. "I am the Commander of this ship and everyone on it! I will walk with him."

"I understand your authority, Colonel. But we aren't part of your crew. We are the Queen Captain's personal guard..."

"CJ," Gunnar said quietly, taking her hand. "It's ok. It'll be better if you don't come. Remember to act when the time is right." He raised her hand to his lips and gently kissed it then turned and started walking, guards in front and behind.

Tears welled in CJ's eyes, partially obscuring her vision as she watched Gunnar turn a corner and disappear. Trying to hold back her emotions, she stepped back into the elevator.

"Command lounge!"

It took only a couple of minutes before the elevator car opened up in the lounge high above the bridge levels. She paused a moment after the doors opened, trying to control her emotional disposition. After wiping her eyes and taking several deep breaths, she stood erect and walked into the lounge, sitting down across from Drax.

"Go ahead," Drax said quietly.

"...I hate you," CJ said, trying to remain calm.

"Tell me how you really feel, Sis."

"I hate you with every fiber of my being." CJ sat stiff, her lower lip quivering. She fought to hold the tears back, but it was becoming difficult.

"Hate is a strong word. Are you sure that's what you want to use here?"

Suddenly it didn't matter if she cried or not. CJ looked straight at her sister.

"Yes, it's what I want to use. I hate you! I hate what we've become. I hate myself for allowing this to happen." Tears flowed freely now, but she let them go, not caring.

"What could you have done differently that would have changed this outcome?"

"I could have done something."

"No you couldn't. I had this figured out as soon as he was brought on board the first time. As soon as they pumped Kodiac Blu into him; this was going to be the outcome."

"Hate isn't a strong enough word," CJ cried, burying her face in her hands.

"You'll be interested to know, I'm having San Sann synthesize more Kalinite."

"He won't be able to have it ready in time."

"Were Gunnar human, I would say you're right." Drax shrugged. "He should have terminated shortly after you two escaped the first time. Clearly he's a lot more resilient than I gave him credit for."

"It doesn't matter," CJ sniffled.

"CJ, I need him alive. His death is not in the Empire's best interest."

"Can't forget about the Empire's best interests. What about his best interests?"

"I just need a couple of things from him and he'll be free to go. I tried to tell you that from the beginning, but you wouldn't stop long enough to listen."

"I know what you have in mind, and it doesn't involve letting him go." CJ jumped up and started for the elevator.

"I assume he told you why he's here."

"You were at dinner." CJ stopped at the elevator.

"The maintenance bay is a great place for a private conversation."

"He just wants to go home."

"If that were the case, he wouldn't be running around Hadrian stirring up trouble." Drax stared at her sister, waiting.

Still angry and hurt, CJ lashed back in a calm, calculated manner.

"Yes, he told me exactly why he's here." Without another word, she turned and disappeared behind the elevator door.

I have the Power

"I hope you know what I'm doing," Tiana grunted.

"What's that?" Dakota responded forcefully.

"The designers of this bird didn't take into account the female anatomy when they were putting it together," Tiana complained. "I feel like I'm going to suffocate."

"I don't think they figured anyone would have to get down here to do this," Dakota responded. They were both squeezed in tight under a mountain of machinery in the bottom recesses of the *Constellation's* engine pit. With only a couple of portable work lights and a few personal wrist lights to see by, they struggled to access the deep internal systems of the complex Starbird machine.

"Freakin' cold down here," Tiana gasped. She pulled her hands up to her mouth and breathed into them. The environmental systems were working again, but only on a limited basis and many parts of the ship had not recovered from the loss of power.

"Could be a lot colder," Dakota grunted. His hands were cold too, but the urgency of their situation pushed him to complete the repairs they were working on. "It's amazing you've learned all this stuff in such a short amount of time."

"I'm a quick study and Billy was a great teacher."

"A Starbird is about as sophisticated as ships get. Most of this crew took a lot of time and experience qualifying to be here. I can't think of a single person who could just walk in and know what to do."

"We're out here in the middle of nowhere about to either get our butts blown off or be taken captive by aliens to have who knows what done to us and you think I was breaking down your door to volunteer for this?"

"Then you've decided to pursue a career in engineering instead of executive officer?"

Tiana stopped what she was doing and looked over at Dakota.

"In the last day or so, I've lost track of time..."

"At this point, I don't think it matters."

"I've seen what an exec has to go through and thanks, the job is yours."

"Dang it, I was hoping I could get back to fighter duty..."

Both worked for several minutes in silence. As Dakota finished what he was doing, he noticed Tiana had become still and worked himself around in her direction.

"Hey, you ok over there?"

"I'm fine," she finally responded. "I just wish Billy were here."

"Gotta put it out of your mind," Dakota said, working his way back to the access shaft.

"Easy for you to say," Tiana mumbled as she followed him. She sat up next to Dakota as he prepared to ascend the pit ladder, then noticed he wasn't moving up.

"I knew Billy a long time before he was assigned to the Starbird program," Dakota said. "I knew him before I was assigned to Colonel Conrad's Teak squadron. He was a line crew tech in charge of thruster operations. The guy had long hair back then. I was just a snot-nosed punk of a kid who thought he was the hottest thing to ever grab the controls of a Thumper. The Colonel was just a Captain back then, but he was so good at flying that after I got over myself and really watched him, I was amazed at how gifted he was. I guess my point is, I knew Billy all through my assignment with the Colonel and his advancements through ground crew status and his assignment here as Chief Engineer. He was a great friend... and I had to watch my friend die in space, with a missile in his hand. That's what being an executive officer is all about."

Tiana fingered the Chyropaz around her neck, remaining motionless as Dakota ascended the ladder to engineering. After a moment of thought, not just for Billy, but Dakota as well, she climbed out. Allowing the engineer's mate, Scott Brandon to move out of the way, she stepped up to the large glass engineering console.

"Where did Lana and Nigel get off too?" Tiana asked, looking around.

"I sent them back up to the bridge. Did we do any good?" Dakota asked, looking at the readouts.

Tiana kept her eyes glued to the screen in front of them as she poked her fingers at several spots on the glass. BachTL stepped up next to them.

"Any improvement?"

Tiana remained silent.

"Not sure yet," Dakota sighed heavily. "Even if we get partial power back, where are we going to go? We're in the belly of a beast."

"True," BachTL agreed grimly. "Even if you had firepower available to blow the doors open, doing so while this destroyer is at Quadra-light would be catastrophic for everyone."

"How so?" Tiana asked quietly, still focused on the controls in front of her.

"Let's say you get enough power back to maneuver the ship," BachTL said, looking around the engine room. "You're going to have to deal with the four tugs holding you in place. The *Dominator* doesn't have tractor technology and even if it did, well, that isn't much of a threat to anybody really. Let's say you get this thing turned around or use your rear guns to knock the doors out. Here's where things get interesting. The decompression storm that'll ensue from a hangar bay of this size is not only likely to blow you out, but anything else not tied down. Then, let's have a discussion about an uncontrolled exit from a Quadra light environment. Even at full power, I'm not sure your shields could protect you from the inherent tidal forces that are sure to develop when exiting a Quadra-light envelope you didn't produce. The energy released from that kind of chaos will tear anything apart and the stress on your ship's hull would likely be overwhelming. I haven't even mentioned what the door breach would do to the *Dominator*. This particular destroyer has no shields, so it would likely lose directional stability and tear itself apart. Now, if you did somehow survive the exit, nothing else that came out with you, would. In which case, you'd be bumping and rattling through all that junk, probably ripping this ship to pieces in the process. Have I left anything out?"

Dakota and Tiana stared blankly at BachTL for a moment, until Tiana went back to her controls.

"No," Dakota finally said. "I think you've covered everything."

"The environment of outer space is impossible," BachTL said. "The human machine wasn't designed to be out here."

"And yet, here we are," Tiana mumbled, looking bewildered.

"Yes, here we are," Dakota echoed. "What's the matter?"

"Not sure," Tiana finally responded, trying several controls. "The power we've siphoned from that torpedo has recharged our main batteries enough that we should be able to restart most of the subsystems."

"Enough to do an engine restart?" Dakota asked.

"Barely," Tiana continued preoccupied with the information in front of her. "But I'm missing some things here and have gibberish in places I need to get to. The Atheon control matrix should be right here, but the fields are blank, and the access portals have symbols I've never seen before. I've tried the normal access codes, but they're not working."

"What are Atheons?" BachTL asked, looking closer at the jumbled maze of pixels.

"The Atheons direct the channeling matrix to send energy to the propulsion and weapon systems. They're sort of a controlling regulator, directing the power demands through the power couplings to those systems; a power management system. When the weapons systems need more power than the propulsion systems, the Atheons automatically adjust and carefully regulate the load through the couplings. It's what allows us to use the weapons systems while we're doing complex or high speed maneuvers."

"Where are they supposed to be?" Dakota asked, looking closer.

"Right here," Tiana said, tapping the glass.

"Is it possible they've gotten moved to a different sub-menu?" Dakota asked. "There are a lot of portals here showing the same symbols."

"I've looked in all the menus, a couple of times. They all seem fine until I drill down into the sub-menus."

"I recognize these symbols," BachTL announced. "They're Colonian."

"Are you sure?"

"I'm somewhat of a linguist. This is definitely Colonian script."

"How did this get in here?" Dakota asked. After a moment of thought, both Tiana and Dakota looked over at BachTL.

"Hey, wait just a minute," he defended, reading their suspicious looks. "I only pressed what you told me to."

"He's got a point," Tiana agreed.

Dakota looked at BachTL for a moment, then motioned for Tiana to follow. He looked back at the Castellian as they huddled next to one of the Interceptor entry elevators.

"Remember all the trouble we were having vetting crew replacements?" Tiana asked in a hushed tone.

"BachTL didn't get vetted at all," Dakota pointed out. "But then again, he's been sitting in the cooler since we landed on Aster."

"It's possible someone else got through the vetting process," Tiana said. "And if we have a spy onboard, it's likely the replacements for the *Athena* are suspect as well."

"So what are our options?" Dakota asked, keeping an eye on BachTL. "What about a reboot of the entire system? Maybe resetting the whole system will bring the menus back?"

"I'm not too excited about taking a system down that's in this condition," Tiana responded slowly as they made their way back to the console. "It could make things worse."

"Will a reset of the system bring everything back up?" Dakota asked again.

Tiana considered for a moment, thinking of why the controls were messed up and how to get them working.

"I don't know," she finally said. "I've never had to do something like this before, and certainly not with a Starbird. I've watched Billy reset subsystems before, but nothing on this scale. If we do an engine restart and then try to reset the system, it will likely shut the engines back down. Then there's no guarantee we'll have the power to restart again."

"How long would a reset take?" Dakota asked.

"Under ideal circumstances, a couple of minutes. But these circumstances are anything but ideal," Tiana replied. "If this is some kind of a viral subroutine, the systems could come back up with everything messed up, if they come back up at all."

Dakota considered a moment, but saw no good options.

"Restart the engines," Dakota ordered. "Let's get the batteries recharged as much as we can before we try a reset. Maybe you won't even have to do that." Dakota turned to BachTL. "Any idea where they're taking us?"

"No way to be certain, but I've got a pretty good idea," BachTL answered. "Calliope."

"What's Calliope?" Dakota inquired.

"My home world. It sits in neutral space between Albia and Tintee. The Rite of Pintar is scheduled to take place there shortly and knowing Drax, she'll want to showcase this ship in front of all the factions so they can see she delivers on her promises. It'll solidify her power as top Queen Captain."

"Sounds like we've got to figure out how to break loose before we reach this Calliope," Dakota said.

"Something to consider," BachTL said. "If the Hessons can't enter in our current configuration, we should be able to just sit tight while the engines recharge the batteries. We can keep trying to clear the systems while we wait. If we end up at Calliope, they're going to need to transfer us to a docking port for processing, so for a brief time it will be just us, those four Ratronian tugs and a couple of escorts. If you're going to try a reboot, that would be the time to do it. Depending on where we end up, it could take anywhere from fifteen to twenty minutes before they have you in a docking bay and trying to pry you open with something a little more substantial."

"I'd love to be a little more operational before that happens." Dakota looked around the engine room, then back at BachTL. "Can you decode this?"

"Pretty sure I can. The real question is, can you trust me with this?"

Dakota looked at Tiana, then back at BachTL.

"We sort of don't have a choice, do we?" Dakota asked, unhappy with the decision. His attention was drawn to the front of the engine bulkhead as the Asium crystal assembly lit up and a low rumble reverberated through the frame of the ship. He looked back at Tiana who passed him a quick smile. "That's a wonderful sound," he grinned gazing at the humming engine bulk. "We gonna need to keep the straw in that torpedo?"

"For a while longer," Tiana responded, enjoying the feeling of the engine restart. "It's channeled directly into the shield system."

Tiana's fingers tingled slightly, listening to the soft hum of the Kalamarion power plant in front of her. She wondered if this was what pilots experienced when they controlled the Starbird during maneuvers. For several minutes she checked the systems that were fed by the batteries as the engine began to breathe life back into the ship. Finally, she turned back to the locked out controls and started working with BachTL.

A feeling of hope seeped into Dakota as he turned to Hayden Hunter, perched above on a lifter next to a softly whistling Mark V torpedo.

"How does it look up there, Lieutenant?"

"I'm gonna need a little help with this," the turret gunner responded, peering through a tangle of cables.

"Now?" Dakota asked, turning to the engineering console.

"Captain," Hayden spoke up. "It's not technical."

Dakota gave Hayden a double take and glanced back at the engineering console. Tiana was deeply focused, trying to figure out how to correct the current problem while BachTL looked on curiously. He finally turned and stepped on to a lift platform and floated up next to Hayden.

"I'm a fighter pilot, Lieutenant, not an engineer."

"I'm a turret gunner, Sir..." Hayden grinned broadly. "Can you maneuver around to this side?"

"Why didn't you have Mister Pippin help you with this while he was back here?" Dakota adjusted the controls on the lifter and floated around the back side of the weapon next to Hayden.

"He looked preoccupied with his assignment."

"So why do you need three hands?"

"I need expertise is what I need. I have almost no idea what I'm doing here. Would have been nice if Command had allowed more of us to know how these scary things work."

"Well, I'm not going to be much help. I went to the briefing on their operation and slept through most of it. These are a little too big to be mounted to a fighter, so I didn't see how any of it was going to apply to me. Can you tell how much power is left in this thing?"

"From what Tiana was able to figure out, this display right here shows how much energy is being drawn from the torpedo engine cell and this display indicates how much is left."

"I believe you," Dakota said, not really wanting to figure out the information in front of them; that was Hayden's job. "So how much is left?"

"A couple of hours."

"So what do you need from me?"

Hayden lowered his voice and peered through the cables down at Tiana and BachTL.

"Did you give Mister TL access to our ship's systems?"

"Yes, we had to in order to have everyone working on getting our systems back up. He was making adjustments while Tiana and I were down working on the pit regulators. What's the problem and why are we speaking like this?"

"Captain, I've been up here the whole time you and Tiana were down in the pit and I noticed Mister TL seemed to be doing more than just making the adjustments you and Tiana were calling out."

Dakota eyed BachTL closely as Scott Brandon worked to pull a large umbilical from an auxiliary access hole.

"I'm sure it's just curiosity," Dakota said, trying to shake his suspicions. "He has very limited access to specific systems."

"I'm sure you're right. I couldn't see clearly from here."

"Did anyone else touch the console while we were down in the pit?"

"Only Engineer Brandon," Hayden said. "But it was only for a moment as he was switching from coupler access one over to two."

"Thanks, Lieutenant." Dakota continued to eye BachTL through the hanging cables as his lift began to sink back to the floor.

Hayden shrugged and went back to what he was doing as Dakota stepped off the lifter and headed for the engineering doors. He watched BachTL carefully as he passed, then turned and headed to sickbay. Dakota nearly collided with Doctor Fuji Yamoto as he tried to enter the medical bay.

"Captain, I was just coming to see if you needed Alder and me for anything else."

"And now you see me," Dakota said, a focused look on his face. He glanced through the open door, seeing Alder Gantrie, the doctor's MedTech, working with something. "May I have a word with you," he motioned to the darkened hallway leading toward the bridge.

"Certainly," Fuji said, glancing back and then starting to walk with the Captain. "I felt the engines restart. Do you have main power back up?"

"We're getting there. Tiana has the engines up and the main batteries are recharging, but we're locked out of the portals controlling the power flows. It appears to have Colonian encryption locking it up."

"Colonian code in Kalamarion software? How is that even possible?"

"That's what I wanted to talk to you about," Dakota said. "Have you noticed anything out of the ordinary with Alder? Things she does or talks about that seem inconsistent with a MedTech background?"

"You're thinking sabotage?" Fuji asked in a whisper.

"How else could the encryption have gotten into the system? And why is it just now starting to manifest itself? Seems a little convenient."

Fuji considered for a moment, looking back down the hall into engineering.

"No, I haven't noticed anything, but to be fair, I haven't really been looking." Fuji tapped a finger to her chin. "Mr. Pippin had nothing but high praise for her performance on the Carolon base, and I've been very impressed with her abilities as a MedTech. I'm strongly considering making a recommendation to the Colonel that we do away with the nurse or physician's assistant and just have a MedTech; more versatility. What about the engineer's mate, Scott Brandon? He has higher access to all the ship's systems."

"I don't believe he has the mind for it," Dakota said, shaking his head. "He's about as basic as they come."

"What about BachTL? When did the encryption show up?"

"Right before Tiana restarted the engines," Dakota said, nodding grimly.

"Certainly points the finger at him, doesn't it?"

"Until we can clear the system encryption, we can't move. I've got Tiana and BachTL working on it right now."

"Isn't he your prime suspect?" Fuji asked.

"Yes, but I don't have much of a choice. He seems to be the only one onboard that understands the encryption."

"Careful, Captain," Fuji cautioned. "BachTL isn't the only newcomer on this ship."

"I like to think the Colonel was on to something when he suggested that the best place to hide something is right out in the open."

"Tiana's under an extreme amount of pressure right now. How's she handling things?"

"Well, that's the good news. Once she was able to refocus, she's been on top of everything. I have no doubt that if anyone can figure this out, she can."

"What can I do to help?" Fuji asked.

"Keep your eyes and ears open," Dakota said, turning for the bridge doors. Once inside, he noticed the outer windows were dark, but the bridge consoles were somewhat alive. Lynette Starman sat at helm control working with several of her displays while Nigel and Doran examined something at the navigator's station. Mister Pippin and Lana Nevall sat close together at the science station working with several glowing glass panels.

"Mister Pippin, a word?" Dakota and Pip stepped across the bridge to the engineering station. "How are things progressing here?"

"We're ready to make a break for it as soon as we have control power."

"Are you noticing anything odd?"

"Odd, Sir?"

"With your systems."

"No, not really," Pippin said, wondering if he was missing something.

"What about your manual backups; are they working?"

"Yeah, they'll work, if we need them." Pip looked around the room. "It won't be pretty, but we could maneuver. The hope is we don't have to use them."

"Agreed," Dakota said looking around. "When you were down on the Carolon base retrieving the Asium, did you notice anything about Alder's behavior?"

"Other than she's real handy with weapons for a MedTech, no. Why?"

"It may be nothing."

"Then why the worry?"

"The engineering portals to the Atheons are locked out. Some sort of Colonian encryption."

"Colonian encryption?" Pippin looked down at the engineering console next to them. "How can you have Colonian encryption with Kalamarion technology?"

"Yeah, that's just what Fuji said."

"No, Captain. You don't understand," Pippin said, touching the glass in several spots. "It's not possible to put the two together. The alphabets aren't even close." Pippin studied the readouts Tiana was working with.

"Nevertheless," Dakota said. "BachTL identified them as Colonian and is trying to help decipher them." Dakota checked the ship's chronometers. "Make sure everyone is ready. We could get Atheon control back without notice. If you and Lana get a chance, have a look at the encryption Tiana's trying to work around."

"Understood," Pippin replied, stepping back over to Lana at the science station.

Dakota watched him go, then turned forward, watching Lieutenant Starman working with her consoles. Hesitating for only a moment, he finally turned and left the bridge, ending up back in engineering. Instead of approaching Tiana and BachTL, he kept back, just inside the Mallory doors, against the wall. From this position he could see everything happening in the engine room. Hayden remained on his floating lifter next to the suspended torpedo and the clustered cables hanging from the warhead. Scott Brandon's head bobbed in and out of one of the smaller access pits next to the engine bulkheads. Dakota watched the assistant engineer for some time, then glancing at BachTL, made his way over to where Scott was working. The young technician was struggling with several large cable casings he had opened, their contents visible through the cooling gel inside.

"What's this mess all about?" Dakota asked, dropping to a knee.

"Power couplings, Sir," Scott grunted, tugging on the outer casings to expose more of the insides. "I have to repair the plasma heliax couplers and fusion splices in all these fiber optics."

"What's running through those couplers?"

"Plasma energy produced directly from the engines. These feed the maneuvering controls and the weapons.

"So, if they leak?"

"You don't want that to happen. Everything in the engine room will be contaminated with Plasma radiation. No one would last very long in here."

"That's a slimy mess." Dakota commented watching him work.

"Yes it is, Sir," Scott agreed, trying to keep his grip on the inner cables. "This is a cooling gel that's pumped through most of these coupling cables during operations."

"Do you have to reinject that slime back in once you're done?"

"No, Sir. The gel fills back in when the pumps are reactivated."

"Where did you learn to do all this?"

"Engineer Moon gave me the tutorial. Some of the newer Hadrian ships have crude versions of this type of coupling system. I've served on several Ratronian and Colonian ships that have them, including the *Realistic* and the *Trax*."

"How long is it going to take to get this done?" Dakota asked, looking closer.

"Not long, maybe ten to twenty minutes."

Dakota glanced up at the whistling torpedo overhead and nodded.

"Good work. Let us know if you need help with this."

"Yes, Sir," Scott smiled, feeling a little better about the compliment coming from Captain Abrams. Engineer Moon was always getting on to him about anything he tried to do.

Dakota stepped over to the turret elevator shaft and looked back at the two figures standing at the engineering console. BachTL and Tiana remained focused on the glass displays in front of them. Tiana was madly working her fingers around the glass while BachTL gave her instructions. He watched them for several more moments, then ascended the ladder to the turret and climbed in. Activating the turret's control mains, he touched several points on a small glass display to his right and looked up at the targeting window. The dark shade in the glass slowly lightened, just enough for him to see outside at the surrounding hangar bay. He touched several more points on the glass and the turret started to move. He adjusted the controls so that the movement was almost imperceptible. Sitting forward, he noticed movement outside.

*   *   *   *

A bustle of activity surrounded the glowing hull of the Kalamarion vessel as several technicians worked with mounds of equipment near its locked boarding hatches. Despite their best efforts and using every piece of equipment they had at their disposal, the Hessons were unable to gain access. As they continued to work, a small army of other technicians quickened their efforts to service the Ratronian tugs holding the Kalamarion craft in place. Several officers entered the hangar bay, flanked by Commander Dalton SoKnack. As they made their way to the Starbird, Dalton studied the ship carefully, making his way around to where the Hessons were working. As he approached, a voice boomed over the hangar bay sound system.

"Standby all hangar personnel. Payload departure in eighteen minutes."

"Anything?" Dalton inquired, stopping behind the engineers. He turned his gaze up at the mid-section of the ship.

"Not enough, Sir," one of the technicians responded. "We make a little progress and then the shield energy seems to strengthen."

"Have you seen any movement from inside?" Dalton looked toward one of the windows, but it was still dark.

"Nothing since it was brought in, Sir."

Dalton turned his gaze up at the ball turret on top of the aft section. He shifted from side to side until he noticed his aides giving him funny looks.

"Does it look like that's moving?" he asked, pointing at the large ball. Everyone looked up and stared at the turret. After several moments of long looks, the other officers shook their heads. Irritated, Dalton walked over to one of the tugs, being careful to remain clear of the tractor beams holding the Kalamarion ship, buzzing overhead. "Is *Warbird II* ready?" he asked looking around. There was nothing in this hangar bay but the Kalamarion ship and the four Ratronian tugs.

"Yes, Sir," one of the aides responded quickly. "In the Alpha bay."

"Are we cleared with Calliope approach?"

"They know we'll be following the escort corvettes in with *Warbird II*. The Queen Captain has instructed that we not follow into the executive docking bay, but remain on a close patrol outside Calliope."

Dalton rolled his eyes. He had already gotten this stipulation from the Queen Captain when he had made his report. He chocked it up to middle management reassuring him that they knew what was happening. He scanned the perimeter of the hangar bay again, stopping on the Kalamarion ship.

"Ok," he mumbled to himself. "This just about wraps things up. All we have to do is get this thing into Calliope's executive bay and I can put my papers in for retirement."

"What was that, Sir?" an aide asked.

"Nothing." Dalton turned and marched out with his entourage close behind him. "Get those Hessons out of the way so we can get this operation wrapped up. The sooner we get that thing delivered, the better."

"Yes, Sir," another aide waved toward the waiting Hessons who quickly worked to clear all their equipment.

As Dalton made his way down the hall toward the bridge, he noticed a young officer trotting toward him.

"Commander SoKnack, orders from the *Tarzana*," he puffed, handing Dalton a small device. "You are to report aboard immediately."

Dalton quickly scanned through the orders, then handed the device back to the officer.

"Have a shuttle ready in ten minutes immediately. I'm transferring command of the *Dominator* back over to Captain Knolls." Feeling a weight being lifted, Dalton moved quickly to retrieve his personal gear and get to the shuttle that would be waiting for him.

May we join you?

Sitting with his legs crossed in the middle of a dark, closed off hangar bay, Rick kept his eyes closed, meticulously listening; moving slowly about in his thoughts. He had reached a point in his Thane meditation where he could master his thoughts and awareness of the environment around him. After spending hours in the quiet of the dark, he probed the air and the elements surrounding him. Reaching deep into their molecular structures, he concentrated on understanding all their dynamics.

When he had trained with Ona Tusk some time ago, it had struck him somewhat odd that the manipulation of the molecules in a given material felt like a violent exchange between consciousness and the molecular makeup of any given atom.

Why do I have to force a molecule to move with such velocity in order to move another object? Is there a way to command at the molecular level that isn't as chaotic as what I've already experienced? Can I not orchestrate change in matter by flow instead of force? On Alvadore, every exchange Blinda and I came up with was explosive in some way, perhaps that's why there was no clear victor to the encounter? Ona taught me to ask the molecules, but have I really done that? Can gas molecules really provide lift or is there more to it than that? Perhaps, it's several things working together? Gas, gravity, environment, and my own existence are all a part of the equation.

Rick drew in a slow, steady breath and let his hands twist slowly up. Sensing his body weight vanish, the darkness of the hangar bay evaporated. The walls, floor, and ceiling transitioned to a ghostly transparent, revealing the internal structures of the great ship. Beyond, he perceived the vacuum of space around *Calypso*. Shifting his gaze, he sensed himself moving within the confines of the empty hangar bay. Wherever he wanted to go, he was there in an instant. He examined every part of the hangar enclosure several times, finally stopping at the enormous outer bay doors. The doors waivered, dissolving away to reveal the emptiness of space. *I wonder...*

Carefully unfolding his legs, he stretched his hands out. Several moments passed and slowly, the infiniteness of space opened up before him. He turned around, looking at a vast wall of clear metal.

Backing away, a broader view of *Calypso's* hull opened to his mind as the ship moved at lightspeed. Marveling at the experience, he backed away further until he could see the front and back of the massive traveling leviathan. Gliding along the side, the excitement churning within him continued to swell. Wherever he turned his gaze, the metal skin became transparent. As he floated alongside the great ship, he noticed several crew members inside the hangar bays where the Starbirds and freighters were housed. *Wouldn't it be a hoot if I just appeared out of nowhere in front of those guys?*

He spotted Caidin and Dãsha Mantose walking together between two of the Starbirds. Backing away from *Calypso* a little further, Rick watched them as they moved around the back side of one of the parked ships. As he continued to observe the activity inside, he realized he had inadvertently moved further away from the traveling giant. Turning around and facing the infinite star field, he detected a glow rushing past him. Looking back toward the thruster ports, everything began to elongate, stretching further back into an empty void of blackness that followed *Calypso* as it traveled faster than light.

Fascinated, he suddenly became aware that he was having difficulty keeping pace with the great ship. Fighting a streak of fear trying to pulse from within, he struggled to control his thoughts and bring himself back to *Calypso*. Managing to move closer, the great ship appeared to be moving past him at a faster rate. Tinges of panic continued to creep into his thoughts as he perceived the hull in front of him starting to solidify. Trying to remain calm became more difficult as the enormous thrusters of the ship appeared in his peripherals. Closing his eyes, he focused on the hull in front of him and stretched toward a narrowing transparency in the metal. A moment later, he dropped to the floor in a darkened room echoing with the rumble of machinery at work. Laying still for several moments, he wound his thoughts back, trying to comprehend what he had just experienced.

After several moments, he got to his feet and looked around. *Calypso's Starboard engine room*. As this part of the ship was typically unoccupied, there were no lights, only winking panel displays flashing information. There was enough light to give him a good idea of how large this engine was. Amazing that this was only one of two moving this great bulk.

As he moved toward the outline of the engine bulkhead, he raised a hand up and spun a finger above his head. Light instantly dissolved much of the darkness, revealing the massive stacks of machinery dwarfing him. He slowly turned completely around in awe at the size of the room and its contents. Completing his scan of this engine room, he noticed a set of open doors different than the others dotting the walls. Several small displays inside drew his attention. Bringing his hands up, he floated through the doors, just because he could. The room lit up as he entered, revealing a curious pedestal in the middle of the room with a console directly beside it. Examining it closely, he was unable to determine what it was. The characters displayed on its tiny readouts were foreign to him. The information scrolling across the nearby displays were equally befuddling. Even using rudimentary deduction provided no clues as to its purpose. There were no real input or output conduits or cabling to indicate how it would transfer its workings to anything external. He noticed several horizontal antenna arrays evenly spaced around the device. Even these arrays seem to have no purpose as there was nothing attached to them. After several minutes of trying to discover its internal workings, he checked the time, deciding he had better start back to the hangar bays.

It took Rick some time to make his way out of the engine room and to a set of elevators. After some searching, he found a pneumatic transport pod where he recognized the hangar bay locations on the control display. Even with the speed of the pneumatics, the trip back to the familiar parts of the ship felt longer than it should. Finally arriving back, he found Jayda and Dãsha standing at one of the hatchways to the empty hangar bay he had been meditating in.

"How did you get out?" Dãsha asked, looking down at the door control. As per Rick's instructions, it was still locked.

"This isn't the only door," Rick answered, taking Jayda's hand.

"I thought you said you'd be in there longer?" Jayda commented, wondering.

"Longer than I should have been," Rick said, taking a deep breath and turning down the corridor for the main hangar bay.

"Thane meditation is a deeply personal ritual," Dãsha said, still looking at the control. "Caidin has spent days at times. How did you get out?" Dãsha finally turned and followed them. "None of the other personnel entrances are operational."

"It took some time to figure it out," Rick said. "How long before we cut back to sub-light?"

"Mister Dacey has just informed me it will be about five minutes," Dãsha replied. "He and Lieutenant Gillespie are quick studies with *Calypso's* controls. You are very perceptive of their abilities to learn to pilot this ship."

"It's a gift," Rick said heading for the *Athena*. "Let's get these birds ready to launch, we may need them."

*   *   *   *

The Nulark system was finally still again; almost. A buzz of activity remained in orbit around Carolon as a few light cruisers patrolled the surrounding areas and a swarm of rescue ships searched a field of broken vessels for survivors. Just a short time ago, a large cluster of Albion ships had been engaged, but now most had departed the area, leaving only the cleanup forces. The tranquility of the cosmos had returned and all was as it should be...

A flash of light and then an enormous bulk appeared within proximity of Carolon. From an age past, the leviathan was a giant among giants. *Calypso* glided silently toward the orange-brown planet of Carolon. As an orbit was established, one of her large side bay doors opened up and two smaller white ships exited, flying separate patterns around the mammoth ship. The Kalamarion Starbirds were dwarfed by the size of the older, larger ship. As the *Athena* moved closer to Carolon, the other Starbird remained in close quarters to *Calypso*.

Rick Niker carefully studied the readouts over Toby's head, listening to the science officer calling out information.

"It's a wrecking yard out there, Sir," Toby said, looking at his scopes.

"How many?"

"How many what? There's wreckage all over the place. Not just on this side, but all along the horizontal axis."

"Any of it ours? Any transponders?" Rick asked.

"Not so far. All that junk looks Albion. There are a few corvettes and light cruisers. A whole mess of what looks like rescue craft. What happened here?"

"Give you three guesses and the first two don't count." Rick turned forward. "Captain Dayton, take us in a little closer, but keep your distance from those corvettes. Chances are, they've got a bloody nose and are a bit on edge. Lieutenant Habba, get me Lieutenant Commander Niker," Rick ordered.

As Captain Lisa Dayton maneuvered the *Athena* into a closer orbit, clusters of wrecked ships appeared ahead of them, stretching out along a large debris field. Several corvette class ships paralleled their flight path, giving everyone cause for nervousness, but to their relief, none of the ships seemed too interested in them and turned in the other direction after a short time.

"If only all engagements were like this," the Starbird pilot commented, glancing at the empty targeting scopes at the weapon's station. "We'd be out of a job."

"Lieutenant Commander Niker, Sir," Laura announced quietly.

Rick turned his attention back to the scopes over Toby's head as he activated the comlink on his armrest.

"How does a brand new Starbird feel?" Rick asked, glancing out the forward windows.

"It's like trying on a new jacket," Jayda responded over the hidden chair speaker. "It looks good in a hangar, but it feels so much better when you're actually in it."

"You could say that about any Starbird," Rick responded, smiling. "How's your crew taking to it?"

"Steep learning curve for some, old hat for others. I think we've got the right people in the right places. I'm sure she'll operate even better when she has Asium loaded in her chamber. Maneuvering thrusters can only give you so much.

"That's our bluff for now. Those four ships will move and look operational and for now, that should be enough."

"How's the *Athena*?"

"Feeling like she just came out of space dock. Nice to have her back under us. There's a lot of junk out here."

"I see that," Jayda responded. "Clearly a pretty intense fight."

"The good news is we're not picking up any transponders or seeing any Starbird debris," Rick said looking back over at the scopes.

"Those Albion ships don't seem to care much about us either," Jayda observed. "What's up with that?"

"I'm thinking the *Constellation* gave them a pretty good spanking." Rick thought for a moment, then abruptly let a chuckle go.

"What's so funny?" Jayda asked.

"Gunnar really kicked the crap out of them."

"I almost wished he were here somewhere floating around disabled. That would answer a lot of questions."

"How's your pilot taking to her temporary duties?" Rick asked as the *Athena* emerged from her orbit around the back side of Carolon.

"Audra is running the other pilots through some practice maneuvers. Docking approaches and close quarter piloting with other vessels. She's using the back side of *Calypso* as both a shield and a surface for reference. She seems to be doing pretty good."

"Good," Rick said, glimpsing the massive shape of *Calypso* ahead of them. "Stick to the plan. We'll mill around out here for a little while longer and see if any of these Albion ships have the stomach to at least inquire who we are or what our intentions are." Toby suddenly turned and pointed at the glass in front of him. "Hang on," Rick said, noticing several targets appear out of nowhere. "Targets, bearing one thirty five point three; on your port side. Looks big."

"We see them," Jayda responded. "Big cluster of ships. Want us back inside?"

"No, but stick close to *Calypso* and have Caidin ready to jump to lightspeed just in case." Rick stepped over to the science station for a closer look. "Captain Dayton, get us back over to *Calypso*. Toby, what have we got?"

"Still more ships dropping out of light speed on the other side of *Calypso*. This looks like an organized fleet. A couple of battleships, cruisers, frigates, carriers and support ships."

"Whose are they?"

"I can tell you they aren't Colonian, Albion or Ratronian. These ships are too pretty."

"Too pretty?" Rick repeated.

"These look like they just came out of a ship builder's yard. Looking for their insignias... here's a match. They're from Cross."

"Cross," Rick mumbled looking closer. "Administrator Vandmire? Lieutenant Habba..." he motioned.

"What would he be doing here?" Jayda asked over the comlink.

"He's probably asking the same thing about us," Rick postulated, watching the last of the Cross ships drop out of light speed.

"General Niker," a voice boomed over the bridge com systems. "I'm relieved and surprised to find you here at Carolon with the Albions milling about."

"Good to hear from you, Administrator," Rick smiled. "I don't think the Albions have much fight left in them. If you look a little closer, it's a wrecking yard here in orbit. My friend threw quite a party."

"That's a big ship," Diord commented.

"Very big," Rick replied. "*Calypso*. Picked her up in the Spartus. Love to tell you all about it."

"General, when and where can we meet?" Diord's voice was suddenly quite serious.

"Have you got room in a landing bay?" Rick replied readily.

"I'm in the *Merc*. The big fancy one. Sending landing instructions now. Will you be the only one joining us?"

"After we're secured from landing, I'll have a few of my officers brought over."

"We look forward to greeting you."

Rick sat back down as Captain Dayton steered the *Athena* toward one of the larger vessels, following the landing instructions appearing on the glass displays in front of her.

"Jayda, I want you and Audra to teleport over as soon as I get the *Athena* settled in their hangar bay."

"Are you sure it's a good idea to let them get so close to it?"

"Diord has already seen the *Constellation* up close and personal. Take your ship back to the hangar bay and close the doors. They won't be able to scan the interior of *Calypso*. But have Constance and Manx standing by in their ships, then call me when you're ready to come over." Rick got up and stepped forward to watch their entry into Diord's battlecruiser. It was as big as the *Tarzana* and appeared to be similar in design, but with a different paint scheme; far brighter in color. While this battlecruiser was still quite formidable, it wasn't as ominous to look at.

As the *Athena* entered the brightly lit landing bay of the *Merc*, Rick marveled at how well the interior was kept; shiny and bright. He recognized several fighters and transport vehicles parked neatly in several rows on the main floor. Many others were stacked in their own compartments lining the walls of the hangar bay. As the *Athena* touched down, CORA 500 came alive.

"Richard, I am picking up Kalamarion telemetry."

"What?"

"It's Alex 7001. He is somewhere onboard this ship."

Rick looked over at Toby who shrugged. Laura had a puzzled look as well as she checked her channels. Surprised, Rick pushed the feeling quickly aside. He looked back out at the surrounding hangar bay, observing a figure in a blue cloak flanked by several others making their way toward the ship.

"If Alex is here, then Colonel Conrad has got to be here as well," Rick affirmed.

There came an air of optimism to the bridge as the expectation that their search was finally drawing to a close. Before everyone got their hopes up too much, Rick came to his feet.

"Everyone remains at their posts until I get back," Rick ordered, heading for the door. It took him only a couple of minutes before he was standing in front of Administrator Vandmire, who greeted him warmly.

"I am so glad to see you," Diord said, putting his hand to Rick's shoulder. "I was certain you were dead."

"Why would you have been certain of that?" Rick asked as they began to walk toward one of the open interior doors. He scanned the bay carefully, looking for any sign of the *Constellation* or his friend.

"Lone astronauts don't last long in the Spartus quadrant."

"Can't argue with that."

"Clearly you were successful though," Diord said looking back at the shiny white hull of the *Athena*.

"It was... an adventure," Rick said, smiling. "So what brings you to Carolon?"

"Probably the same thing that's brought you here. We're looking for the *Constellation*."

Rick stopped in his tracks and turned to Diord.

"We've only been here for a little while and haven't been able to find any trace of her; only what's left of an Albion incursion."

"But there is a second Starbird," Diord said. "We assumed it was the *Constellation*."

"No, that's a different ship," Rick answered quickly.

"The *Realistic* reported an Albion ambush over the planet when the *Constellation* arrived. They must have bugged out at some point."

"Evidently Gunnar really kicked the crap out of them," Rick said, flashing to what the fight must have been like.

Diord's expression suddenly shifted.

"What?" Rick asked.

Diord looked away, then motioned for Rick to follow.

"What?" Rick insisted. "Diord, we picked up Kalamarion telemetry from Colonel Conrad's AI droid when we landed. He's onboard this ship?"

"Forgive me, General. Yes, Alex 7001 is here. Come on, I'll take you to him."

"What about Gunnar? Where is he?"

"Please be patient, General." Diord quickened his pace toward a small elevator that whisked them high up into the command decks.

"Either he's here or he isn't," Rick bristled. "Which is it?"

Exiting the elevator, they were joined by several guards in front and behind them. Diord turned several corners without slowing, giving Rick cause to become more impatient. Finally stopping at a set of guarded doors, Diord pressed a device to the security pad and stepped inside, flanked by Rick.

"General," Alex 7001 exclaimed, becoming airborne. Diord dropped immediately to a knee and bowed to a young woman seated on the far side of the room. Rick looked at her curiously as an older woman took a position beside her.

Is that...? "Janox?" he whispered.

Jana Oxlind sat poised in the chair, a pleasant look forming. She was dressed in a radiant outfit, far from what he remembered of her on Aster.

"You are required to bow in the presence of the Princess," Tonnie instructed, resolutely.

Rick looked around as Alex 7001 took a position next to him. He bowed as Diord rose to his feet and, giving the administrator a glance, looked back at Janox.

"Janox? What's this all about? Where's Gunnar?"

"You will address her majesty by her proper name," Tonnie demanded, stepping between them. "Princess Jana Oxlind."

Rick blinked.

"Jana? I don't understand. Who is this woman?" Rick asked.

"I am Tonnie, the Princess's Au Pair." Tonnie looked at Diord. "Administrator, why do you bring this man before the Princess?"

"Forgive the intrusion," Diord apologized. "This is General Richard Niker, Gunnar's friend and protector."

Jana came to her feet, but the royal nanny remained between Rick and the Princess.

"Princess Jana, where's Gunnar?" Rick asked again, trying to see around Tonnie.

"It's all right, Tonnie. General, please have a seat. I'm sure you have a lot to tell us of your adventures to find your ship and crew."

"That's an understatement," Rick said. "Do you mind if I have a couple of my officers teleported here?" he asked, touching a control on his wrist communicator and taking a seat.

"That would be acceptable," Jana nodded. "You'll have to forgive my Au Pair. She's a little protective."

"It's nice to see she's real," Rick said.

"Quite real, General," Tonnie snorted.

"Tonnie," Jana scolded. "General…"

"Rick, please call me Rick," he said as two figures materialized in the room directly behind him. Tonnie backed up closer to the princess. A firm hand from Jana brought the android down to her side. Rick smiled and stepped over to Jayda.

"Princess Jana, Tonnie, this is Jayda Niker, my wife and first officer."

"You are the reason Rick left Aster," Jana said, rising and taking Jayda's hand.

"Well, supposedly it was to find his ship, but I like the *just me* idea." Jayda smiled as Rick helped her find a seat.

"Commander Atlanta," Alex 7001 exclaimed, springing back into the air. The little white droid circled Audra with scanning antennas extended.

"Audra," Jana whispered, astonished. The young princess gave her nanny a reassuring gesture as she stepped toward Audra.

"Yes?" Audra looked tentatively around the room and smiled as Alex 7001 continued to circle them. "Do I know you?"

Jana raised her hand to Audra's cheek as she moved closer.

"Gunnar told me so much about you," she breathed, looking deep into Audra's brown eyes. "You are what has been missing from him for so long."

"Commander," Alex said, softening his voice. "I am at a loss for an explanation."

"That would be a first," Audra smiled. "Do you know where Gunnar is?" Audra asked, feeling a little uncomfortable.

Jana stared at Audra with a sad, distant look, then slowly shook her head.

"He is not here," she answered quietly, turning back to her seat. "General Niker..."

"Please, call me Rick."

"Rick... No doubt you have come to Carolon to find Gunnar and the *Constellation*. It is my understanding that Alex 7001 has most of the information of our adventures stored in his memory banks, as does Tonnie," Jana said, gesturing to the Au Pair. "If you will be patient, we will try to explain what has happened to your friend in your absence." Jana gave an anxious Audra a quick glance. "I stowed away on Gunnar's ship when he came to Carolon in search of Asium and forced him and Alex to take me back to Reako to find Tonnie."

"I'm not sure I'd call it forced," Alex 7001 interjected. "Although you did stun him a couple of times."

"You stunned Gunnar?" Rick burst out gleefully. "I would have paid good credits to see that."

"It was wrong of me to do so," Jana responded, embarrassed. She looked back at Audra. "He offered to take me after they had finished their mission here, but I did not want to wait."

"Wait a second," Rick said, holding up a hand. He paused, developing an odd look. "Tonnie has memory banks?" Rick looked over at the Au Pair. She raised an eyebrow.

"She's an android," Diord said, leaning closer to Rick.

Rick turned back to Tonnie.

"May I?" he asked, taking her hand.

Tonnie nodded silently, allowing the Thane a closer examination. He touched her skin and hair, then looked into her eyes, gave her face a quick scan.

"May I look into your access portal?"

Tonnie pulled her hair back and lowered her head, exposing the back of her neck.

"At the base of my skull. You should feel a tiny bump."

Rick carefully probed just below her hair line until he felt it and carefully applied pressure. A small panel gently opened, revealing her internal workings. Tiny colored lights danced about in a mixture of circuits and mechanics.

"Amazing," he breathed, closing the access. "Nearly perfect."

"Perfect enough that my cohort, Lou Aurora, wants to marry her," Diord smirked.

"It wouldn't be the first time I have been confused for a human female," the android admitted, furling her hair back into place.

"Who was your manufacturer?" Rick asked.

"I don't know," Tonnie revealed quietly. "There are parts of my memory and programming that I am unable to access. I only know my history with the Princess."

"Can we let the Princess continue please?" Audra complained impatiently.

"Yes, of course," Jana agreed. "My apologies. Gunnar and Alex 7001 helped me find Tonnie in Reako at the bottom of our well."

"How did you...?" Rick started, but Tonnie continued with the explanation.

"When Jana was but a child, we were imprisoned in the Reako biosphere by agents of the Albion Secret Service. Sometime later, I received a command to kill the Princess by drowning. As this was in direct conflict with my primary directive to protect the Princess, I threw myself into the well where my externals were rendered inoperative and unable to activate the flooding mechanisms. At great risk to themselves, Gunnar and Alex 7001 were able to pull me back to the surface. It's unclear whether my directives became active as a consequence of being pulled out of the well or if something else triggered the flooding mechanism of the biosphere."

"When we returned to the surface," Jana added, "we hid in Reako's asteroid belt while Alex 7001 reactivated Tonnie's systems and helped her recover."

"How long were you down there?" Rick asked.

"The Princess was just a young girl when I received the directive to flood the biosphere," Tonnie said.

Rick sucked in a deep breath and nodded.

"Once I was able to function somewhat," Tonnie continued, "I revealed the Princess's true identity to her and it was decided that if she wanted to take her rightful place on the throne of Hadrian, we had to retrieve her genealogy and royal tokens from their hiding place in a repository on the Albion moon, Dither."

"I don't like the sound of this," Rick mumbled.

"It's got Gunnar written all over it," Jayda whispered back.

"We were able to land and retrieve everything Tonnie had stored in the repository," Jana continued, her voice starting to crack. "But somehow, we were detected and detained before we could get back to the Interceptor." Jana opened her mouth to finish the account, but nothing came out. Jayda increased her grip on Rick as the realization of Jana's emotional state became apparent. The young Princess looked around, embarrassed, her gaze settling on Audra.

"Alex 7001 brought the Interceptor in close as a skirmish developed," Tonnie continued, detecting the Princess's emotional state. "I did my best to shield the Princess from small arms fire while Gunnar ran interference as we made a break for it. He almost made it to the ship, but was injured by a grenade. I helped him the rest of the way, but both of us were taking heavy fire. He ordered Alex 7001 to take off while he tried to get aboard. Gunnar pushed me in and I tried to help him inside, but we were already airborne. I held onto him for as long as he would allow. At the altitude we were at when he let go, it was not possible for him to have survived the fall."

Tonnie glanced around the room, looking back at Rick and Jayda. Rick's head was down and Jayda's eyes were sparkling with moisture. Tonnie looked at Audra sitting on the other side of the room, staring blankly at Jana's pooling tears.

"So what are your intentions now?" Rick finally asked, looking at Tonnie.

"We've been working with an operative in the Albion High Command," Diord said. "We're going to take the Princess to the Rite of Pintar on Calliope. If we can present her before the Hadrian council of factions, they will have no choice but to accept her royal status and place her on the throne."

"Who's your operative and how are they going to help with this?" Rick asked, skeptical.

"CJ Barker," Diord answered.

"Colonel Barker?" Rick burst. "Isn't she in command of the Aster base? How is she an Albion operative?"

"That's a bit of a tangle as well," Diord responded.

"I'm sure we're all going to love the explanation," Rick grumbled angrily.

Jayda squeezed Rick's arm, feeling his frustration with the unfolding news.

"Apparently, when Gunnar was taken from Alvadore by Blinda Koss and Seelix Monroe, he was injected with a derivative of a drug called, Kodiac Blu. Its original design, among other things, was to make prisoners manageable for processing. Improperly counteracted, the drug will kill it's subject within a short time. The dose Gunnar was given would have killed a normal human almost immediately."

"I flew wing with him during that escape," Rick flared, still visibly shaken by the blossoming revelations. "He was totally in the zone! He was fine during the buildup on Aster too."

"It didn't last," Jana interjected quietly. "Shortly after you left to find Jayda and the *Athena*, he started to deteriorate. It was speculated that something had happened on the *Tarzana* that no one knew about. Gunnar's doctor could not identify it."

"I delivered a private message to CJ on Aster," Diord began again, "from the Albion Queen Captain, Drax Blair. We decoded it shortly after CJ left her command. Drax has blackmailed CJ with the Kodiac Blu antidote. She was required to return to service with Drax in exchange for the antidote." Diord shook his head grimly, pulling something from under his cloak and fingering it. "Once she was settled onboard the *Tarzana*, CJ contacted me and delivered this antidote to me at a neutral location just outside the Jurass system." He held up a hypo filled with a dark blue fluid. "She insisted that Gunnar get it immediately when he returned from the Carolon mission. Sometime later, the Princess showed up in the Interceptor with Tonnie and Alex."

"I have made the request of Administrator Vandmire for his services to repatriate my royal blood line to the throne of Hadrian," Jana announced to the room. "He and his friend, General Lou Aura of the Tomplie have pledged their fleets along with the Aster force to getting Tonnie and I into the Rite of Pintar on Calliope."

The room was silent for a long moment, everyone looking around the room wondering what was next. Rick stared at the floor, shaking his head slightly.

"General?" Jana coaxed.

"Janox... Jana," he corrected himself. "Princess Jana," he restarted. "When I left Aster you were a rock climbing wild girl that could barely put a sentence together. What makes you think you can rule an entire galaxy?" Rick came to his feet, visibly agitated. "You want us to believe a socially awkward orphan girl, who's been imprisoned in a subterranean biosphere by herself for most of her life, is supposed to rule a galaxy on the brink of civil war? Maybe you were able to talk Gunnar into taking you where you wanted to go, but I suspect he was more under the influence of this Kodiak Blu than willing to stick his neck out and get himself killed on account of the off chance..." Rick fumbled with the words. "No," his voice elevating. "That's not why I'm here... why we're here! I've gone through hell and back to bring my wife and what's left of my crew home. Now, with Gunnar gone, I have no reason to stay here other than to find the *Constellation* and bring it home, if it's still in one piece."

Jayda gently pulled at Rick, grasping his hand tightly. Another awkward silence prevailed until Alex 7001 floated out into the middle of the room. He turned his head assembly completely around, stopping at Audra.

"While the Colonel was often illogical in his behavior, I believed his reasoning in this matter to be sound. He and Tonnie went to great pains to explain all the options confronting the Princess while we were still in orbit over Reako. If I may quote the Colonel. To the Princess, *You don't have to accept any of this. We can take you to Cross and you can live your life there in the forests and jungles of Alvadore.* Tonnie agreed with the Colonel, saying to the Princess, *You are under no obligation to pursue any of this. There will always be rulers of Hadrian, good and bad.* But the Colonel thought it important enough to risk and ultimately lose his life in the attempt to retrieve the royal documents and tokens. His last orders were for me to leave Dither without him and get the Princess to safety."

Everyone waited for the tiny droid to deliver more of a lecture. Instead, Alex floated across the circle and out into the hall.

Jana studied Rick's demeanor for a moment.

"I don't pretend to know what it is to be a royal or how to govern a galaxy, but I feel like it is my duty to try. Not to punish or enact vengeance on anyone for what happened to my family, but I believe my parents would have wanted this of me."

"Rick?" a shaken voice spoke up from across the room. Audra's eyes were pooling as she fought to push her words out. "You said

Gunnar had an important role to play in repairing the rift to this fragmented galaxy…"

"Gunnar's gone," Rick said emotionally. "I've failed as a protector and as a Thane."

"As I understand it, a Thane isn't just a protector of persons or things, but of an ideal they believe in. If Gunnar gave his life trying to do what he believed to be right, then don't you think it's something that's worthy of a Thane? Especially his best friend?"

Audra abruptly got up and stepped quietly out into the hall, a hand to her mouth. Jayda followed as the room fell silent.

"General Niker is quite right," Diord finally said. "This is not their fight. They didn't ask to be here. Only circumstance has brought them here. You and your people have been through so much already and we have no right to expect you to sacrifice any more. I, for one, would like to offer our services to help you find the *Constellation*. Then once all this business with the Rite of Pintar is over with, whether we succeed or fail, we'll try to help you get back home."

Diord came to his feet and bowed to the Princess. As the occupants dispersed, only Rick and Jana were left in the room. Tonnie lingered in one of the doorways leading to the Princess's private chambers.

"Rick," Jana began, emotion still heavy in her voice.

"Please, don't," he responded quietly.

"I know he was your best friend."

"He was so much more than my best friend…"

"He was my best friend too," Jana defended. "His only thoughts were for my welfare and safety. He understood who I was and helped me to see it. I couldn't help but love him."

"He certainly had a way of growing on you." The two sat silent for a moment longer before Rick finally came to his feet. He looked at the opposite wall and let a heavy sigh go.

"What will you say to Audra?" Jana asked.

"There isn't much I can say." He looked at the Princess as she came to her feet. "I think now I have a better sense for what Gunnar has been going through."

"I am so sorry," Jana sniffled, burying her wet cheeks in his shoulder.

Rick remained silent, holding the young Princess as she wept. He glanced at the main doorway, noticing the shadows of Jayda and Alex 7001 pass by.

"I need to find the *Constellation*, then figure out a way to get home."

"Best way to do that is to come with us," Diord said reentering the room, his breathing elevated.

"Pretty sure I've made my intentions clear..." Rick responded.

"I just received a transmission from CJ."

"How?" Rick looked disbelieving as Jana pulled away to listen. "You said she's onboard the *Tarzana*. Wouldn't Drax be monitoring her?"

"Probably, but she knows how to get around her sister."

"Sister?" Rick and Jana exclaimed in unison.

Diord froze, realizing he had let something out.

"Oops... Uhm... Yeah, I probably shouldn't have said that," Diord said regretting the flub. "What I mean is the *Constellation* was here and they did get into a huge firefight with a lot of Albion casualties."

"Clearly," Rick smiled, his attention diverted for the moment. "Captain Abrams has one heck of a report to give."

"But the Albions, with the help of the Ratronians did manage to disable it and get it onboard one of their heavy destroyers, the *Dominator*. It's being taken to Calliope. Drax wants to showcase it in front of all the factions at the Rite of Pintar."

Rick's expression shifted, giving Diord a stern look.

"We're coming with you," Rick demanded. "Can you guarantee *Calypso's* safety?"

"That old tub?" Diord asked, perplexed. "Why not just leave it here? The Albions have nothing that's going to bother it. I've got room here for your Starbirds."

"There are four other Starbirds," Rick replied.

"There's four?"

"Six if you count the *Constellation* and the *Athena*. No, I won't leave *Calypso* here by herself. Besides, your Hadrian ships can't seem to scan her interior. We can hide in plain sight. Plus, I don't want to risk someone getting a hold of any more tech."

"What, some kind of prime directive or something?" Diord asked.

"Or something," Rick agreed. "What about the Aster force and your friend Lou?"

"Lou went to Aster to retrieve CJ's command personnel and ships. He was supposed to meet me here, but was delayed to investigate a distress beacon. He said he would go directly to Calliope once he had completed any rescue operations."

Rick considered for a moment as Jayda stepped back into the room.

"I appreciate you helping us with this," Rick said, gripping Diord's shoulder.

"I was going that way anyway." Diord handed Rick the Kalinite hypo. "Here... It doesn't do you much good now, but you might as well have it, for... I don't know." Diord turned to Jana, "Your highness..." he said, bowing and leaving the room.

Rick held the hypo up, looking at the curious blue liquid.

"I heard you talking about the *Constellation*," Jayda commented, her demeanor somewhat reserved.

"It's been captured and taken to Calliope," Rick responded, quickly pushing the hypo into his shirt pocket. "Princess..." Rick nodded and ushered Jayda back into the hall. "I want you guys to teleport back to your ships, but remain in *Calypso's* hangar bay. Tell Caidin we have a new heading. I'll make sure Diord sends over the coordinates for a light speed jump."

"What about you?" she asked.

"The *Athena* is going to jump right along with everyone else."

"But you haven't tested her new engines yet."

"She'll work. Have Dãsha keep *Calypso* behind Diord's ships." Rick noticed someone step silently to a window at the end of the hall. "We leave in thirty minutes."

"That's not much time," Jayda said looking over her shoulder.

Rick drew in a deep breath, watching Audra stare out at the blackness of space. Jayda kissed him and started in the opposite direction. Rick watched her go, then stepped up behind Audra. He had no idea what to say. The cruel irony of their circumstance stabbing at him, he wished his Thane powers could do something to help.

"We talked about retiring someday," Audra began, fighting to get the words out. "He wanted to build a house... from materials provided by the land." Audra tried to chuckle through her swelling emotions. "Materials from the land? Is that even... possible?"

"Where do you think our metals and composites come from?" Rick replied quietly, trying to hold onto his composure as well.

"He loved the idea of some place remote, at the edge of a wooded area with rolling hills of long grass and big trees. I loved it because it got him out of the cockpit. He wasn't interested in becoming some grand facilitator for change." Audra's anguish suddenly burst. "He only wanted... to do what he loved... and what he was good at." She abruptly turned, pounding her clinched fists against Rick chest. He put his arms around her and held her close as she sobbed quietly.

"If only I could have been there," Rick said, trying to find words that would help.

For several stretching minutes, he held her as her grief flowed freely. Rick felt tears flowing from his eyes as well and for a moment, he felt mortally wounded by the loss of his best friend. Finally crying herself out, Audra looked back at Rick's pained expression.

"You're a... good friend," she sniffled, barely audible.

"Maybe we can make arrangements... to have his body located and returned."

"I'd like that, if we could manage it. I think it's the right thing to do."

Rick smiled slightly, trying to push his feelings back and focus on what had to be done.

"Ok, Commander," he said holding her back, looking into her eyes. "We need to go find the *Constellation*."

Audra wiped away her tears, gathering her composure. She finally took a step back and brought her communicator up, but hesitated.

"That reminds me, these ships haven't been christened."

"I'm not much for ceremony," Rick answered. "You give them good names and make them great ships." Rick gave her a wink and a quick salute, then turned for a nearby elevator as Audra activated her communicator.

Thanks for the lift

Logan cracked an eyelid. The ceiling he was looking at was a stark white with diffusion lights illuminating the entire surface and down the walls. He slowly turned his head and stared at an oddly colored wall. There were bright vertical strips in a regular pattern lining an otherwise unremarkable surface. After several moments of trying to process, he realized he was hearing a buzzing sound similar to what he had experienced outside. *Outside… I'm not outside anymore!!*

"Now you have an idea of what I went through when you pulled me inside the ship."

It was a familiar voice, but he could neither see who it was nor put a face to it. Logan tried to sit up, but found himself too stiff and weak.

"Take it easy there, Turbo. It'll take a while for that crap to wear off."

Logan was finally able to follow the sound of the voice and turned his head far enough to see a figure sitting on the other side of another set of bright vertical lines. He tried shaking his head clear, but it hurt too much and he had to lay back.

"Where am I?" Logan grunted.

"A guest of some crazy hussy that I've only seen twice."

"And what does this… Did you call her a hussy? Billy, is that you?"

"Hey, you are all there," the Starbird engineer chuckled.

Logan forced himself to sit up. He swayed a couple of times, wondering if he wasn't better off still on his back. Blinking his eyes again, he looked over at Billy sitting against a far wall eating something. Getting to his feet, Logan staggered toward the engineer.

"Hey, best to stay away from…"

Logan suddenly found himself lying on his back, stars boiling through his vision and his head feeling like it was going to explode.

"You all right there, Captain?"

"What the… Holy buckets, what hit me?"

"You walked right into it."

"Into what?"

"Energy bars. Hussy lady's version of force fields or like in the olden days, metal bars. We appear to be in a holding cell. Probably want to stay as far away from those beams as you can."

"Yeah, I'd say they're pretty effective." Logan got back to his feet and stepped up to the buzzing beams separating his cell from Billy's. "You ok?"

"Yeah, I'm fine. It takes a while for the effects of that crap she shot you up with to wear off. Then someone or thing will be in with something to eat and drink."

"...or thing?"

Billy pointed at a large droid sitting motionless next to a set of double doors.

"That brought you food?" Logan gestured toward the droid. "Looks more like a battle droid."

"I don't know if it brought the food in or not, but the food was here when I woke up and it's the only thing I've seen so far. You've got some too, if you're hungry." Billy pointed at a container on the floor just inside the force field.

"So what happened?" Logan asked as he picked up the container and started eating.

"I was deep in the Ion exciter manifolds when out of nowhere, this ship appears."

"Where is the ship anyway?"

"If you don't have it, I have no idea," Billy continued quickly. "So, I'm thinking we're saved or someone here on the planet saw us come down and has stopped to help us out. Wrong! This blonde chick comes out with a couple of these battle droids and about a dozen guns pointed at me. I had no choice but to do what she says. You took the only pistol; not that that would have helped." Billy pointed at the armaments on the droid. "She marches me up into her ship and then pops me in the back of the neck with some kind of tranquilizer dart. Hurt like the dickens. Next thing I know, I'm waking up here. Seems like it's been about a day or so."

"Sounds about right," Logan agreed. "I saw your mystery ship disappear out here in the middle of nowhere."

"Where's the middle of nowhere?"

"That wide valley we saw out in the distance. You were looking at it with the binocs?"

"So if you don't have the ship, then how did you get all the way out here? Did hussy girl come get you?"

"I hiked it."

"You did what?"

"Seems like a long way to walk just to get captured."

"So you really have no idea where the ship is?"

"I assume your hussy lady has it."

"Notice our com bands are missing?" Billy pointed out as a set of double doors popped open.

"Doe-mon tee dayo," a voice said as a medium built woman with short blonde hair stepped into the room holding up two wristbands. "Wanbee tou kydic ma su pikey dulonee." She was flanked by two more battle droids. The mechanized machines were formidably armed similarly to the idle droid already in the room. She tossed the bands to Billy and Logan, who caught them and quickly snapped them back on. "Yes, I have it, and I don't fancy myself as a hussy; provided it means the same thing to you as it does to me." The woman stepped up to the force field and looked through at Logan. "So, where did you come from?"

"I would have been happy to discuss it with you if you hadn't shot me up with whatever that stuff was you blasted into me," Logan said rubbing his chest.

"I don't get many visitors out here in the middle of the Placid. You caught me off guard by jumping my fence and snooping around. What are you doing here on Moyle, what's your business?"

"No business," Logan answered. "We were having some problems with our ship and needed to set down for repairs."

"That's what I was doing when you came and shot me in the back of the neck," Billy cut in, sitting up.

"The toxin takes effect faster there."

The woman looked at the two men behind the force fields for several moments, then pulled out a device and touched a control. The vertical beams instantly disappeared as the woman turned for the door.

"If you'll follow my droids, they'll keep you from getting lost." She started off with one of the droids directly behind her while the other two waited for Billy and Logan. They looked at each other and followed, making their way up a brightly lit hallway and sweeping ramp that curved into a massive, lavishly decorated room.

"I could have just shot you both," the woman said, stepping over to a decanter and pouring herself a drink of orange liquid.

"You did shoot us both," Billy said, stopping next to a large, beautiful couch.

The woman lifted her eyes through her glass as she took a drink.

"I'll rephrase; I could have killed you both."

"That sounds less to my liking," Billy responded.

"Please," she directed as her droids took up positions around the room. "Make yourself comfortable. Can I pour you a drink?"

Billy wasn't interested, but Logan fancied exotic drinks. He hadn't had anything since they had left Kalamar two years ago. He didn't drink a lot, but he did like the stuff and being without for that long gave him reason to volunteer for a glass. The woman poured him a little and handed the Captain the glass, then sat down in one of the big chairs.

"You really shouldn't be drinking this. You're still a bit dehydrated."

"I'm hoping you have more to offer," Logan said, taking a sip.

"I don't operate on benevolence," she countered. "I want to know where you came from and why you're here." She leaned forward a bit. "And where did you get that ship?"

"I thought we already covered why we're here," Logan said, carefully trying to dodge the other questions. "Wherever here is?"

"The planet Moyle; the Placid Basin. The largest dry ocean bed of its kind in the surrounding systems. Your ship is not from around here; where did you get it?"

"I've had it for a little while," Logan answered after a brief pause. "I bought it off a group of, shall we say, procurement artists, beyond the Jurass system."

"Tomplie?"

"I promised I wouldn't give out too much information. They like to keep a low profile."

"It wasn't in as good a condition as we'd hoped," Billy spoke up. "We were at light speed and started having trouble, so we came here. Getting through your outer atmosphere just about brought us down into a hole. But my friend here got us on the ground in one piece. I was just getting a handle on the problem when you bushwhacked me."

"That's what you get for being so gullible," the woman said. "It could have been a lot worse. Moyle is surrounded by several layers of Oort. Makes for a great natural shield from unwanted visitors. Takes a tough ship with strong shielding to get through that muck. Even the big cruisers have a rough go of it. You're lucky you're in one piece. So how is ole' Treed? Still hiding out in the Gudulka system?"

Logan gave Billy a quick glance.

"He's not in the Gudulka system anymore. Got chased out of there by the Colonian 3rd fleet."

"Really," she mused, swallowing down the last of her drink. "Well, you can call me Seelix Monroe. You realize you're in a lot of trouble."

"How so?" Logan asked, glancing at Billy again.

"Because I hate being lied to." Seelix picked up a pistol from an end table and discharged the weapon at Billy. The blast pitched him back on the couch, motionless. Logan came to his feet, but held his ground as the guard droids instantly raised their weapons. "Now, perhaps you could amend your story a little?" Seelix suggested quietly.

Logan looked anxiously at Billy for some sign of life, but the engineer remained motionless, slouched on the couch.

"Not much to amend," Logan said cautiously.

Seelix turned her pistol toward him.

"What?" Logan raised his voice a little and tried to shuffle out of the line of fire. "You want me to make something up? We told you!

Our FTL engines were having difficulty. We had to drop out and land to make repairs."

"Where were you going?"

"We were trying to get to Alvadore on Cross."

"Keep talking," Seelix hissed. "You're getting closer to the truth."

"If you already know the truth, then why are you blowing holes in us?"

"What? That?" Seelix motioned with her pistol. "Your friend isn't dead, just unconscious. I put a stun charge into him. He'll be out for a couple of hours and be super sore when he wakes up."

"Not to mention pissed off."

"Yes, certainly a little upset. Your travel path doesn't add up, and there is no Treed. I made the name up. Now, please amend or we'll start this all over again in a couple of hours."

"We got it from a Lou Aura; a Tomplie pirate. We came out of the Nulark and are headed for Cross to see Administrator Vandmire about repairs."

"There's nothing in the Nulark."

"We were prospecting."

"For what? There's only a small garrison of Albion flunkies and a bunch of dirt."

"Asium."

"Never heard of it. Make up something else."

"I'm dead serious."

"Poor word choice."

"Carolon has a high concentration of the stuff."

"What do you use it for?"

"It's a channeling mineral used to direct energy flow to light speed drives."

Seelix considered for a moment, then shut her pistol off and dropped it on the table. She motioned for Logan to sit, then poured herself another drink.

"Well, I'm still not convinced, but it'll do for now. Go ahead and check on your friend. You'll find he's just fine."

"That seems like a poor word choice," Logan said, checking Billy's pulse.

"Had to get your attention somehow." Seelix put her feet up and took a sip. "Tell me more about your ship. What do you know about the markings?"

"Not sure whose they are, but they look pretty awesome," Logan said checking the still smoking hole in Billy's shirt. "It would look great on a shirt, don't you think? Geez, what did you shoot him with?" He carefully probed through the burn hole in the plain flight suit.

"An electron stinger," Seelix responded.

Logan pulled a tiny bead from the smoldering hole in Billy's shirt.

"This is a big hole for such a tiny pellet," Logan complained.

"It had a zenite coating. An ion charge ignites the zenite when it hits the target. It is specifically designed to burn through clothes because it doesn't have enough velocity to penetrate fabric otherwise. The bead then soft lands on the skin and releases its energy. The design is to not permanently injure."

"Who are you?" Logan asked, dropping the bead in his pocket and trying to make Billy a little more comfortable.

"I suppose what you're really asking is, what do I do; my vocation." Seelix smiled and looked around. "I like to hunt things."

"Is that all you do?"

"No, but it's one of the things I take the most pleasure in. I make a good living as a commodities broker. I buy and sell just about anything. I suppose that's all a part of being a hunter. To make a good living at it, you have to know how to hunt for what you're looking for."

"What determines that?"

"Opportunity," Seelix said after taking a drawn out drink. "Also, knowing everything there is to know about what I'm after." Seelix gazed at Logan for a long moment, then set her glass down and got up, moving to one of the walls. "I've seen the symbol on your ship before," she said, casually looking at several of the trophy heads mounted all around the room. Logan looked up from Billy, watching Seelix examine a ghastly alien head.

"With what you're boasting about, I wouldn't doubt you've seen a lot of them."

"Private entities and their vehicles don't generally have markings, except for something nondescript. Only a military craft carries something large and recognizable. I pride myself on being well connected to the goings on in most parts of Hadrian. That insignia goes with a ship rumored to have been produced on Alvadore. Curious that you're on your way back there."

"Isn't it though?" Logan agreed readily.

Seelix strolled slowly along her trophy wall with her hands clasped behind her back, Logan watching her. Once she had reached the end, she crossed to the other side of the room, looking at the trophies mounted there.

"Do you know how long it took me to acquire these?" she asked, continuing down the wall. She stopped and looked at one of the larger animal heads. Logan had started to count the trophies, but couldn't focus.

"I don't care?"

Seelix chuckled, then looked up at another animal head.

"This is a Murdulian Crank. It's considered one of the most ferocious aquatic predators in Hadrian; certainly on Tomain. They have anywhere from eight to twenty-eight tentacles with life draining suction cups and fangs around each one. When it decides that it likes you and wants to have you for a snack, it will wrap itself around you and hang on. Try to pull one of these off a human once it's gotten a hold of you and it will tear your flesh right off your bones. Once it has you immobilized, it bites you with all its fangs and injects a low grade acid. By itself, it's only an irritant to other life forms, but it's considered the predigestive process. It softens up organic tissue for easier digestion. But it has a chemical reaction to the saline mixture in human blood. The mixture changes from a low grade acid, to a highly corrosive one. You get to watch as your flesh disintegrates in a matter of minutes."

"Told you I didn't care," Logan replied. He had heard enough of that story.

"Not pleasant at all."

"How did you get it up there on your wall?"

"It took a lot of time and patience." Seelix continued to gaze at the animal. "There aren't very many of them in the universal realm of things and since they're aquatic, that makes them even harder to track. Fighting something like this in their own environment is a challenge, to say the least. Getting permits to hunt them is a real trick. Finding someone willing to risk their neck as a guide is almost as tough, but if you're willing to pay enough, people will forget about their own mortality. Then there's the chore to get properly outfitted."

"I think I'm getting the picture," Logan said.

"I can be patient when it's needed," Seelix stated after a long pause.

"I'm not sure I believe that," Logan said looking back down at Billy. "You shot him to prove you mean business. You haven't shot me yet..."

"Yet," Seelix repeated.

"Because you think I'll try to problem solve my way out of this situation by giving you more information."

"Is it working?"

"Since you have the droids and the guns, you'll have to make that determination."

Seelix smiled and nodded, leaning against one of her droids.

"Do you like my battledroids?" She wrapped her knuckles against the heavy armor plating and stepped back, making a presentation pose.

"Heavy duty."

"Vivitar 700 series."

"This is Colonian hardware," Logan said, stepping toward one of the droids.

"Very good, you know your equipment. Yes, as I have no political affiliations, I have a pretty good cross section of things from all the factions here in Hadrian. You should see my collection of ships."

"I would love to," Logan said intrigued.

Seelix instantly perked up. She considered for a moment then moved to a set of double doors.

"Let's go have a look."

Logan glanced back at Billy.

"Oh, don't worry about him. If he wakes up, my droids will see that he doesn't go anywhere."

Logan gave the unconscious engineer one last look, then followed Seelix through a set of security doors and onto a small ground transport. They were whisked down a dark tubular hallway that stretched around a distant bend.

"So, what are we talking?" Logan began. "Speeder models, maybe a few retired military transports?"

"Yeah, I have a few of those, but I don't want to boast," Seelix smiled, touching a control on the panel in front of her.

"If you didn't want to boast, why are you showing me?"

"Well, I fancy myself as a pretty good judge of character," she said, giving Logan a glance. "I think you can appreciate a good collection when you see one."

"So, what's a nice girl like you doing out here all alone?"

"HA! I don't think I've ever been called that before."

"It's got to get lonely."

"Who said I was alone?"

"All I've seen is military hardware. Just a bunch of droids. Hardly something to use for recreation."

"If I need recreation, I know where to go to get it. I have everything I need here. I don't need anyone else."

"So this planet is deserted?"

"There are a couple of settlements on the other side in the upper hemisphere, a few loners like myself scattered around the globe and even one or two floating villages. But, nothing large enough to attract attention. There is rarely any interaction as they're here for the same reason I am. We Moylians like our seclusion and we're far enough out of the way from the other factions that no one takes notice. I go off planet when I need something or have business."

"I guess to each their own."

As they approached the end of the tunnel, a large set of doors swung open. Logan took a deep breath as the snouts of several ships appeared. He couldn't see where there wasn't a ship either sitting on the expansive floor or parked in elevated storage cubicles all around the structure. It seemed every smaller class ship imaginable was housed within the enormous bay. As the transport glided deeper into

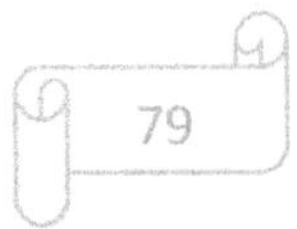

the hangar, Logan could scarcely keep track of what he had already looked at.

Seelix finally brought the transport to a halt and sat back, watching her guest's expression. She smirked a bit recognizing the childlike look of wonder on Logan's face. It took him several minutes to take everything in. As he turned back to the Moylian Hunter, he noticed a ship near the main bay doors with its cargo doors wide open. Neatly tucked just inside was the Interceptor. All of its external access doors were still open. Logan stepped off the transport and started toward the Kalamarion fighter, but froze in place when he heard the sound of whistling power generators coming from the walls of the hangar bay. He looked up, seeing several blaster turrets swinging around in his direction. Slowly turning around, he looked back at Seelix.

"It's all there," she finally said with a shrug. She acted as if she wasn't worried. "I haven't even touched it… yet. But you already know it's your ship. You came to see mine."

Logan let his eyes wander around the ships closest to him.

"You mean to tell me all these are yours?"

"Well, not all of them," Seelix admitted. "About half. The rest are renting space. I'm not as well off as the Administrator of Alvadore and rent pays the bills."

"What could you possibly do with all these?" Logan asked moving carefully back to the transport.

"Fly them, of course; special purpose and utility. Some of them, just because I can." Seelix pointed to a tiny ground vessel that looked as though it had been designed for air flight. "That one there serves no practical purpose."

"Looks like it's missing some, essential parts."

"It'll seat four," Seelix said pointing at the cockpit glass. "But I don't think I've ever had more than two in it."

"What's its purpose?"

"Speed."

"You can get that from any ship and probably be far more comfortable doing it."

"True," Seelix admitted stepping over to the twin engine machine. "But how often can you fly a foot off the ground, and still be on the ground? Ships these days all have antigrav systems, so us humans can withstand the stress of high speed space travel. This thing puts you in touch with your senses. No anti grav and no slip stabilizers."

"You can adjust an antigrav. They're made to do that, you know? Looks like it was built to fly," Logan commented stepping around the stubby winglet mounted over the fuselage. The large underslung engines were mounted only a foot on either side of the cockpit. The wheels were hidden beneath aerodynamic fairings on the underside and the vertical fin swept back sleekly behind the fuselage. "What happened to the wings?" he asked, trying to see how it might have been modified.

"It's called a Telladega. Standard jet engines igniting mixed air and liquid fuel for propulsion. It uses sonic buffers to keep it stabilized once you're up to speed," Seelix explained. "It's a little tricky to get used to, but once you get the hang of it, it's a riot. I think I run this thing more than anything else in here." Seelix looked around at all the ships. "That's why I keep it here next to the doors." Turning the other direction, she raised her communicator and mumbled something then looked back toward the Interceptor. "So, what kind of problems were you having with your ship?"

"FTL stabilizers," Logan responded, turning back to the Interceptor just inside the transport doors.

"What's the propulsion?" Seelix asked, taking several steps toward the open doors.

"Asium Ion induction. It's a miniature derivative of... How did you get my ship in your hold? You've got no one here except your Vivitars."

"Short range tractor hauler," Seelix answered, pointing at an arm mechanism just inside the doors. "It's good for picking things up, but that's about all. Looks like that bird has seen a lot of action," she commented, as they both walked leisurely to the transport. "Recently," she added.

"Yeah, it's called, your atmosphere," Logan replied.

"You've got shield generators," she countered.

"Stuff still gets through." Logan kept his focus straight ahead at the Interceptor.

"I know battle damage when I see it," she cut back quickly. "Don't suppose you'd like to give me a tour?"

Logan glanced at Seelix, then back at the Interceptor.

"I suppose it wouldn't hurt to explain a few things about the outside."

"I've already seen the outside..." Seelix ramped.

"Would you let me in the cockpit of one of yours?"

"The answer is yes, but touché."

Logan stepped up to the Interceptor and gave the open panels a quick examination. All seemed to be just as Billy would have left it.

"So, what are we looking at here?" Seelix asked, peering into the open panel next to Logan.

"These are some of the associated subsystems that make the ship's FTLs function." Trying not to give away his ship's identity or its technology, Logan fought the urge to close all the panels. Not that Seelix could understand what she was seeing, but he got the feeling she was hiding something.

"Just the FTLs?" she asked. Moving toward the thruster ports, she put her hands on a protruding knob near the rear of the ship. "What's this?"

"Shield generator node." Logan pointed to various parts of the fighter. "They're mounted in several places for overlapping coverage."

"Effective," Seelix said with folded arms. She scanned the entire surface, taking notice of the large red phoenix insignia on the underside of the grey V-wing. "And these?" she asked, pointing at several lumps of machinery near the thruster ports.

"Asium inverter manifolds."

"This Asium of yours; is it a liquid or solid fuel?"

"Neither," Logan smirked, randomly closing a few of the access panels.

"Neither?" Seelix repeated. "Didn't you say you guys were in the Nulark looking for this Asium on Carolon?"

"Yes, we were prospecting. It's a mineral."

"So what can you do with a rock in an FTL engine?"

"It's used to channel FTL energy." Logan was growing weary of the passive aggressive inquisition. "Let's cut to the quick here. Are you going to let my friend and I continue on our way or do you have something else in mind?"

Seelix considered a moment, looking out at movement in the main hangar. She turned her gaze up at the underside of the V-wing and the insignia. The moment several Vivitars started rolling up the ramp, she carefully stepped to a small compartment in the hold of the ship and produced a small blaster pistol.

"By all means, let's cut to the chase and get on with it."

Logan looked at the Vivitars carrying Billy's limp form, then back at Seelix.

"Is this where you shoot me like you did my friend there and take us to your leader?"

"Depends on if you're willing to make yourself comfortable in my crew quarters."

"And if I don't, you'll stun me like you did him."

"No," Seelix said, stepping up to Logan, recognizing his expression of defiance. She put the muzzle right to his chest. "This one doesn't shoot stun darts."

Logan looked down at the barrel of the stubby pistol resting on his chest.

"I don't really need you," Seelix said confidently. "I can't access your ship and I doubt you're willing to give me access, but I'm pretty confident I can convince your partner there to open it up."

Logan slowly raised his hands and tried to back up, but bumped into one of the Vivitars.

"Put your hands down," Seelix flapped disgusted. "You look like a dang fool. You've got nothing. Follow my boys here and they'll make sure you're comfortable during our little trip."

"Where are we going?" Logan asked, turning to follow the other droids carrying Billy.

"I've got some wheeling and dealing to do on Calliope."

Logan raised an eyebrow, wondering. He hadn't heard of Calliope. Looking back at Seelix, he saw her turn for a hatch at the other end of the hold as the side doors began to close.

Time to Go, Wait!

Dakota watched curiously from the darkened window of the ball turret as a small army of men and women began to scatter, taking their equipment with them. *This can only mean one thing.* Leaving the turret, he scampered down the ladder and burst into engineering.

"Tiana, tell me you've got this all figured out," he said, skidding to a stop next to her. BachTL looked up from the glass console with a worried look.

"What's going on out there?" the Castellian asked, twisting to a darkened window.

"I think we've arrived. They're clearing everything away from the ship."

"We've got maybe fifteen minutes or so before they start offload procedures," BachTL confirmed.

"I'm almost there, Captain. I've made it to the submenus, but they're locked out by a different set of encryption symbols."

"These symbols look Ratronian," BachTL mumbled.

"Are you sure?" Dakota asked, looking at the displays.

"They appear to have the correct cyphers and hyphens in all the right places, but something still isn't right," BachTL grimaced. "These symbols aren't in the right order for normal Ratronian. It's almost like they were translated from another dialect."

"What do you mean?" Tiana asked.

"Sort of like a slang," BachTL said. "The spellings and use of common characters can shift from one region to another, creating a codec that's difficult to decipher. Trying to trace it back to its origins would be problematic." BachTL fidgeted, trying to recall where he recognized it. "Think, think, think," he repeated, closing his eyes. "I've seen this form of *slang* somewhere before, but where?" BachTL mumbled uncertain. He focused harder, trying to achieve recall. "It's a specialty codec," he groped. "Somewhere, but..." Suddenly snapping his fingers, he looked around and smiled. "Of course! It's an engineering dialect."

"A what?" Tiana responded.

"Engineers use these codecs to speed up programing subroutines when they're tracing out a problem. It's basic in nature and a bit sloppy, but when they're in a hurry, they use them and then go back

when they have more time and clean it up to make it more fluid with the rest of the system." BachTL developed a nervous look. His eyes shifted from Dakota to Tiana and then around the engine room.

"So how do you know about them?" Dakota asked, giving him a wary eye.

"I'm the personal aide to the Queen Captain," he stumbled a bit getting the announcement out. "There are lots of times I've been in engine rooms on different ships."

Tiana gave Dakota a glance as the Captain shook his head, sucking in a deep breath.

"BachTL," Dakota said, taking a step back. "I'm gonna have to ask you to step away from the console."

The Castellian looked down at Dakota's holster. The Captain's hand was resting on the handle of his pistol. He quickly ran the scenarios through his mind and carefully brought his hands up while taking a couple of slow steps backward.

"Now wait a second, you've got this all wrong," he stammered. "I said I recognized the cyphers. I didn't say I knew how to write them."

"I think we're down to your word against the evidence," Dakota said, motioning the Castellian further back. "The fact that you've saved this ship and crew a couple of times is the reason I don't have this pistol out and powered up right now."

BachTL read the determined looks staring at him. Were this an Albion ship, there would be little left to the imagination. He would have been shot without any dialog whatsoever.

"I understand your suspicions and appreciate the fact you aren't going to shoot. I still believe I can help with this, if you'll allow me."

"First law of troubleshooting a problem; eliminate as many variables as you can. Please step out into the hallway. Doctor Yamoto," Dakota said, activating his communicator. "Can you meet me in the hall, please?"

"Yes, Captain. What's going on?"

"Just need a little backup."

"Backup?"

BachTL and Dakota slowly backed up toward the sickbay door as Fuji stepped outside, stopping short when she noticed Dakota's posturing.

"What's this all about, Captain?" Fuji asked, giving BachTL a glance.

"Pretty sure we've found our problem child."

"Dakota, please give me a chance to explain," BachTL complained.

"We don't have time," Dakota said, pulling his pistol from his holster and handing it to Fuji. "Doctor, I need you to keep an eye on our Albion aide."

"Not liking this, Captain," Fuji objected nervously.

"I suppose I could have him slammed against the wall with a couple of pistols shoved in his face."

"It's ok, Doctor," BachTL reassured. "This makes sense. The Albions would have acted much sooner, if they had let me onboard at all. Where would you like to have me?"

Fuji frowned and motioned to the closest crew quarters door. After the door had closed behind the Castellian, she instantly turned to Dakota who was already heading back to engineering.

"I don't like this any better than you, Doctor," Dakota said knowing she wanted to complain. "But the evidence is pretty overwhelming. He all but admitted to it."

"But what about all the help he's given us?"

"The Colonel once told me that sometimes the best way to hide is right out in the open. I get the strong impression that applies here. For all we know, getting us to Calliope was his plan all along."

"You could always ask him."

"I could, but right now, I have this ship and crew to worry about, besides, do you really think we can trust anything he says now?" Dakota turned for the Mallory doors.

Fuji watched him pass through, then turned back to the crew quarters.

Approaching the engineering console, Dakota noticed Scott Brandon emerging from the number two coupling access. The young engineer stepped over to the console next to Tiana and tapped on the screen, looked at several readouts, then expanded several windows. After touching a couple of points on the glass, several more areas on the displays began to flash information and a few more readouts turned green.

"If I read this right," Scott said, "the power couplings are working. We should be able to disconnect from that torpedo now."

Tiana shifted from what she was doing and examined the displays Scott had been working with.

"Engines and main batteries are feeding the shields now; nice work," she smiled, seeing the readouts stabilizing. "Hayden, you can go ahead and pull the plug on that thing."

"Don't toy with me woman," Dakota perked up. "I'm in no mood."

"This isn't a joke, Captain. Main power is back online. Well, most of it is anyway." Tiana paused a moment as she checked all the systems on the console.

"What's left?"

"Atheon control, but this last cypher is in the way." Tiana rubbed her eyes and shook her head. "Afraid I haven't got a clue."

"What about you, Mister Brandon?" Dakota asked, turning to the young engineer's mate.

Scott froze, a bewildered look solidifying on his face.

"Sir?"

"You said you served on a few Ratronian ships. Do you know anything about these?" Dakota asked, pointing at the engineering console.

Scott leaned closer and examined the displays.

"This looks like complete gibberish to me, Sir."

"So you've never seen anything like this before?"

"It doesn't even look like Ratronian script."

"What does it look like?"

"I could be wrong, but it looks Castellian."

Dakota turned back to Tiana leaning against the console with her hands and head down.

"What can I get you?" Dakota asked.

Tiana looked at him for a moment, then shook her head.

"BachTL."

"No," Dakota ramped frustrated. "Clearly, this is his doing."

"You don't know that for sure."

Exasperated, Dakota glanced down at the flashing readouts on the console.

"Mister Pippin?" he called out into the com system. "You and Lana are looking at this too. Any ideas?"

"None, Sir," came the response from the com panel. "Maybe, if we had a couple of hours."

Dakota looked down at the battery indicators. With the power couplings back online, most of the systems were coming back. Even the auxiliary batteries were starting to recharge.

"Not a lot of time before they start offloading," he said looking up at Tiana.

Without a word, she moved back to the glass display and continued working. Dakota turned to Hayden as he stepped off his lifter and reached for several cables still strewn across the floor.

"The turret batteries should be charging now," Dakota said quietly. "If we can't get this thing working, you may be the only guns we have."

"A weak charge isn't going to go very far," Hayden pointed out as they stepped over to the elevator.

"Until we have scanners and sensors operating, you're our eyes to the outside." Dakota looked after the turret gunner as he ascended the ladder. Heading toward the Mallory doors, he paused a moment, looking into the still open number two access pit. He glanced around the engine bulkheads at the open number one access.

"Is there a reason you've left these open?" Dakota asked as Scott hurried past with an armload of cabling. The engineer looked down and continued around the Captain.

"The doors close automatically after the diagnostics routines on the power couplings have been confirmed. Tiana said she'd take care of it."

Dakota watched after the young engineer, then looked back into the open access pit. He turned back to Tiana and watched her work for several moments. He finally made a quick exit, heading directly to the door beyond sickbay. He stopped only long enough to unlock it. Inside, he found Fuji sitting across the room from BachTL, quietly chatting.

"Captain," Fuji started. "I don't think you fully appreciate…"

"Will you give us a minute?" Dakota interrupted.

"Certainly," Fuji got up. "But I'm sure you're gonna need this," she said, thrusting his blaster pistol back at him.

Dakota holstered it as the door closed behind him. He walked across the room and touched a control. The outside window immediately lightened, allowing the light from the hangar bay to filter in. He leaned against the wall looking out at several tiny one man utility craft moving into position in front of the bay doors. Another figure leaned against the wall on the opposite side, looking out the window with him.

"They're in a hurry," BachTL said quietly. "I think all the trouble you guys have caused since you got here has them on edge."

"You're saying they've moved up their schedule?" Dakota asked as several warning beacons started flashing all around the bay.

BachTL counted down on his fingers. The giant bay doors suddenly started to open, revealing a starfield cluttered with nearby ships. "I'm pretty sure I saw Commander SoKnack outside the ship before we got enough power back to darken the windows."

"SoKnack is the commander of the *Dominator*?" Dakota asked, noticing the *Constellation* starting to rise. He looked forward, seeing the tug on this side pulling them up.

"He might be now. When I was still Drax's aide, he commanded the *Tarzana*. My guess is she put him in charge of the Torag that attacked us."

"So why the hurry?"

"Probably a combination of a couple of things. Drax has a low tolerance for failure and wants results instead of explanations. As Colonels Barker and Conrad have been a proverbial thorn in her side, no one wants to take any chances during this maneuver. Clearly they've expended a lot of assets to capture this ship. Plus, if I know Drax, she's gonna want to make a show of things to assert herself. Presenting this ship before all the factions of Hadrian is just the kind of spectacle she'll want to create."

"I have every intention of disappointing her," Dakota said as the ship began to move toward the open bay doors.

"I can help, if you'll let me."

"Why?" Dakota looked at BachTL. "Aren't you the reason we're here?" Dakota pointed out the window.

"I admit it was my assignment. Drax sent me to Tintee to try to retrieve this ship. A tall order, but you don't refuse Drax Blair. She can be somewhat overpowering. My hope was to at least gather intel. Never in my wildest imaginations did I think I would ever get close to it, let alone onboard."

"Well, here you are." Dakota's tone was less than understanding.

"Yes, here I am, facing the same dilemma as you."

"Bull! You can step off this ship anytime and be hailed a hero. Then your precious Queen Captain will welcome you home with open arms."

"Think it through, Captain. The Albion military has been explained to you and you've seen it in action. I've been gone for a considerable amount of time, and now I'm onboard an alien ship being sought by nearly every faction in Hadrian. I don't know how your military works, but in mine, people have been shot for far less. Let's not forget the fact that I've been helping you. Now, I can tell them I wasn't helping you all I want, but when they interrogate your crew, they'll find out otherwise. In the Albion military, that's considered treason. They don't do courts; they do accidents or outright executions."

Dakota watched as the interior of the *Dominator* made way to the black of space.

"I assume you know where we are then?"

"The last information I had on the Rite of Pintar was that it was being held on Calliope. It's my home world, such as it is."

"What do you mean?"

"Long ago, my planet, Castell, was decimated by an intergalactic plague. It nearly wiped out the entire population and rendered my world uninhabitable. Rather than let our unique culture be swallowed up by merging with another populous on their world, the factions of Hadrian came together to figure out a solution. Headed up by the engineers of Cross, they converted a suitable asteroid to house the Castellian race. We are somewhat isolated as we are off the beaten path from other cultures, but not so far distant we can't interact and be part of the other factions. Since the other factions can't ever agree on a place they can all meet, it was decided that Calliope would serve as neutral ground for most of Hadrian's intergalactic conferences and governmental events, such as the Rite of Pintar."

Dakota caught a glimpse of Calliope as the *Constellation* was pulled away from the Albion destroyer. The myriad of lights and structures built up on the jagged, uneven surface gave its stark, dead look, beauty. The asteroid was hardly round like a normal moon, but resembled more of a broken chunk. As there was no external atmosphere, it appeared to be just a big lump of rock, with the exception of all the external structures and lights. As it was naturally tubed, there were a myriad of portals everywhere. Every exterior opening had shield generators installed so merchant and military ships alike could pass freely into an internal, breathable atmospheric environment and dock in massive hangar bays or docking ports.

Dakota took a deep breath, watching a variety of ships hold their positions well clear of the *Dominator* and the *Constellation*.

"This is against my better judgement, but Colonel Conrad used to say that sometimes you have to go with your gut. I need you back in engineering. If we can get the engines and weapons back online, we stand a chance of getting out of here."

"There are a lot of faction ships out here," BachTL pointed out as they passed a large flotilla of Ratronian vessels. "Most of them are armed to the teeth with itchy fingers on slippery triggers. There's the *Tarzana*," he said, pointing to the enormous dark shape of the Albion battlecruiser surrounded by her support ships. "We just passed the *Maxell* and let us not forget the King Commander's personal dreadnaught, *Indominable*." BachTL pointed at a long bulky ship, its size rivaling that of the *Tarzana*."

"I've considered that," Dakota said, sighing. "But I can't just let them take this ship without a fight. It's just not in my blood. Besides, we have regulations..."

"They're likely to find out what all our blood looks like," BachTL said, turning to Captain Abrams. "I promise I won't try anything. You can stand next to me with your gun if you'd like."

"No, I think I'm just going to trust you," Dakota said, gesturing toward the door. He followed BachTL out and back down to engineering, stopping next to Fuji Yamoto.

"Nice to see common sense win the day," she commented with a smile.

"We'll see if it wins the war," Dakota said, motioning BachTL to join the others. "I'm counting on you to do the right thing," Dakota said, handing his pistol back to Fuji.

"I really don't like these things," Fuji said, frowning.

"Buck up, Doctor. You know where you are." Dakota spun around and exiting engineering again, trotted back up to the bridge. Upon entering, he found every console alite with information dancing across each display.

"Report," he ordered, watching Lieutenant Starman work through her preflight checklists.

"System power is back online," she said without looking up.

"Thank you, Tiana," Nigel grinned gleefully from the navigator's station.

"I'm still red over here, Sir," Doran spoke up from weapons control. "I can't fire anything, but I do have telemetry and targeting."

"Same here," Lieutenant Starman announced. Dakota leaned down a little closer. "Everything is working…" She tapped several controls on her glass panel. "Just nothing available on the thruster sequences."

"Tiana is still working on the Atheons," Dakota said, turning to the science and communications stations. Everything seemed to be working here as it had before they were attacked over Carolon. He scanned the glass displays, looking for anything that might be out of place.

Turning to leave, he noticed a flash in the top right corner of the science station's glass panel. Sharpening his focus, he took a half step closer for a better look. He had to get right up on the display in order to recognize what the readouts were conveying. This was the Personnel Transponder tracking monitor. He gave it an odd look as it was flashing erratically, showing an intermittent signal. *Probably a malfunction in the boot-up; maybe stray data clearing from the matrix buffers?* Noticing a shadow casting over the bridge, he looked up. They had just slipped inside Calliope's outer shield layer. Glowing rock walls revealed dwelling structures built into the jagged walls of the consuming cavern. Dakota glanced back at the science station, noticing the external environment scanners registering an atmosphere. Indeed, he recognized a slight wobble as the ship passed through the surrounding air. Ahead of them, a bright city appeared. Enormous buildings, business districts and residential areas glowed with lights of every color and intensity.

"I'm guessing that's our destination," Lieutenant Starman said, pointing at a large glowing slit opening in one of the cavern walls.

Dakota stepped up behind the helm officer. Before them lay an enormous complex filled with large windows and stumped docking ports. A large hangar bay door was pulling to one side revealing its lighted interior. As they drew closer, Dakota noticed several Albion corvettes pulling alongside. Another one took up a position right in front of them.

"Keep your windows dark for now, Lieutenant," Dakota spoke quietly, as if trying to remain undetected.

"Captain Abrams, this is Lieutenant Hunter."

"What do you see back there, Hayden?"

"Pretty much surrounded on all sides, Captain," Hayden replied. "I saw two corvettes swing in under us and I'm looking at the belly of two more right on top of us. Oh yeah, there's one behind us too. You want me to cut them down to size? I've got a good charge in my batteries. Would make them think twice about getting too close."

"Negative, Lieutenant. Keep your windows dark and your guns powered down. We're not quite ready to play our cards yet. We need to give Tiana a little more time." Dakota sank into the command chair as the tugs made a sweeping turn toward the open hangar bay door. He studied the surface all around the docking bay, seeing no defensive weaponry anywhere. There were docking port extensions spiking out in a couple of different areas. No doubt designed for ships too large to fit in the hangars. As the doors loomed closer, he touched one of the buttons on the armrest.

"Engineering, please give me some good news," he said, still trying to take everything in as they reached the threshold of the landing bay.

"Thanks to BachTL, we're almost there," Tiana answered.

Dakota sat back, noticing a long line of fighters and gunships parked in a neat row inside the hangar bay. Most of the interior of the landing bay was barren of windows except toward the upper portion on one side. It appeared to be an observation area as the windows were quite large and long. In every corner, turnstile laser cannon emplacements sat unmoved, but pointed right at them. Dakota noticed the tugs slowing and settling to the bay floor. As he felt the *Constellation* touch down, Lieutenant Starman jerked around.

"Captain, I have control!"

"Bless you, Tiana!" Dakota jumped forward, looking at the displays in front of the helmsman. Information began to populate the blank fields on the displays. He glanced at the weapons console as Lieutenant Cartwright tapped at his controls.

"Weapons are coming online, Sir," came the excited announcement.

Dakota noticed Lieutenant Starman reaching for her engine controls and touched her shoulder.

"Steady, Lieutenant," he said quietly.

"But you said as soon as..."

"Yes, I know what I said, but let's not give ourselves away just yet. Let her idle for a bit," Dakota said cautiously. He scanned the hangar bay. "I've got an itchy feeling. Keep the windows dark for now. Lieutenant Hunter, what's it look like behind us?"

"The corvettes peeled off when we came in, but a couple of them came back around and are sitting just outside. The doors are still open. I can see gun emplacements in the corners pointed right at us and a squad of troopers on the far side. There's also a couple of gunships hovering in the corners just inside the doors."

"Hold on for a little longer," Dakota responded, touching a button on his armrest.

"Doctor Yamoto, will you please escort BachTL to the bridge?"

"Sure he won't mess something up there too?" the doctor replied snidely

"On the double," Dakota said, ignoring the doctor's attitude.

Several minutes later, Fuji and BachTL entered the bridge, moving quickly to the engineering station to watch.

"What do you make of this?" Dakota asked, turning to BachTL.

The Castellian stepped up next to the helm. He studied their surroundings for several minutes, occasionally looking back at the displays over the science station.

"They certainly have you surrounded," he commented, stepping back to the science station.

"What's their next move?" Dakota asked.

"If they follow hangar bay protocol, they'll hold you right where you are until called for. If Drax is calling the shots, she'll want to present you in a static position, probably like we are right now, then get all fancy and have you moved up closer to the observation deck up there, so everyone can see her prize," BachTL said pointing to a long line of windows near the ceiling.

"How long before that happens?"

"There is no way to know," BachTL responded. "I have no idea what her schedule would be for something like this. If you're thinking of trying to get out, I would advise against it... for now."

"Why?"

"Is this ship ready to do heavy combat with every gun emplacement and fighter you see in this hangar bay, not to mention whatever's going to be waiting for you outside?"

"In a word?" Dakota frowned looking at the readouts at the engineering station. "No."

"Sir?" Doran spoke up. "We could do some real damage with our Pin missiles."

"And what would happen when they fired back with the amount of shield energy we have?" BachTL pointed out.

"He's right," Dakota said. "We could really tear this place apart provided no one was firing back at us. Sounds like we may have some time to sit tight and let everything charge back up. Drop us to yellow alert, Lieutenant," Dakota said, turning to Lana at communications. "Everyone stays ready at their posts though. We'll wait for an opening, whenever that is."

Arrival

Rick watched stoic as the *Athena* dropped out of light speed operations and a mass of clustered ships surrounding a bright asteroid came into view. Glancing at Toby, he could see the target returns of *Calypso* and the Alvadorian ships on the scopes in front of the Science officer.

"Better hang back here before they get a chance to scan you," Diord called from the comlink. "Everyone will be curious about *Calypso* and my ships, but considering what most the factions have been after for the past two years, if they detect the *Athena,* it'll be too tempting."

"Captain Dayton," Rick said, still steely. "Pull us back behind *Calypso* and nose the ship between her main thrusters. The wake turbulence should give us plenty of cover. What's the next move?"

"I'll take the *Merc* in closer and identify the delegation from Cross. Make sure they're still going to let me participate."

"Aren't you required to be here?"

"Technically, yes," Diord responded. "I have the invite in my hand and I'm transmitting my clearance codes. But in the last year or so, things have gotten a little tense between Cross and some of the other factions. Namely the Albions and Colonians. They seem to think we developed the Starbird and are holding out on the other factions. They're accusing me of trying to take the upper hand and present Cross as the rightful faction for power in Hadrian."

"Well, they sort of have it right," Rick smirked. "You have Starbirds behind you and the rightful heir to the throne onboard."

"I feel so much better being reminded of that." Diord sounded a little worried. "I wished my nut job friend, Lou, would get here. It would look a lot better if his fleet cluster were sitting in position."

"Fleet cluster?"

"Each faction's representative fleet is assigned a position around Calliope."

"And orbital position?" Rick asked.

"No, no orbits," Diord responded. "Calliope is a man-made oddity. The Cross engineers built it for the Castellian race a very long time ago."

"They built an asteroid?"

"Not exactly. They found a suitable asteroid and modified it to make it habitable. There is no rotation or outer atmosphere."

"There's one heck of a magnetic field around it," Rick observed looking at Toby's displays. "Makes it hard to look inside. I bet communications are a nightmare."

"Yes," Diord agreed. "But those magnetic fields create some added benefits for the inhabitants. There's standard gravity inside and a breathable atmosphere. The main ore, Corvantium, reacts with the other minerals, producing a luminescent glow. We installed flux generators in strategic places all around the asteroid to control the magnetic lines."

"How did the air get there?"

"It was pumped in, then vegetation and air scrubbers were installed to maintain it. Passive shield generators are operating at every opening to keep the atmosphere from escaping."

A flash from one of the displays between the communications and science station caught Rick's attention. He gave the ship identifier readout a double take. Whatever the flash was, it was gone now. *Just a reflection...*

"So back to Lou. Where did you say he was?"

"En route, he encountered a distress call from a transport in the Lieastan sector and insisted that I continue on and he would catch up with me here. While he loves to profess his benevolence, more than likely, he's just hoping for some kind of payload salvage. Normally, I would just as soon have him chasing off after something else, but in this instance, it'd be helpful having the *Executioner* here."

"Why?"

"More of a distraction for the other factions. Right now, the Cross delegation has everyone's attention, but throw in the Tomplie and all the factions are going to be looking over their shoulders."

"Understood," Rick agreed. "Would it be better if I took the *Athena* back into *Calypso*?"

"I don't think so," Diord answered. "You're going to need to act fast if an opening presents itself. We can't even detect you right now."

"Very good, we'll wait for your signal."

"Thanks, General. I'll give you some kind of indication when everything is ready. I don't expect it to take more than a couple of hours."

"We'll be ready," Rick said, turning to Lieutenant Habba. "Patch me into the other ships." Rick looked up at a sea of metal as Captain Dayton finished positioning the *Athena* between the thruster ports of the great ship. He didn't have to imagine what it would be like to fly in such close proximity to such enormous ships. He and Gunnar had used these tactics many times when attacking Valkyrie capital ships. Flying through wake turbulence during thruster operations was a tricky business under controlled circumstances, but he and Gunnar had learned where the sweet spots were. Thankfully, *Calypso* was not using her thrusters at the moment.

"General, I have the other ships tied in," Laura announced as an overhead display flashed on, revealing several faces tiled together.

"Commanders," Rick started. "Are you staged for deployment?"

"We're ready," Jayda and Audra answered.

"The *Phalanx* and *Paladin* are staged directly behind the *Intrepid* and *Montego*," Captain Fowler spoke up.

"Nice names," Rick smiled. "You think like Starbird commanders."

"Remember the whole, *You guys may be good at what you do, but we make you look good doing it*?" Jayda stated.

"In this case," Audra cut in. "We look good doing what we're doing."

"Just be careful," Rick cautioned. "We don't want to tip our hand until it's absolutely necessary. I'd hate to see one of those shiny new birds wrecked."

"Think how we feel," Manx said. "Are you sure deploying us is a good idea? We're painfully undermanned and with no Asium in our chambers, we're essentially pushing these things with our bare hands."

"Your batteries will give you what you need for this. The hope is you don't have to launch, but if you do, you're meant to be a bluff. I'm betting that word of what the *Constellation* put the Albions through over Carolon gets back to the other factions."

"General," Toby said, focusing. "Several faction ships are turning in our direction. Nothing armed, yet. But they have definitely noticed *Calypso*."

Rick studied the displays on the scanning equipment, then looked up at the tactical information overhead.

"I'm sure they're wondering what she's all about," Rick mused. "I don't see anything out there that could be a threat except the *Tarzana* and that dreadnaught." Rick studied the displays further, seeing the Ratronian heavy cruiser *Maxell* moving in their direction. "Since *Calypso* is facing them, I doubt they have an accurate perspective of just how big she is. Lieutenant Habba, get me *Calypso*."

"You're on, Sir."

"Richard," came the voice of Caidin over the com.

"How are you handling things alone in that big ship?" Rick asked.

"Your Mister Dacey and his friends are doing all the work, I'm just supervising."

"While we're waiting for Diord's partner to get here, I wonder if you guys could turn *Calypso* ninety degrees to Calliope? I think the other factions would like a better view of her."

"That's not a problem at all, General," Dacey called out. "This thing drives like a dream. Hard to believe it takes so little effort to maneuver such a big ship."

"Yes," Rick said looking around the bridge of the *Athena*. "It sure is nice when everything works as it should. Just hold her in a trolling mode after you make your turn. I'm sure we'll get their attention."

"Are we clear to move with the *Athena* parked where she is?" Dacey asked.

"We'll be fine. Remember, I have the best Starbird pilot in the fleet."

"I'm sure Lieutenant Starman is better than I am, Sir," Captain Dayton piped up from the helm position.

"We'll have that debate when we have them alongside again," Rick replied as *Calypso* began to pivot. "Caidin, I'm going to have you transferred over here. I get the distinct feeling I'm going to need you."

"Whenever you're ready," Caidin responded.

"CORA 500, can you bring Caidin over?"

"I'm on it, Richard," the AI responded.

Captain Dayton carefully manipulated the *Athena's* controls, holding the Assault Corsair in position, tucked tightly between the mammoth thruster ports of *Calypso*. As the maneuver continued, Rick studied the scanning scopes closely. He could feel his pulse accelerate a bit as several primary targets maneuvered around the Cross fleet. To his dismay, the ships that had started toward them when they had first arrived, were still moving in their direction. As *Calypso* finished her turn, letting her shape stretch broadside to the approaching ships, her enormous size became apparent to everyone. Rick waited impatiently; hoping.

"Instructions?" Captain Dayton asked, watching the *Maxell* make a fast circle around *Calypso*.

"No doubt they're sizing things up," Rick said, watching several other destroyer class ships swing in close to the dwarfing hull of *Calypso*. "They would have stopped back here if they were able to detect us. Just hold your position."

"What about the other ships inside *Calypso*?"

"With her doors shut, they won't be able to scan what's inside." Rick watched the small swarm of ships maneuvering around the great ship.

For several more moments, everyone held their breath until, as if all the curious onlookers had been given a signal at the same time, they turned off and headed back the way they came. Everyone let out a collective sigh of relief as the space around *Calypso* cleared and they were once again alone, outside the fleet clusters surrounding Calliope.

"Signal from the *Merc*, Sir," Lieutenant Habba announced. "She's signaling all clear to bring *Calypso* in. Coordinates for positioning coming through now."

"This ought to be a lot of fun," Rick said looking back up at the tactical display. "Mister Dacey, are you getting this?"

"Yes, General. That seems pretty close to the asteroid."

"That'll probably work to our advantage. Just take it slow and easy. Captain Dayton, keep us snuggled up close," Rick ordered as the stars began to move. "How long will it take to reach our parking spot?"

"Barring any problems," Toby said without looking up, "twenty eight minutes, thirty five seconds."

Rick waited impatiently as the seconds slowly ticked down. He kept glancing at the tactical information flashing across the display over Toby's head. As *Calypso* moved into its assigned position next to the *Merc*, Rick caught a flash out of the corner of his eye and looked over at the scopes in front of Toby. There seemed to be nothing out of the ordinary, other than the multitude of targets the science officer was tracking on his main scopes.

"I'm receiving another voice message from the *Merc*, Sir," Lieutenant Habba said, putting her finger to her earpiece. "Administrator Vandmire and his delegation have left for Calliope. The *Executioner* is delayed a bit longer."

"A bit longer? How much longer?" Rick asked.

The com officer only shrugged. Rick turned forward, but gave the transponder display a double take. He was sure he had seen another flash toward the upper right quadrant of the display.

"Lieutenant Habba," Rick said looking back up at the tactical display. "Request the *Merc* and her sister ship take up positions on both sides of *Calypso* and their support ships directly above and below."

"Sir?" the com officer asked, looking at him oddly.

"Helm, standby to pull back."

"Aren't we supposed to wait for Administrator Vandmire's signal?" Toby asked, turning.

"We're just getting ready," Rick cautioned. Again, there was flash on the display to his right. He turned and looked closer, focusing on the upper right quadrant of the monitor. Under normal operations, all the surrounding transponders would be filtered out so as to not trigger false readings. Looking directly at the display, it flashed again, bringing him to his feet. Toby and Lieutenant Habba were so focused on what they were doing, they hadn't noticed the sporadic flash. Keeping his eyes glued to the display, Rick stepped over to the science station and leaned in close, bringing Toby's attention to his commander.

"Sir?" the science officer inquired quietly, looking at the display.

"What is it?" Lieutenant Habba whispered, diverting her attention.

All three held their breath and waited. Suddenly, there it was; a flash.

"There..." Rick uttered, stabbing a finger at the glass. The readout froze under his touch. "It's the *Constellation*!"

Toby went to work, his fingers sweeping across his glass at a blur. Displays changed, information scrolling past them at a dizzying rate. The readouts on the transponder monitor flashed again, this time a little faster.

"She's here," Toby mumbled, continuing to work with his instruments. "But she's really hard to get a hold of. That asteroid is causing a lot of problems with our scans."

"Can you pinpoint her?" Rick asked anxiously.

"Maybe if we were closer," Toby said, still working.

"Can you get me anything?" Rick asked.

"The asteroid has no shields, just an energy barrier at the main opening holding the atmosphere inside, and..."

"And what?" Rick asked, glancing at the screens in front of Toby.

"A very large ship, sitting right at the opening." The science officer tapped a target on his glass and one of the overhead displays drew out the tactical information of an enormous ship, sitting right in front of the opening to Calliope's interior.

"Albion dreadnaught," Rick settled back in his chair, inhaling.

"The *Indominable*," Toby finished. They stood back and examined the tactical specifications of the enormous Albion warship. The information kept coming. Toby forced down a gulp as Rick focused. "Look at all those Colonian ships on the other side."

"The gang's all here. Lieutenant Habba, signal the *Merc*. Tell them we're going in."

"Going in?" the com officer repeated. "Going in where?"

"There," Rick said, pointing at the far edge of the asteroid. "Get me Mister Dacey again and tell the *Montego, Intrepid, Phalanx and Paladin* to hold tight behind *Calypso's* doors. They'll remain undetected if they stay inside.

"*Calypso*, Sir," the com officer announced.

Rick touched the comlink button on the command chair.

"Mister Dacey, can you pivot around one hundred and eighty degrees?"

"It's feasible," the Drake pilot answered.

"Can you swing the tail end around within a couple of hundred Kaldants from the asteroid?"

"A couple hundred Kaldants?" Dacey sounded a bit apprehensive. "I'm not that comfortable at the controls of this thing."

"It'll be a piece of cake," Rick tried to reassure him.

"General," Lieutenant Gillespie interrupted. "We can barely understand the clearance parameters this equipment is giving us. Trying to get that close without actually hitting anything is a precision maneuver."

"No worries," Rick responded again. "We'll be back here on your tailpipe guiding you through it."

There was silence for a moment.

"Ok," Dacey agreed skeptically.

"The asteroid will mask our departure from *Calypso* and entry into Calliope," Toby commented, reading what Rick had in mind.

"Better that no one knows we're inside," Rick said looking back up at the scanning displays as the giant ship began to pivot. It took nearly ten minutes before Calliope came into view. Only a couple of maneuvering adjustments were necessary to bring *Calypso's* thruster ports within the distance parameters Rick had requested.

"Mister Mavis, scan the surface of Calliope. What do you see?"

"A whole lot of nothing, General," the science officer replied. "Lunar dust sediments, meteor craters and tunneled vents with air shield generators around them."

Rick remained silent, waiting as the science officer worked his systems. Captain Dayton glanced at Rick as Toby looked forward.

"The vents..." the science officer repeated.

"The vents," Captain Dayton echoed, immediately starting to work her controls.

"Suitable entrance, bearing one thirty-five mark two," Toby called out.

"Helm, let's go." Rick said as the back end of *Calypso* continued to pivot around.

Captain Dayton didn't wait to be told twice, nor was she too worried about any residual wake turbulence from *Calypso*. Letting her fingers sweep across her glass controls, the *Athena* pulled straight back while pivoting and blasted away from the huge bulk of *Calypso*.

Following the navigation telemetry toward the surface, Lisa quickly found the opening and drove directly into the glow of Calliope's interior.

A closer look

Gunnar sat up and carefully stretched. At least his bed was comfortable, but that did nothing to alleviate the raging headache and the pain firing through the back of his neck and spine.

Kind of weird having a bed this nice in a cell block, or infirmary; whatever this place is.

The room was equipped with various medical equipment built into the walls and hanging from the ceiling. There was a place setting with food on a table a few steps from his bed. He checked his band for the time. His communications band was gone, but the one he had taken from Captain Bridger was still around his wrist.

"And how are you feeling?" A tall man with jet black hair and lab coat walked in.

"Apparently, Ditairian wine doesn't agree with me." Gunnar really wanted to shake his head, but didn't dare; it hurt too much.

"Splitting headache and extreme pain down the back of your spine?"

"You've had it before..."

"It's not the wine. You have the final stages of Kodiac Blu poisoning. If you can manage it, you can sit down over here and have a bite to eat. Then we can get on with the rest of the examinations."

"So are you trying to fatten me up first?" Gunnar turned his nose to the aroma coming from the food. It looked really good.

"Not at all. I was instructed to do everything I can to make you comfortable."

"And you are?"

"San Sann; You can call me Doctor Sann."

"Should I just use your first name or your last?" Gunnar stepped over to the table and sat down.

"Never heard that one before."

"When do you start dismembering me while I watch?"

"Who said anything about dismemberment?" Doctor Sann picked a couple of items from Gunnar's plate and popped them in his mouth. "Not bad."

"Isn't that what Drax has in mind for me?"

"Certainly not. What do you think we are, barbarians?"

"Well, one was led to believe..."

"This is good stuff. Wished I got food this good. No, I need to find out everything I can about you. Would be nice if you could provide the detail, as you're a very different specimen."

"What about the antidote for the Kodiac Blu?"

"What about it?"

"You're going to give me some, aren't you?"

"First of all, Kalinite is very difficult to synthesize. The Queen Captain took my only supply a couple of days ago. I'm in the process of making more, but like I said, it's a long process. Curious, I haven't received any orders to give you any."

"Well, I would think if the Queen Captain plans on keeping me alive, I'm going to need some pretty soon. Look at me, I'm a mess."

"I can see that. Your scans show an exponential growth on your upper spine, working up inside the back of your lower cranium."

"You could have just said the bottom of my skull and I would have gotten the message. When did you do the scan?"

"While you were asleep. You've got some kind of micro electronic device embedded in your head behind your left ear. It didn't seem to be affecting anything, so I just left it. I repaired several projectile wounds and removed a couple of large chunks of shrapnel. Got the DNA sample we've been after too. Like I said, you are a unique specimen."

"So I've been told. If you have all that, what else is there to do here?" Gunnar continued to eat as Doctor Sann worked to get the rest of his equipment set up.

"Well, powerful people are interested to know what you know and what effect certain stimulus has on you."

"I'm not sure trying to figure those kinds of things out is a very good idea. It could prove painful."

"No, no, there shouldn't be much pain at all. We'll try a host of different stimuli designed to trigger different levels of biological responses."

"In my condition, just say something to piss me off and you'll get a response."

"The idea is to see if there is a way to control it without more drastic measures." Doctor Sann turned and leaned against the examination table to wait. "Whenever you're finished, we can start."

"Feel like I'm eating my last meal," Gunnar said.

"To be honest, this is highly irregular. As I said, I don't even get food this good. But it's a direct order from the Queen Captain herself."

"I suppose she's there behind that wall?" Gunnar asked, nodding in the direction of a wall with obscured glass.

"No one there right now, but I think if we find anything of interest, we'd have an audience soon enough." The doctor gestured to all the cameras mounted around the perimeter of the room.

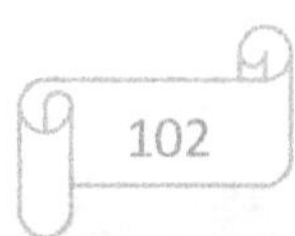

Gunnar took a drink and sat back, considering.

"Ok," Gunnar finally agreed, getting up on the table. "Let's get on with this."

"Perfect, now, I'm going to restrain you..."

"What makes you think these are going to do any good?"

"Well, under normal circumstances, these would be more than adequate, but as I understand it, there's nothing normal about you, so while I know you can pull right through these, I'm hoping they will give me at least a moment to get out of the way."

"I suppose I could always tell you before that happens."

"My experience has taught me not to count on that."

"Suit yourself," Gunnar agreed as the table back moved into position. Leaning back, several mechanical appendages hovered above him, one directly over his head. He noticed several lights blinking on the inner surface of a half-moon shaped paddle assembly. *This is like something out of the dark ages,* he thought. "Where you going?" Gunnar asked as Doctor Sann moved towards the far end of the room.

"The controls are over here," he said, pointing to a console stand. "So, there are no surprises, here's what's going to happen. I'm going to put you through a series of visual stimuli and read your reactions."

"What's to keep me from blowing this place apart?"

"I'll monitor everything and if things start to escalate, I'll change the course of the tests."

"How long is this going to take?"

"You've got somewhere else to be?"

"Yeah, there are a lot of other places I'd rather be."

Doctor Sann looked up from his console, shook his head, then went back to work. For a while Gunnar didn't feel any different. He could see the lights blinking and changing color, but that was all. They bothered his eyes a little, but nothing like his head and neck. After an endless battery of testing, Doctor Sann stepped up and touched a control on the panels above Gunnar's head, then positioned the movable panels directly to his forehead.

"So far, so good," the doctor said, making some adjustments to the apparatus on Gunnar's forehead. "This next test is designed to measure your control capabilities. This could take a while, so keep doing what you're doing."

Gunnar looked back up at the device and tried to relax.

"Tell you what's gonna get my hackles up; sitting here for hours on end looking at your idiotic blinky lights. Give me something interesting to look at."

The testing continued on for some time before Gunnar found his mind wandering in and out of recent events. The dinner last night, then his capture in the launch bay. The look on Malina's face as she

was leaving in her shuttle. Their adventures in the Administrative Command suite and the trouble with the ID band.

How odd, he thought. *Everything seems to be running backwards. The detail was amazing.* Gunnar looked past the apparatus for Doctor Sann; he was gone. As the images continued in reverse, there were the conversations with Malina Cass, as well as convincing the guards they were married and before that, the meeting in the hallway where she slugged him. Boarding the ship and back to where he finally left Eldon and Rayna Grant at the main station in Zeppelin. Through his entire time with them at their home. His mind replayed the crash and crawling through the open cargo bay to the cockpit while inflight. As the scenes continued, he found he could skip over some of the events, even black out faces. Some were easier to skip over than others. Many of the images were halted and viewed from varying angles, giving him a different perspective. Fascinated, he let the memories flow backwards. Making his way through the jungle after falling from the Interceptor. The feeling of free falling when he let go of the hatch. Getting shot several times trying to escape. The fire fight outside Repos Three and the brawl inside where he killed Captain Bridger. Walking the great halls of the Repos facility and traveling to Dither. Getting out of Reako, finding Tonnie and being taken hostage in the Interceptor. As Gunnar became increasingly enveloped in the images rolling through his mind, he became aware that he was experiencing the emotional effects associated with each memory.

He relived all of the events that followed his meeting with CJ Barker. The images slowed down to the long tedious work of repairing their damaged ships after the wormhole. Gunnar became aware of tears rolling down his cheeks as he passed through these images, the emotional pain of Audra's death as vivid and as raw as when he had originally experienced them. His senses heightened as he watched a burial pod materialize from the cosmos, growing larger to his view, then dematerializing, only to reappear in the teleporter room of the *Constellation*. He could see his fingers on the control TAs, several crew members lifting the pod into place on the teleporter pad. He could feel the intense pain and the drain of the energy it took to remain composed through the entire funeral proceedings. The images continued to roll backward with an intense gush of emotion as he stumbled out of sick bay right into Rick and Jayda's arms, then images of holding Audra in his arms and their last moments together. The images sped up again, but then stopped as they tumbled through the wormhole. For several passing moments, he looked around at the images surrounding him in his mind. He stood in the engineering vestibule, looking into the darkness of smoke reaching out at him, a form emerging from the smoke and a bright flash developing just behind the figure. He carefully studied the scene from every angle, moving around inside the scene with his mind's eye. The scene suddenly wound back a little further, this time as Gunnar was just entering the vestibule and hitting the air control.

No, he tensed up in his mind. *Please no,* he thought. *Please, I'm begging you, don't make me relive this!* He concentrated, trying to pull away from the scene with all his might, but it felt like the memory was being torn loose from his mind's grasp.

The scene started, moving slowly forward at first, then in just under real-time. Gunnar watched it all happen again, this time as a bystander, but feeling everything as it happened while reacting to the surroundings and conditions. He crashed into the engineering door and it opened, releasing a toxic cloud of smoke. Barely able to roll sideways, he hit the air control and the smoke was instantly pulled back. Shards of composite pieces launched in every direction as the groaning of twisting metal rang familiar in his ears.

"Get out of there!" he yelled as a figure stumbled through the smoke toward the engineering door. Gunnar tried to move; to somehow step through the smoke and pull Audra through sooner. She would avoid the falling ceiling that was coming down the moment she came through the door. But he could only watch himself stand aside as Audra made it through the archway only to be crushed by a cascade of heavy debris. The instant she was buried, the image stopped and rewound, taking him back to just before the explosion, then it started again in real-time. This time the scene played out all the way through to the point where Audra passed away in Gunnar's arms. The scene rewound a couple of times and played through at regular speed, stopping where the burial pod faded to a pinprick in space.

Finally, the images melted back into the back of his mind, fading away into memory and leaving him feeling exhausted. Darkness filled him, but as he blinked through his blurred vision, he detected a short figure moving toward him. It took him a moment to recognize the form of Blinda Koss stepping up next to the examination table. She carefully wiped his tears and took his hand.

"You're full of all kinds of interesting things," Blinda said, leaning a little closer and gently stroking his hair. "You've got some serious emotional baggage. Was she your wife? Tisk, tisk, that's some rough stuff." She wiped Gunnar's face with a cloth. "I could have gone back further; found out all sorts of fun things about you and your culture, but there are some more recent events and people I need to find out about first. Doctor Sann seems to think too much of the memory grab can do more harm than good. Personally, I think you can handle far more."

"Miss Koss," Doctor Sann said, stepping up behind her. "He needs rest. The memory grab process is extremely taxing on the human mind. He needs time to recover."

"As you well know, he's not human."

"I must protest. You just had me put this man's mind through five times the normal emotional trauma a person endures under stress of this kind. He needs time to decompress."

"He can decompress when I say he can," Blinda hissed. She gave Doctor Sann a quick glare. "Now get out."

Doctor Sann stood firm until he felt the pressure in his ears suddenly build up. Wincing painfully, he finally turned and disappeared through one of the other doors. Blinda turned back to Gunnar and leaned back down close.

"I imagine you want to tear me apart," Blinda said, looking into his tired eyes.

"It's not because I'm still restrained," he rasped.

"I believe you," Blinda said, slowly working to remove the restraints. "However, you're welcome to give it a go when you're ready."

"Give me a minute or two."

"While you're working yourself up to it, I'd like to ask you some follow up questions to what we just saw."

"Why don't you just keep using your mind thingy on me?"

"Because, while it's very good at recalling all your deepest memories, you can subconsciously manipulate the images, causing them to blur and lose detail. Besides, this gives us a chance to get to know each other a little more... intimately."

Gunnar shook his head, smirking.

"You don't get out much, do you? You're just here doing Drax's dirty work."

Blinda reset herself and leaned in even closer.

"I can see you misunderstand my relationship with Drax. You're under the impression I serve the Queen Captain; that I'm her protector."

"Unless you're willing to tell me otherwise, what am I supposed to think?"

"I serve the Empire, not Drax."

"So what detail are you hoping for?"

"Who was the girl and the other woman with her, and what was so important about Reako that you went back?"

"The girl's name is Kara... Americ. She was shipwrecked there. We went back for some of her personal effects."

"Seems a little risky for a bunch of clothes."

"It was dicey, but when you've got the skills..."

Blinda grinned playfully, rolling her eyes.

"You're certainly full of yourself. But to your credit, you and Colonel Barker have an uncanny knack for getting out of some of the tightest predicaments imaginable."

"We're kind of cool that way."

"What were you doing on Dither and who were the others with you?"

"Just a couple of lady friends. We were on a nice field trip until you and your boys showed up and started shooting up the place."

"Nice try."

"Just trying to help a couple of friends get back to where they belong."

"Why risk capture to get onto the *Tarzana*? Bold move; something I would have done. Who would think of looking for you so close to home? Taking Captain Bridger's identity and seducing his widow. That was really smooth. I wonder if you can be just as smooth with me?" Blinda slithered, drawing in closer. "Do you suppose Malina is as good a lover as I am? With how I can manipulate matter, can you imagine the pleasure we could have blending our DNA?"

"I think your mind is a little warped," Gunnar responded, pushing her away. Blinda pulled back, visibly frustrated.

"Where did Miss Bridger make off to? I'm certain you and Colonel Barker discussed it."

"Yes we did. It was a delightful conversation in the Black Tiger maintenance bay."

"She and Diord have something in mind at the Rite of Pintar."

"You'll have to ask Diord and CJ about that."

"Come on," Blinda pressed him.

"Think about it," Gunnar said. "CJ and I can plot and scheme until we're blue in the face, but unless she was able to take the *Tarzana* to Cross or get a message off in the amount of time that's passed since I last saw her, there's no way for me to know what they discussed, so you'll have to fill in the blanks and figure it out when we get to... Did you say where we're going?"

"Calliope."

"I thought the *Tarzana* was your big bad ship? What's Calliope?"

"It's a space station of sorts." Blinda frowned, thinking. *He's mostly correct. CJ's been monitored constantly since she delivered Gunnar to the Tarzana's infirmary. It's been impossible for her to send anything to anyone without me knowing it. Besides, CJ isn't that stupid.*

"I'm glad I could fill you in on all the details." Gunnar tried to sit up a little in his bed, but the pain in his head and neck prevailed on his ability to move.

"Perhaps I need to have a short interview with Colonel Barker," Blinda suggested.

"You think you're going to get more out of her?"

"Properly motivated," Blinda countered.

"You don't think her big sister would have anything to say about that?"

"What she doesn't know... You forget, this is the Albion military. People have accidents all the time and sometimes they disappear. At this point, both of you have about out-lived your usefulness."

"I'm getting bored..."

"So am I," Blinda growled, sticking one of her long, sharp fingernails into the underside of Gunnar's chin. Getting no response, she pressed harder. Gunnar only smirked.

"You didn't consult with Doctor Sann before you came in, did you?" He stared blankly into her dark, mascara painted eyes.

Blinda suddenly found herself spinning through the air and slamming against the wall behind Gunnar's bed. Her wrist smarting and head spinning, she looked around the room. Dazed, it took her a moment to figure out why everything was upside down. As her wits suddenly came to her, she realized that she was crumpled against the wall on her head. Before she could right herself, she was grabbed by the leg and sent flying across the room. Still trying to regain her wits, she watched the room spin madly, only catching a glimpse of her attacker just before she slammed into the far door. Blacking out momentarily, Blinda lay motionless, hearing approaching footsteps. Just as she regained consciousness, a vise-like hand gripped her throat and picked her up. Gasping, she grabbed at the hand, trying to pull the fingers loose as they began to constrict. Her eyes bulging, she looked down at Gunnar. His countenance on fire, his eyes flashed different colors as he narrowed them at her.

"I don't respond well to... external stimuli."

Blinda flailed her legs at him, but he held her safely out of reach. As Gunnar tightened his grip, he suddenly felt his ear drums compress, massive pressure squeezed at his already tortured head and back. There came a jet blast of air and something else that knocked him off his feet, sending Blinda tumbling. Coughing uncontrollably, she rose to her hands and knees holding a hand to her throat and looking through her jet black bangs for Gunnar. He was just getting to his feet and coming at her again. She had just enough time to twist a wrist at him, sending a vicious swirl of air at him. He instantly upended and slammed violently into a wall. Still coughing hard, Blinda staggered to her feet and using both hands, clawed at the air directly in front of her. Gunnar felt his arms sink into the wall. Shaking off the collision, he looked at his new restraints, then at Blinda who was rubbing her neck.

"Really? You're gonna go with this?" Gunnar pulled his shoulders in a little and his arms popped back out of the wall. In an instant, he was back on his feet. Trying to hold him back, Blinda threw another blast of air at him, but he was ready for it and leaned into it, still coming right at her. The moment before he reached her, she jumped. Gunnar nearly ran into the door behind her, but came to a sliding halt. He no sooner turned around than he found a razor-sharp half disc only inches from his throat. Breathing hard, Blinda gulped carefully while holding the other half of her glowing Balkrum back, ready to strike.

Gunnar was acquainted with the Balkrum and what it could do. Rick had acquired one during his adventures in the Nulark system. Demonstrating how the strange weapon worked, Rick had reinforced the need for extreme caution when it was in use. Just being this close to the popping radium field around the blade was burning Gunnar's tough skin layers. He looked across the snapping energy field at a wild eyed Blinda.

"I should cut you in two where you stand," she sneered, her face half covered with hair.

"I have nothing to lose," Gunnar snarled defiantly.

Blinda flexed to strike.

"Blinda!" Drax grabbed her forearm and pulled the glowing Balkrum away from Gunnar's neck. Blinda glared at a visibly infuriated Queen Captain. "Stand down!"

The Thane resisted Drax's restraining hand for a moment, then looked at Gunnar and backed away. Drax released her hold on Blinda, but kept her eyes on the Thane as she retreated.

"You need to keep your girl on a tighter leash," Gunnar remarked.

"Shut it!" Drax gritted angrily, producing a small hand grip from under her cloak.

Gunnar saw a tiny flash come from the end of the device, then found his vision spinning and going black. He felt himself hit the floor before he lost consciousness.

Fuming, Drax watched Gunnar drop and motioned for several medical attendants, who quickly loaded him onto a levitating med cart and rushed him from the room. Doctor Sann met them in the hall and checked his patient. Drax then turned to Blinda, who still had her Balkrums activated, but lowered to her side.

"Let me guess," Drax started with controlled anger. "You just happened by and he almost ran into your blades?

Blinda touched a control on both Balkrums and the snapping blades fell silent.

"I came for information."

"And did you get it?"

"No, he doesn't know anything more than what he's already told us." Blinda rejoined her weapons and clipped them back into place.

"So what were you gonna do, cut him in two?" Drax did her best to hold control.

"He became belligerent." Blinda started for the door, but found herself pinned against the wall, an angry Drax burrowing her fingers into both her arms.

"I don't know what you think you're up to here, but stay away from CJ and Gunnar. I will deal with them in my way. Have I made myself clear?"

Continuing to breathe hard, Blinda glared back at Drax.

"I serve only the Empire."

"You used to, but I'm not sure what or who you serve now." Drax squeezed harder, making it clear who was in charge. She finally released the Thane, letting her drop to the floor. Drax burrowed her eyes into Blinda for a moment longer, then exited the room, leaving the Thane to regain her composure and follow a short time later.

What is it?

CJ Barker watched with careful anticipation from the bridge of the *Tarzana* as the *Merc* and *Valdez* maneuvered into their assigned positions near Calliope. Similar in class to the *Tarzana*, these Alvadorian ships distinguished themselves with vastly different paint schemes. Watching the Alvadorian support ships weave around several fleet clusters, she detected a figure emerge from the command elevator.

"How many factions are left to arrive?" she asked one of her aides.

"The Tomplie is the only major player," the aide responded quickly. "There are three or four other smaller factions coming from the Spartus quadrant."

"Where's the *Executioner*?" she mumbled, turning to a short burly man coming to a halt in front of her.

"Another ship further out, Commander," the aide announced.

"Get me a tactical readout." CJ developed a broad grin looking at her friend. "Dalton." She was genuinely glad to see an old friend.

"Was this your doing?" Dalton asked, saluting and standing at attention.

"Enough of that," CJ ordered good-naturedly. "Of course it was my doing. You don't think Drax would have purposely put the two of us back together again?"

"Does she even know I'm here?"

"Right now she's up to her eyeballs in galactic brass, so what she doesn't know isn't going to hurt her." CJ tipped her head away from the command aides. Stepping some distance off, she found a comfortable wall next to the command lounge elevator. Dalton parked himself against an opposing window.

"I believe congratulations are in order," she said, folding her arms.

"I don't know about that," Dalton responded. "Have you seen the debrief and asset report from the Carolon mission?"

"That bad?"

Dalton paused, looking around, then let a heavy sigh go.

"You've seen these Starbirds up close and personal?"

"Yeah, been onboard one. They are an impressive piece of tech."

Dalton's tone turned cold.

"I sent in four heavy destroyers and umpteen corvettes, plus four squadrons of Black Tigers and Starhoppers. We lost nearly every fighter and they barely even had a chance to engage the Starbird. Instead, they were all tangled up with one of its fighters; one fighter! Most of our corvettes were either destroyed or disabled. Two destroyers were blown apart and one heavily damaged. The only reason the *Dominator* wasn't hit was because I kept it out on the Nulark's rim with Colonian security. Oh, and I should add the Torag is a complete write off. We managed to get it back to the *Dominator* before our life support completely failed." Dalton leaned a little closer. "CJ, one fighter against four squadrons!" Dalton's voice was elevated, but still hushed. "One fighter!"

"I thought your intel painted a different picture?" CJ suggested, smiling slightly. "You and Blinda engaged one of them at point blank range."

"Clearly they weren't very functional at the time. If we try to go toe to toe against more than one of those things..."

"Well, now Drax has one."

"Yeah, if she can even gain access. Their shields, even at low power, are impossible to get through. I bet the King Commander could order the *Indominable* to fire on it and have a hard time causing any real damage."

"Well," CJ said, trying to calm her friend. "While they are certainly impressive, they are from a different place and I don't think we have any business trying to take something that doesn't belong to us. Nor do I think we have the wisdom to use it responsibly if we did."

"Then tell me why we tried so hard to get our hands on one?" Dalton elevated in a whisper.

"You know the answer to that," CJ reminded him. "What condition was the Starbird in when you left the *Dominator*?"

"Commander?" an aide interrupted.

"Yes?" CJ and Dalton answered simultaneously.

CJ smiled at Dalton's embarrassed look as she turned to the aide.

"Yes, Ensign, what is it?"

"Here's the tactical you ordered on that ship," the aide said, handing her a display device.

CJ gave the display a double take and motioned for Dalton to follow as she hurried to the control pits for a closer look.

"That's not the *Executioner*."

"What is that?" Dalton asked, looking closer.

"Commander Barker," the scanning tech called. "There is no registry for that ship."

"Have Delta patrol take a closer look," CJ ordered. "Certainly is big," she commented leaning down next to Dalton.

"The Ratronians already have scouts on the way," Dalton said, pointing at the main tactical display.

"No doubt Admiral Shipley has the *Maxell* turned in that direction. Have Delta make a wide swing and come back." CJ started back up to the command deck.

Dalton lingered at the tactical station for several moments, then followed CJ.

"Any idea what this is all about?" he asked, catching up with her.

"No idea," CJ said thinking. "It came in with the Alvadorian delegation." CJ took a hand display from one of the aides and fingered through several screens.

"That's Administrator Vandmire's group." Dalton said.

"Yes, really odd," CJ said, studying the display. "He's only cleared for the *Merc, Valdez* and their support ships. That isn't a support ship by a long shot." CJ let a half chuckle go. "Diord, what are you doing?"

"You don't think he's planning on making a challenge for the throne of Hadrian, do you?"

CJ considered Dalton's question silently.

"If anyone were to do it, he'd be the one I'd finger."

"Doesn't the Queen Captain already have plans for the throne?"

"You know as much about her plans as I do."

Dalton looked out the bridge windows at the enormous ship in the distance as CJ continued to study her display.

"What would you do?"

CJ looked up.

"Dalton, you know I have no political aspirations."

"That's not what I asked."

The two started walking together, away from the aides.

"If you had the chance to restore the rightful heir to the throne, would you?" Dalton asked.

"The contention is that there is no rightful heir."

"CJ, I've known you for a very long time. I've served with, and under you many times. I've gotten to know how you think and what your true convictions are. Your somewhat irrational departure has really given me cause to wonder what you're up to. The Queen Captain appears to have only one thing on her mind. My guess is she's after the throne."

"She seems hell bent on getting it through strength."

Dalton looked away from CJ, carefully trying to put his thoughts into words.

"After what I witnessed over Carolon, any attempt at trying to harness the power of those Starbirds would only end in disaster. We can't possibly survive this conflict by escalating our military might with super-weapons and soldiers we have no understanding of. This will only consume us."

CJ stepped silently over next to Dalton and put a hand to his shoulder, both looking out at the large ship now turning its profile to Calliope.

"You know where I've been," CJ started quietly. "What you may not know is what I've been doing. I've been onboard the Starbird and like I said, it is a true marvel, but as you've pointed out, it's not ours. We didn't invent it and we have no idea how to use it. But more important than the tech, are the people who run it. They're just like you and me. They have hopes and aspirations. They have personal triumphs and deep tragedies, just like you and me. Have you noticed they haven't tried to start a fight with any of us?"

Dalton thought a moment, then looked at CJ.

"They've only defended themselves," CJ continued. "They come from a very far place and they just want to get back home, but we keep getting in their way. Did Drax tell you what they were after on Carolon?"

Dalton shook his head.

"A mineral called Asium."

"Never heard of it."

"No one has, but apparently I've been kicking it around that base the whole time I was there. It's what powers their ships."

"A rock?"

"Apparently a very special rock. Mill and shape it a certain way and it has the capability of producing immense power."

Dalton thought back to Carolon. *If one small Corsair class assault ship has the capability of inflicting that kind of damage on Hadrian vessels, what would a fleet of such ships be able to do? And what if they had even larger ships; Frigates, Destroyers or Battlecruisers?* He looked around the bridge of the *Tarzana*. "We don't even know how to use what we have…"

"No," CJ echoed. "Apparently we don't."

The two remained silent in their thoughts for a moment.

"How do we get out of this?" Dalton finally asked. "Where does it end?"

"We're either going to implode and break into all-out war that will destroy most of Hadrian as we know it, or…" CJ paused. "A few of us will do the right thing and sanity will rebalance itself."

"But what is the right thing?"

Dalton realized the alien ship had finished turning broadside of them and was now showing its entire profile. As his attention was consumed by its size, CJ's wrist band made a noise. After fumbling with it for several moments, she leaned back against the rail to watch as Ratronian and Albion scout ships buzzed around the enormous vessel.

"Commander?" An aide approached them slowly. "I have a message from Calliope Approach Control. The alien ship has been given clearance to enter fleet space. It will be posting with the Alvadorian cluster."

"Who submitted the request?" CJ asked, looking at Dalton. His gaze was still fixed on the alien ship.

"Administrator Vandmire."

"Where's Delta patrol?"

"The far end of their sweep."

"Bring them back in."

"Commander?"

CJ gave the aide a quick look, enough to get her point across and he hurried off to carry out the order.

"I'm transferring command of the *Tarzana* over to you," CJ said, turning to Dalton.

Dalton considered for only a second.

"You're doing what?"

"I need to be on Calliope."

"For what?" Dalton asked, turning to her.

"The right thing," CJ muttered. She looked to another aide close by. "Have my Flightstreak ready and clearance to Calliope approved through Approach Control." She turned to leave but hesitated next to Dalton. "The Queen Captain has ordered the *Tarzana* to remain on a fixed patrol on the outer perimeter of Calliope space. But you are the Commander of this ship now, and you decide what the right thing is." She grasped his shoulder and looking him in the eye, smiled slightly, then hurried off to the command elevator. As she disappeared behind the sliding doors, Dalton felt a sense of foreboding. Looking around the bridge of the great ship, he wondered if this would be his last command.

Going My Way

Billy's head felt like it was splitting wide open. *Maybe if I keep my eyes closed and don't move... Something is turned on... What is that?*

"Are you going to wake up or what?" a voice spoke quietly.

"If you can fuse the two pieces of my skull back together," Billy groaned. "I might."

"I think you've had enough sleep."

"How long this time?" Billy winced, cracking an eyelid.

"About twelve hours."

"Crap! I thought for sure it would be your turn to have the ray of death used on you."

"Wasn't a death-ray. You'd be dead if it had been. Hence the term, 'death... ray'."

"Says you." Billy rolled over and sat up tentatively, blinking as he looked around. "Glad the lights are off in here," he grimaced, putting his hands to his head. "The mean hussy lady shot me again, didn't she?"

"Yep," Logan replied quietly. "Here, this'll help."

The Interceptor pilot held out a gel pill. Billy took it as he worked his way to a small freight container behind him.

"I really don't like her very much."

"Me either," Logan admitted.

"So what's our situation? Isn't that the ship?" he asked, catching sight of the Interceptor in the dim light.

"We're on our way to a place called Calliope," Logan nodded while standing up and looking out one of the portal windows at the whirl of stars outside. "I suspect we're about to be the focus of some very interested people that don't have our best interests in mind."

"I'm not so sure there's anyone around here that has our best interests in mind. Any idea how long before we get to wherever we're going?"

"From what I've been able to decipher from the display over there, we should be just about there. After that, who knows."

"We can't let anyone get a hold of that bird," Billy insisted, pointing at the Interceptor.

"I know," Logan agreed. "I'm just not sure how we're supposed to get out of this one. The amount of power left to run the guns on that thing would never be able to blow those cargo doors open."

"They've got to open them up sooner or later," Billy pointed out. "We just have to be ready with the engines running."

"Yeah, and every gun muzzle they have, big or small, will be pointed right at us."

"At least we'll have a fighting chance."

"You said it yourself, no one can get their hands on it. If we try to fly it out of here, maybe we'll get clear of the doors and maybe we can get past whatever they might have trolling around outside, but in its current condition, there's no way we could ever outrun whatever they have."

"Then I'll fix it so we can outrun them," Billy said, getting to his feet and staggering toward the Interceptor.

"You can barely stand," Logan said leaning against a nearby airlock hatch. The stars outside suddenly shifted and became still. He turned and peered through a small window in the hatchway, noticing a cluster of large ships sweep past his view.

"Welcome to Calliope, gentlemen," Seelix's voice echoed through the cargo bay com system. "Looks like we've arrived right on schedule. After I've secured the ship from landing, my droids and I will be down to introduce you to some important clients of mine. For those of you who are interested in all the technical navigational stuff getting through the mess out there, please refer to the courtesy monitor near the outer airlock. The Rite of Pintar has attracted a lot of attention and there are plenty of ships out there. Makes my collection look kind of small. We should be landing in ten to fifteen minutes."

"I can't do anything with this thing in that amount of time!" Billy cried facing Logan.

Logan looked down at the display near the airlock hatch. Ships were clustered everywhere outside. He glanced up, seeing a large, lighted asteroid pass by his view. Looking back at the Interceptor sitting in the shadows, his thoughts began to race.

No one can ever get their hands on this technology. Even if me and Billy could get this fighter out of the cargo bay in one piece, we aren't likely going to be able to fight our way through all these faction ships and it would fall into the wrong hands anyway. No, the Interceptor has to be completely destroyed.

"How hard would it be to set the light speed engines to overload?" Logan asked, still thinking deeply.

"Overload?" Billy asked, trying to see up into one of the open access panels of the fighter. "It was about to overload when we popped out over Moyle."

"How fast can you do it?"

"Take about two minutes. But if you do that it will vaporize this transport and everything on it, including us. There won't be anything left..." Billy stopped and turned back to Logan. "You're not suggesting..."

"Billy," Logan affirmed quietly. "It's the only way."

Billy opened his mouth to object, but stopped, recognizing the logical outcome of what lay before them. The engineer glanced over at the display, seeing the flightpath of the transport as it made its way through the masses of ships surrounding the asteroid. Without any more objection, he turned for the lower hatch of the Kalamarion fighter.

"Billy," Logan called after him. "Make sure she sees it coming," he said pointing at a camera near another door in the cargo hold.

Billy didn't stop, disappearing up the hatch of the big fighter. Moments later, several landing, navigational lights and beacons began to flash, followed by the engines starting to wind up. Another minute passed before the engineer remerged.

"How long will it take?" Logan asked as Billy joined him next to the airlock.

"Five, maybe six minutes."

"That's a long time," Logan complained.

"You're in a hurry to get blown to bits?"

"More worried about what's gonna try coming through that door to stop it," Logan said pointing at an entry hatch.

"Excuse me, gentlemen," Seelix's voice rang out over the com system. "What are you doing?"

"Just making sure everything is working for your clients," Logan called back.

"That's not necessary. Please shut it down. There's nothing to be gained by trying to escape."

"Who said anything about trying to escape?" Billy yelled over the noise of the revving engines.

No response.

Billy tapped Logan's shoulder as he bolted to the cargo hatch and started pushing heavy containers and equipment in front of the door. After several moments of grunting and heaving, they had the door blocked and paused to rest. Billy started for another hatch on the other side of the hold, but noticed Logan wasn't behind him. The Interceptor pilot was working with something near an airlock hatch. Billy started towards him, but came to a halt when he recognized what the pilot was holding.

"Oh no. Not a chance!"

"Come on!" Logan hollered over the noise. "It's our only ticket out of here."

Billy looked back at the first cargo hatchway as the mechanism flashed active.

"Pretty sure I'd rather get blown to bits."

"Just get it on!" Logan yelled as he scrambled to put the suit on.

"If I die, I'm gonna kill you!"

Billy struggled with the suit as Logan grabbed a couple of helmets and worked the airlock mechanism. He looked to the far end of the cargo hold as the other entry door swung open and a bright stab of light streaked into the hold. Moments later, the hold lights flashed on. Logan grabbed Billy by the helmet ring and pulled him unceremoniously through the airlock hatch, closing it behind them. Touching the control pad, the door instantly locked and the airlock started to decompress. He barely got his helmet locked in place and his air turned on when he saw several Vivitar droids surround the Interceptor. As the airlock indicators slowly sank, he turned and helped Billy with his helmet, snapping the lock in place. Logan smacked the air control on Billy's suit just as Seelix stepped into the hold. Walking around the throttling fighter, she finally turned to the airlock.

"Ready for this?" Logan asked as they stood at the outer hatchway.

"No, I'm not ready for this!"

"Come on, Billy. You've done this a million times."

"Not with a ship that's about to get blown to smithereens!"

"Sure you have. You did it with the *Constellation,* and you had an armed missile in your hand. How is this any different?" Logan armed the emergency control on the outer hatch.

"Logan? Billy?" a voice called from the com system. Both men turned. Seelix was standing right outside the inner airlock door. "What are you hoping to gain here? Shut this thing off before it tears up my cargo hold."

"It's not supposed to tear up your cargo bay," Logan said, touching the door control. In an instant, both men were catapulted from the airlock, streaking away from the ship.

Seelix's eyes widened. She turned around observing smoke pouring from several of the open access panels on the Interceptor. Looking down at a suit laying at her feet, she glanced back at the airlock hatch. Turning back to the growling fighter, she noticed sporadic flashes beginning to issue from the back. Shaking her head, a smirk developed just as a brilliant flash of light filled her vision.

*　　*　　*　　*

Audra sat quietly at the helm of the *Intrepid* feeling a sense of Deja Vu. It took her some time to recall why this felt so familiar. This was the exact position she was in during the time the *Constellation* was

stuck in the gravity well of the Oneida Caldron waiting for rescue. Then, like now, all she could do was wait. She glanced out at a sea of steel directly in front of the Starbird. If Rick called, she would be the first ship out. Jayda's ship, the *Montego*, sat directly behind the *Intrepid*. The hope was that they would not be needed. For now, it was best to stay inside the hold of *Calypso*, where they could remain undetected from the Hadrian ships.

Her attention was drawn to Dãsha, squirming in her seat at the science station.

"Problem with the chair?" Audra asked, turning her chair around.

"I'd be more comfortable if I weren't wearing this confining body suit."

"You're going to complain about what you're wearing?" Audra asked bewildered.

"My skirts or shorts would be far more comfortable."

Audra rolled her eyes. She and Jayda had had several conversations about Dãsha's taste in attire and they had both tried their best to help Dãsha understand the principles of modesty.

"If you're going to wear something like that, you might as well wear nothing at all."

"Nothing?" Dãsha responded, settling. "Certainly not. That would be entirely unhygienic."

"Not to mention I wouldn't want your naked butt on my seat," Audra snickered as Dãsha examined a flashing symbol on the vast glass console in front of her.

"What does this mean?" Dãsha asked, pointing at the object.

Audra came out of her chair, arriving at the science station at the same time as Alex 7001.

"Another ship coming in for a closer look?" Audra asked, leaning in.

"No, Commander," Alex corrected. "This ship has just dropped out of light speed and is coming up directly behind *Calypso*."

"How long before it makes contact?" Audra looked closer to the displays watching the target approach.

"Five minutes."

Audra considered carefully, then moved over to the communications panel and touched several points on the glass.

"Jayda?"

"I see it too. You're lead ship," Jayda said. "How do you want to play this?"

Audra sat down and looked at the hovering droid.

"Alex?"

"One moment. Ship coming into view now." There came several flashes on the scanning screens, in the top right quadrant. "Commander," Alex bellowed. "Detecting Kalamarion transponders."

"What!? Whose?"

Alex hovered closer.

"Captain Logan Dalley and Chief Engineer William Moon."

"They're on that ship?"

"Affirmative, and something else," Alex said, studying the readouts in front of them. "Interceptor two from the *Constellation*."

"Maybe they've hitched a ride here?"

"Whatever they're doing, something is wrong," Alex said. "The Interceptor's light speed drive is on a build up to overload."

"How much time?" Audra asked, looking closer.

"Estimate three and a half minutes."

"That'll blow that ship to pieces," Audra surmised.

"And anything close to it. Standby..." Alex paused. The droid shifted his head back and forth several times, emitting a short string of chirps. "The Captain and Engineer have ejected from the ship."

"Ejected?" Startled, Audra looked up at an overhead monitor, watching the transport ship streak toward *Calypso*. "Are they still alive?" Audra touched several more points on the display in front of her. "Mister Dacey, this is Commander Atlanta. Standby to open the hangar bay doors."

"You want me to come with you?" Jayda asked from the other open channel.

"Life signals from both transponders," Alex confirmed.

"Better not," Audra responded, scurrying back to the helm station. "I imagine I'm going to get enough attention, but be ready in case we get into a bad situation. Mister Dacey, get these doors open, now!"

The rear console swung out from under the front as Audra sat down and pulled her headset down into place. Tapping several spots on the console in front of her, she glanced up as the enormous hangar bay doors began to spread apart, revealing the cosmos outside.

"As soon as I'm out, get them closed," she ordered as she brought her ship into a hover. The moment the doors were open far enough, she took hold of the control yoke and pushed the throttles forward. Watching the rear viewer as the doors immediately began to close behind the *Intrepid*, she banked the Starbird hard, skimming along the hull of *Calypso*. "Alex, do we have enough power to teleport them at the same time?"

"Barely," the droid responded.

"Get me trajectory telemetry, then go man the teleporter." Audra glanced back at the scanners and turned forward. "Be ready to bring them aboard on my signal."

As the stern of the great ship loomed ahead of her, she glanced at the readouts in front of her watching a pictograph appear showing the trajectories of the two transponder signals. Fingering the right glass display, her eyes widened realizing where they would end up. As

Calypso's huge thruster bells appeared, she pushed the throttles ahead and swung away from the giant ship.

"This course will take your friends very close to *Calypso*," Dãsha announced, pointing at the flightpath depicted on her screens.

"Yes, it will," Audra agreed, focused on following the flight parameters.

Dãsha looked out at the starfield ahead of them. It looked as though Audra was flying off into nothing. Calliope appeared, streaking across their view as Audra continued to navigate along the calculated flightpath for intercept. She pushed the throttles a little further forward, watching her displays as they converged on the transponder signals.

"This is going to be close," Dãsha warned, watching her screens.

"Alex, standby," Audra said looking out at two tiny specs hurtling toward the hull of *Calypso*. She checked her left display and adjusted the *Intrepid's* flight path. Pulling back on the throttles slightly, she gently twisted the yoke.

"Alex, do you have a lock on them?"

"Ready, Commander."

"Bring them aboard."

Several agonizing seconds passed. Suddenly, the transport exploded, disintegrating into a streaking mass of molten ship parts. The blast was violent enough that the shock wave buffeted the *Intrepid* as Audra banked in a tight circle back the same way they had come.

"Alex?" Audra called. Several more seconds slipped by in slow motion. "Alex!"

"I've got them, Commander."

"As soon as I've got us secured inside *Calypso*, have them report to the bridge," Audra said, relaxing a little as the hangar bay doors came back into view. She looked back at Dãsha. "Does it look like we're going to get away with this without being noticed?"

"If I understand this correctly, there are no ships that are moving in our direction," she replied.

Audra felt her heartbeat relax a little as she set her ship down in the hangar bay and the doors had closed. Looking out the forward windows, she could see the *Montego* now staged in front of the doors. That suited her fine. She had taken a chance and appeared to have gotten away with it. After securing the pilot's station, she stepped to the command chair and sat down, turning it toward Dãsha as she continued to study the readouts in front of her.

"Remind me to transfer to anywhere," Billy Moon croaked looking back at Logan as they stepped through the bridge door. "Away from you!"

"Reporting as ordered," Alex announced, heading to the engineering consoles.

"Thank you, Alex," a voice from the command chair replied calmly. "Gentlemen, I'm sure you have a tall tale to tell." The figure in the chair remained facing the science station.

Billy and Logan developed an uncertain look and took a tentative step forward, but instantly froze when the chair turned toward them. Lieutenant Commander Audra Atlanta smiled as she stopped the chair's turn.

"Welcome aboard the *Intrepid*," she said, loving the looks on their faces.

Billy and Logan gave each other a glance and stepped a little closer.

"This isn't funny, Alex?" Logan bristled.

"No, it is not," the droid replied without turning.

"Is this some sort of hologram or something?"

"No, Sir."

"Come on, Billy, Logan," Audra said, bursting to her feet with her arms open. "It's really me; Audra." She didn't wait for the stunned men to gather their wits, throwing her arms around them and pulling them close.

"You can't be here," Logan said, not believing his senses. "You're dead."

"Do I feel dead to you?" she asked, holding on to them.

"Well, no, but... I don't understand?" Billy asked, realizing she was very real.

"I would love to sit down with you guys and explain everything." Audra let go and stepped back. "But as it's all still a little fuzzy to me, it might be best to wait for a better time. Right now, there are bigger problems to be addressed."

"I don't know," Billy said, still in shock. "This seems pretty big to me."

"What's this all about?" Logan asked, looking around the bridge. He could scarcely concentrate on reality right now.

"First things first," Audra said, turning back to Billy and Alex 7001. "Engineer Moon, effective immediately, I'm relieving Alex 7001 of his engineering duties here and making you Engineer in Command of the *Intrepid*. Alex, you will report to the *Phalanx* and assume engineering duties there. Captain Dalley, you will report over to the *Paladin* and assume the Engineer in Command duties there."

"*Phalanx*? *Paladin*? Engineer in Command?" Logan repeated. "I don't know anything about engineering. I'm a fighter pilot. "

"You know more about it than who they have over there right now," Audra replied, becoming serious.

"Who's they? We need more information here, Commander. I can't fix things," Logan objected. "I blow stuff up." Logan was confronted by Audra's stern expression as she turned and faced him.

"I've only got time for the abbreviated version of this story," she said, folding her arms. "The *Intrepid* is one of four ships found in the holds of two Kendalon freighters General Niker rescued from the factions of this galaxy."

Kendalon? Billy mouthed from behind.

"We have no idea why they were there, but here we are. Right now, the *Constellation* is being held inside that asteroid," she said pointing at the images of Calliope. "The *Athena* went in after it. These four new ships currently have skeleton crews onboard and most of them have never seen a Starbird before. They have little to no idea what they're doing. I need officers to take command." Audra softened a little. "Logan, I need both of you to help get the *Constellation* back."

"It certainly looks new," Logan said, sucking in a deep breath and looking around the bridge. "But you're gonna need pilots for Interceptors if things get crowded outside," he said.

"General Niker has ordered all Interceptors latched and locked for maximum ship defenses if needed," Audra replied. "Time is valuable right now, Captain."

"Understood, Commander," he finally nodded. "I'll keep you to your word about your story," he said, turning for the door with Alex beside him.

"As will I," Audra responded smiling.

As the two left the bridge, Billy scanned the engineering readouts for a moment, then turned to Audra.

"Commander, I know these birds are all supposed to be the same, but would you mind if I did a little snooping around down in engineering? Try to get to know her a little better?"

"I was hoping you would take to her as I have," Audra nodded. "Be ready if things start to happen."

Billy turned to leave, but stopped before he reached the door.

"Do you know if the Colonel is on the *Constellation*?"

"I really don't know where he is for sure." Audra's tone was suddenly subdued and distant.

The engineer looked back at her, trying to decide if this was reality or if he and Logan had been blown to bits with Seelix Monroe. Hesitating for only a moment, he finally disappeared behind the closing bridge door.

"Your friends deserve a better explanation," Dãsha said as Audra stepped up behind her.

"Yes, they do," Audra agreed. "I wish we had the time." Audra leaned closer to the displays in front of Dãsha. "Doesn't look like anyone noticed us outside."

"Perhaps the transport's destruction averted attention away from us?" Dãsha offered.

"Somehow, I don't think we could be that lucky." Audra rubbed the back of her neck and stepped back to the command chair.

"Are you still feeling the effects of your hibernation sickness?"

"It's not that. Gunnar used to tell me that anytime you're in combat, you'll feel uneasy. If you don't, then something is wrong."

"Perhaps you need a couple more containers of Ganaush," Dãsha suggested, turning her chair at Audra.

"I feel like I could use a good stiff drink, that's for sure," Audra said, leaning back and closing her eyes.

Let me in

As the Alvadorian shuttle settled to the floor of Calliope's executive hangar deck, Diord nodded to the com officer and made his way to the back where Jana and Tonnie sat quietly.

"It'll be just a couple of minutes," he said, sitting across from them. "Calliope security has to do a sweep of the ship for explosives, then we can get you moved up to the delegation quarters."

"This isn't going to work," Tonnie objected. "Someone is going to recognize me."

"A nervous android," Diord said, shaking his head. "No one is going to recognize you."

"How could you possibly know for sure?"

"I have delegate appropriate clothing and IDs for you."

"That may work for the Princess but it's not going to stop someone from recognizing me."

"Tonnie, you need to stop," Jana said, putting her hand to the android's shoulder.

"Was she this difficult with Gunnar?" Diord asked.

"No, but to be fair, Gunnar wouldn't have put up with it if she had," Jana smiled.

"Forgive my anxiety," Tonnie said, calming somewhat. "There are so many variables to go wrong in this situation that there is a continual assault on my prime directives. The most glaring? The Princess doesn't have all her tokens. She has her mother's ring, but has lost her father's. To claim her right as the heir apparent, she must provide both."

"Let's recap our earlier discussion," Diord sterned up. "This is our only chance to get her in front of everyone in order to make her case. The fact that we've gotten this far gives me great hope for success. Now, you've either got to back out of this right now, or we go forward at full speed." The Cross administrator gazed at both women for a few moments and was about to get up, but Jana stopped him.

"We have come this far with you, Administrator. Tonnie and I will see this through to the end."

"Sit tight," Diord said, moving to the rear hatch.

After a thorough search of the transport by Calliope security personnel, the Cross delegation was escorted out of the hangar bay

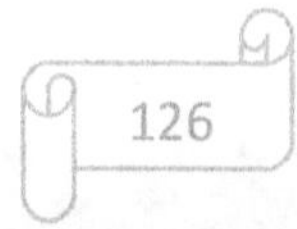

and whisked away to the delegate Multiplex. Once the room and the immediate area were inspected, Diord made his way to the delegate entry portal with one of his aides. Finding their way to delegation services, they secured an audience with the Deputy Grande Wamar.

"Thank you for seeing me on such short notice," Diord said, sitting across from the deputy.

"My apologies, the Grande Wamar couldn't see you himself. He's attending preparation meetings with the other royal delegations. You're cutting it kind of close, Administrator. Trying to get a presentation on the Rite program at this point is normally not allowed, but considering the Cross delegation is making the request, I think an exception can be made. Every faction in the galaxy would like to be heard. But as I said, there just isn't the time, so we have to pick and choose. You'll only be allotted a few minutes, otherwise the ceremony would go on forever. So..." the deputy picked up a viewing device and started scribing something on it with his finger. "What is the subject matter for your presentation?"

Diord's expression flattened, drawing a blank.

"It's a surprise," he fumbled awkwardly.

"A surprise?" the deputy repeated with a chuckle. "No, there are no surprises with the Rite of Pintar. Everything is choreographed down to the minute. Surprises don't work with a ceremony as momentous as this. I need a justification for the presentation and the subject matter. I also need the number of people involved and a resolution."

Diord stared blankly at the deputy. He hadn't counted on needing a rationale for presenting.

"The subject matter is..." Diord fumbled awkwardly.

"Historical Royal procedure," his aide spoke up, looking at the administrator.

"That's right," Diord continued, picking up on the aide's quick thinking. "Royal procedure from Hadrian historical records. I expect we'll have several factions represented during the presentation, so let's say about eight to ten people."

"That's a lot of people for a historical presentation," the deputy said slowly.

"Well, you know how some of them are. If you leave them out, they get offended and then we've got a squabble on our hands."

"I understand completely," the deputy agreed. "Can't have that."

"The hope is to educate all of the faction representatives," Diord said. "Certainly, many will be in attendance that have never participated in a Rite ceremony."

"That's very true. I'm so glad you brought this up. It makes perfect sense. I've got you scheduled on the floor directly after the formal introductions. Make sure you're at the presentation portal by the time marked on this portal key." The deputy handed Diord a chip fob and shook his hand.

Diord and the aide gave a huge sigh of relief as they made their way out of the delegate services suite and back toward the delegate quarters. His aide split off to run several errands before heading back to where the Princess and her guardian were housed. Diord took his time, making his way around the lounge areas, scoping out who was who in the represented political circles. As he started for an exit hallway, he turned a corner and stopped short. Blinda Koss was casually leaning against the wall directly around the corner. She was dressed in her usual dark colored, skin tight bodysuit, her weapons boldly displayed on her hips.

"Do you ever wear anything that hides your figure?" Diord asked, stepping closer.

"Right after I shower." Blinda reached back and flipped her velvet dress cloak across her shoulder. "How's this?" she asked, smiling playfully. She adjusted the clasp holding the material hiding her neck.

"Well, I think you've managed the spirit of the idea."

"So what do you and Drax's little sister have cooked up for this event?"

"Never were much for small talk." Diord started past her, but the Thane stopped him with a hand to his chest.

"I'd like a word," Blinda said, softening her hold on him.

"A word is singular."

"Some place a little more intimate?"

"I like open spaces."

"Come on, you're not going to hold the incident on Alvadore against me, are you?"

"I'm still a little pissed about it, so… yes."

"You were well compensated."

"Compensation wasn't the point."

"It's important," Blinda pressed him.

Diord looked around and seeing an empty, open booth that was somewhat out of the way, gestured toward it. After the two had situated themselves in the darkened booth, Diord leaned forward onto his elbows.

"Ok, so what is this word you want to have with me?"

"Can I get you something from the bar; my treat?" Blinda asked, maintaining the casual air. She touched a control on the table and pressed one of her fingers to a tiny pad.

"Nothing for me, but please, feel free."

"I usually do," Blinda grinned as a drink platform hovered over the table until she had taken her glass. "How was the trip?"

"What do you want?" Diord asked, impatient.

"Calm down. Can't a girl be interested in the wellbeing of her friends and associates?"

"Sure..." Diord said, looking around. "You tell me which girl in this room is interested and I'll sit and listen."

"Oooo, that's cold," Blinda responded unphased. "Tisk, tisk; come, come. You and CJ have something cooked up for the Rite ceremony."

"And you want me to tell you all about it?" Diord was a little annoyed.

"I want you to do the right thing. If not for Hadrian, then at least for CJ."

Diord considered for a moment while Blinda sipped her drink.

"Ok," he finally said. "Yep, we've got something cooked up, for the good of Hadrian. As one of the major factions, Cross will be presenting after the opening ceremonies and introductions."

Blinda eyed the administrator carefully, leaning a little closer.

"And what or... who are you presenting?"

"It occurred to me some time ago that there are very few still alive that have ever attended a Rite of Pintar, and well, you know this younger generation. These days, they don't take the time to do the research and learn about the intricacies of a ceremony such as this. I ask you, what's the older generation to do? So, we've put together a comprehensive presentation on navigating the Rite of Pintar. Sort of an ROP 101 for idiots."

Blinda stared blankly at him. Diord gave her a quick sarcastic grin and went to get up.

"Hold on," Blinda beckoned, grabbing his wrist. "A lecture on navigating the ROP? That's what you're gonna go with?

"Quite frankly, I don't give a flying Hartwig what you believe. I don't even care if you show up. This is a once in a lifetime opportunity and I, for one, plan to be a part of it regardless of who they put on the throne."

"Regardless?" Blinda repeated giving him the eye. "Do I detect a hint of doubt?"

"Come on, Blinda," Diord retorted calmly. "You can't possibly believe the Duchess or King Commander are right for the throne."

"Have you got someone else in mind?" Blinda fired back in a hush. "Come Diord. You and CJ have something more than a public service announcement cooked up. Maybe I can help."

"Is Drax presenting anything or anyone at the Rite?"

"I don't know." Blinda took a sip. "I really don't know what's going on in her head." Blinda paused a moment. "Who was Gunnar traveling with on Dither?"

"Huh? He was on Dither? When did you see him?" Diord let his attention draw away.

"On Dither, he was with two women. The older one looked familiar. Very alien looking; white pupils and hair."

"Well, you'd know more about that than I would," Diord dodged. "How did you manage to track him to Dither? How long ago was that?"

"I was given a tip." Blinda leaned down on one hand, letting her little finger slip between her lips.

"A tip?" Diord repeated, suspicious.

"Who was the woman?"

"Not sure why you're asking me," he said, shaking his head.

"Because you and CJ know who they are," she snapped, slapping both hands on the table.

Diord felt his ears pop, but remained aloof and passive. He looked around to see if anyone was paying attention to them.

"Have you asked CJ about these mystery women?" Diord pushed his words out confidently.

Blinda smiled, pulled her hands off the table and leaned back. Looking around and seeing no one was showing any interest in their conversation, she watched Diord as he melted back into his chair.

"What about these secret alien ships I've been accused of manufacturing?" he asked.

"Drax will hold you responsible for holding out on the Empire, while glorifying the fact that she captured one."

"This great super weapon that she believes will give her the right to rule Hadrian," Diord commented snidely.

"Without a blood heir, the balance of power must go to the most powerful. It's galactic law."

"I don't want to be around when there's a dispute about what faction that is," Diord said.

Blinda shifted, looking at him as a smile formed.

"How does it feel, knowing I know you and your girlfriend are up to something?"

Diord smiled back and got up, looking down at a seemingly confident Blinda.

"Probably the same as us knowing you don't know what it is." As he turned to leave, Blinda spoke up.

"See that information on the Rite is all you present," Blinda simmered, shifting her cat eyes after him. She looked out the large bay window of the lounge area at the crowd of ships clustered all around the asteroid. She could just see the tail end of the *Indominable* positioned near Calliope. The *Tarzana* was trolling space further out. Something was amiss, she felt it, but couldn't put her finger on what it was. *Gunnar's protector has to be here somewhere.* The fact that she couldn't sense him had her on edge. She finally gulped down the rest of her drink and left, stopping at an automated delegate kiosk just outside the lounge. After a moment of browsing, she headed toward the delegate Multiplex.

The Multiplex was a maze of winding hallways and stairs. Some moved from one level to another while others were accessed by escalators. For the guests that didn't want to wait or exert themselves, there were plenty of elevators. Some of the elevators moved horizontal and vertical. In a place like this, if a guest didn't have their information on them, they could easily get lost.

Blinda looked a little out of place as most of the guests were wearing lavish colors and fashions compared to her dark colored velvet bodysuit and high heel boots. Wearing a blaster on one hip and her Balkrum dangling from the other, garnered occasional looks from passers-by as she made her way through the labyrinth of walkways and light crowds, searching. After several minutes, she peeled off the main thoroughfare and made her way toward a large section of lavish apartments. The open foyer to the multiplex was several levels high, with natural lighting. Foliage was planted among engineered streams and waterfalls, giving the enormous room the look of a lush rainforest. As the foyer was void of anyone visible, she summoned the air and blew up to one of the upper levels. Silently touching down, she scanned the identification pads next to the doors lining the walkway. After some investigation, she stopped at an inset door and looked around. A couple of people had appeared down on the main terrace and one or two were moving about on the walkways below her, but her level appeared to be deserted. Looking back at the door, she held a hand up and closed her eyes. After only a moment, the door slid to one side and she moved inside.

Odd, the lights are on. They should be off, Blinda thought, probing the space in front of her. Taking several cautious steps, she entered the first living space. A large room with only a few pieces of furniture, decorated in a generic fashion. Several doors lead away to other areas of the apartment. As there was nothing of interest here, she kept moving deeper into the dwelling. The long hall opened into a mid-sized food prep and eating area with vaulted ceilings shared by another level. Several other lighted hallways were connected to the room.

As Blinda turned to investigate the nearest hall, she caught movement out of the corner of her eye and shifted back. A young woman stepped into the dining area and moved toward one of the dispensaries. After pulling something from an opening in a panel, she turned to a countertop. As she went to eat, she did a double take, seeing Blinda standing across the room. They looked at each other for a long moment, neither moving.

"I'm sorry, I must be in the wrong apartment," Blinda finally said. She stepped to the back of a chair, resting her hands in plain sight.

She looked carefully at the young woman for a moment. "I was looking for Administrator Vandmire."

The young woman hesitated a moment, then took a bite of what was in her hand. Glancing down, she quickly switched hands, lowering the other behind her back.

"Yes, this is one of his apartments," she said with a mouth full. "But he's not occupying this one right now. We are."

"We?"

The woman paused chewing and turned to the entry she had just come from.

"Yes, my mother and I. We are here for the Rite of Pintar, as the administrator's guests."

"Have we met before?" Blinda asked, looking at her carefully.

"I'm sorry; you are?"

"Blinda Koss. I'm an old acquaintance of the Administrator's. I'm here for the Rite as well. I thought I would drop by and do a little catching up. You look..." Blinda paused, considering. "Will you be participating in the Rite?"

"No, the Administrator has offered us his personal box."

Blinda nodded, still considering. The room remained awkwardly silent for a moment until Blinda turned back in the direction she had come.

"Well, I really should be going. It was nice to meet you..." Blinda stopped. "I'm sorry, I didn't get your name."

"Kara," the woman answered quickly. "Kara Americ."

Blinda paused again.

"Kara..." she echoed softly. "It was nice to meet you. Please let the Administrator know I dropped by. I'll try to meet up with him before the ceremony." Blinda turned, looking off for a moment, then disappeared back the way she had come. When the outer door had closed, Jana let out a relieved sigh and sank into one of the chairs close by. After she had settled for a moment, Tonnie stepped to the rail of the upper balcony.

"Is everything all right?" she asked, looking around at the room below. "I thought I heard you talking to someone."

Jana shifted her gaze to the android, then back at nothing.

"Your highness?"

"No, I was just imagining what I might say to the court of Pintar tomorrow if we get in."

Tonnie smiled and leaned forward.

"You'll be just fine. Remember, it's not what you practice to say. It's what you manifest to others from the inside."

"Someday, I hope to understand all the advice everyone has been giving me."

Mad Dash

Gunnar woke to the smell of fresh sheets and a soft pillow. He blinked a couple of times. The room was only partially lit, but enough to see he wasn't in the examination room any longer. He hoped that would be the last time he would have to endure the mind probe. The reality of living through the actual events was hard enough. This was a whole new level of torture and having to relive those memories over and over again, amplified exponentially, was more than he could manage.

"How are you feeling?" Doctor Sann asked, turning up the lights a little.

"I would love to have my glasses back. Other than that, the usual head, shoulders and back pains associated with being poisoned."

"I'm heading down to the lab for the antidote after we're done here. Do you think you can sit up?"

"My head is going to explode whether I'm laying down or sitting up." Gunnar forced himself up on one arm, while Doctor Sann helped him the rest of the way.

"I'd like to do a quick eye scan if you can manage it."

"Couldn't you have done that before, while I was asleep?"

"No, we transferred you to Calliope's infirmary while you were asleep. Besides, eyes function differently when you're conscious."

"Your eyes might. My medical people always had trouble getting anything out of their gear when they'd do a scan of mine. Wait a second... Transferred me where?"

"Calliope," the doctor said, holding a small scanner up to Gunnar's face. "The Queen Captain wanted you here for her big presentation; whatever that is." Doctor Sann mocked the idea of a grand presentation, then put the scanner up to Gunnar's eyes. The entire scan took only a minute or two and the doctor eagerly studied the information. "With the number of layers you have in your cornea, it's odd your eyes are so sensitive."

"It's their composition and function."

"I could spend half my tour of duty on just your eyes alone and probably still not fully understand how they work."

"I've been told I'm a bit unique. So what is this Calliope?"

The doctor looked at his wrist band and turned for the door.

"It's a large asteroid converted into a space station of sorts. Smaller than a moon, and instead of everything being built on the outside, it's all inside. The place is a labyrinth of crisscrossing tunnels and caves. The nice part about it is its internal atmosphere and gravity. If you're up for it, I'll have your breakfast brought in."

After Gunnar had eaten, several guards appeared, one carrying an odd-looking object.

"You've been summoned by Queen Captain Drax Blair. Hold out both hands." All the guards raised their weapons at Gunnar. It seemed pretty convincing to him, so he brought his hands in front of him. The guard then took the object and opened it, revealing a set of handles. "Take hold of both handles." Gunnar did as instructed, and the lead guard closed the top half around his hands and wrists to the bottom half and activated a control on the top, effectively binding his hands inside.

"What, no gag?" Gunnar asked, his voice dripping with sarcasm.

"Give me a reason," the guard said, leading the party out into the hallway where they nearly ran into Doctor Sann.

"Hello, what's this all about?" Once he recognized everyone, his tone changed. "What do you think you're doing? I haven't released this man!"

"Queen Captain's orders," the guard said, sweeping past him.

"This man is not fit to leave this facility yet."

"Take it up with the Queen Captain."

"I will," the doctor gruffed indignantly. He stuffed a hypo into his smock pocket and grabbed a small medical scanner. It took him several running strides to catch up and jump into the elevator just before the doors closed.

"Where are you taking him?" Doctor Sann demanded.

"Special Forces observation deck."

"For what?"

"Queen Captain's orders."

"Do you really believe a Queen Captain needs to tell her Captain of the Guard why she gives orders?" Gunnar asked, giving the doctor a grimaced look.

Doctor Sann pulled the hypo from his pocket and turned to Gunnar, but was stopped by the guards standing between them.

"You're not to touch the prisoner."

"I was ordered by the Queen Captain to give this to him when it was ready."

The guards held steadfast in his way.

"You can remind the Queen Captain of that as soon as we get to the observation deck."

"This man has advanced stage Kodiac Blu poisoning. I must give him the antidote before it's too late." Doctor Sann moved to put the hypo to Gunnar's arm, but was again stopped by the guards; this time a little more forcefully.

"Doctor Sann," Gunnar said with tired eyes. "It's ok. I've been without it this long, I think I can hold out for a couple more minutes."

Doctor Sann looked at the guards, then back at Gunnar's bedraggled expression. The circles under his eyes looked darker and his skin tone appeared to have changed. He recognized the discoloration on the back of his neck as the last phase of Kodiac Blu poisoning. Following would be uncontrollable madness, increased pain and then death.

The doctor backed into a corner, putting the hypo back in his pocket as the elevator continued to move. He brought his scanner up, trying to get a reading on Gunnar, but there were too many people in the elevator. The door finally slid open and the group came out, turning and making their way down a large hall. It was a busy corridor, with lots of crewmen moving back and forth in a frenzied manner. The echo of engines was heavy in the air as they made their way through the bustle, turning into a smaller, less traveled hall and down to another door. After a series of corners, they stepped into a long room lined with large paned windows on one wall. Still surrounded by his guards, Gunnar moved to the windows and looked out across a large landing bay. At the far end, beyond a set of open bay doors, he could make out glowing rock and metal structures pocked with lights.

A large group of fighters sat parked in a long line near the far wall. Another group of fighters were parked against the opposite wall. Several smaller ships were just making their way into the landing bay when Gunnar recognized the distinctive shape of a Starbird just inside the main doors.

Straining to ascertain its condition, his attention was drawn back inside the room as several aides and higher-ranking officers filed in. Some of the uniforms he had never seen before. He did a double take as CJ Barker entered the room, flanked by several more aides and then followed by Drax Blair and Blinda Koss, who were conversing quietly. Drax nodded several times, giving CJ and Gunnar several looks. Blinda gave Gunnar a quick glance, then slithered into a corner as another small group surrounding a single figure came in. A medium, unassuming man in a light colored uniform. The Albion Supreme Commander, King Commander Thoene Dismon had a look of only mild interest in what was happening. As his group settled near the line of windows, another group of similar makeup came in and took a position near the King Commander. Teleknee Duchess, Stephanie Benetar looked around the room, observing Gunnar. Her attention was drawn to the Albion Queen Captain when she caught sight of CJ Barker standing near Drax. After the group had taken a position of some comfort, Drax turned to everyone.

"Ladies and Gentlemen, let me introduce the Albion Empire's next super weapons," Drax said, gesturing to the windows. CJ covered a gasp, recognizing the familiar shape of the *Constellation* sitting motionless under the bright lights of the landing bay. Gunnar glanced back at his ship, taking notice of the blue hue surrounding it. Again, he tried to look for hull damage, but had to close his eyes as his head started to spin. He was finding it increasingly difficult to focus on anything.

Rescued at last, Gunnar thought. *They'll pick up my transponder signal and teleport me out of here. Then... Wait, no, that's not going to work. Ah nuts!* He felt himself starting to lean to one side. He opened his eyes to try and hold his balance, but went into one of the guards anyway. He heard a shuffling as a set of strong arms steadied him, then the voice of Doctor Sann.

"Pardon the interruption, QC Blair."

"Yes, Doctor Sann. How is our guest getting along?"

"Not well, QC. I have the Kalinite ready, but your guards refused to let me administer it."

"Understood, Doctor. As soon as I'm done with my presentation. Ladies and Gentlemen, if you'll turn your attention to the far end of the room, you'll also be the first to see what our genetics labs will be using to produce our highly anticipated super soldiers."

Everyone turned to Gunnar, the guards steadying him so everyone could see. Doctor Sann tried to get to Gunnar, but was unable to breach the gawking crowd.

"This is your super warrior?" Field Mashall Von Hoffen scoffed from the crowd. "Even if you didn't have him all doped up, he looks entirely too old for super soldier status."

Even in his deteriorated state, Gunnar managed to roll his eyes.

"Am I wearing a sign?" he grumbled, trying to remain alert and composed.

"It's not the exterior that we're talking about here," Drax gleamed. "It's what's percolating inside him."

King Commander Dismon pushed through his gauntlet of guards for a better look.

"I thought I had made myself quite clear, QC?" The King Commander barked, scanning the room for Blinda Koss. "How is it you think you can just blow by orders and expect to get away with it?"

Drax turned to the Supreme Commander and bowed flippantly.

"What?" she asked, acting surprised. "I haven't disobeyed any of your instructions. This man came aboard the *Tarzana* from Albia on his own accord. Neither myself nor any of the crew had anything to do with it."

"And what of Colonel Barker?" he asked angrily, turning to CJ.

"Again, I have breached none of the orders you've given me," Drax continued innocently. "My dear sister realized her place was with the Albion Empire and I accepted her back with open arms." Drax turned to CJ. "Although, I'm not exactly sure why you aren't at your post."

"How could I miss my sister's big moment?" CJ shrugged, folding her arms. Drax glared at her.

"What about the Nulark system?" KC Dismon demanded. "I told you to stay away from Carolon."

"And I have," Drax defended. "I didn't go anywhere near it."

"Clearly you did," the King Commander bellowed pointing down at the Starbird in the landing bay.

"You attended the faction delegation conference. You tasked me with the procurement of this ship." Drax leaned against the glass, looking down at the Starbird held to the floor of the landing bay. "Here it is. Commander SoKnack led a small task force into the Nulark with the help of Von Hoffen's tugs and Colonian system perimeter security. I have done everything I said I would do, all for the good of the Empire."

Thoene turned back to the crowd, a cold scowl permeating his expression. He hesitated, passing Stephanie Benetar, and moved in Gunnar's direction. The King Commander stopped in front of him and looked him over.

"This man has Kodiac Blu poisoning."

"Yes," Doctor San called out, trying to push through the crowd. Several guards held him back. "I have the Kalinite right here."

Thoene looked back at the Doctor, then shaking his head slightly, turned back to Drax.

"I assume you have everything you need from this man?"

"Almost," Drax said, taking a couple of steps closer. "I believe I've uncovered a plot that will implicate a number of people in this very room of the highest degrees of treason to Hadrian. This Colonian terrorist is one of a couple of people who have the details."

"Colonian terrorist?" Duchess Benetar croaked nervously. "The Empire has no knowledge of this man or his activities. What evidence do you have against him and how he's connected to the Empire?"

"Haven't you used a mind probe on him?" Thoene asked. "That should have exposed everything he knows."

"As he's alien, his resistance to the mind probe is considerable. While we were able to extract some information, his biology is just different enough that he's able to mask many crucial sequences."

"So how do you propose to uncover this plot and expose these traitors?"

Drax smiled confidently, ignoring a quick look from CJ.

"Since he is unwilling to give a full accounting of all his terrorist activities," Drax said, pulling a blaster pistol from under her dress cloak. The crowd quickly moved away from the windows, compressing against the opposite wall. Surrounding the King Commander, Gunnar's guards left him teetering by himself near the landing bay windows. "Let's see how well he responds to reality, shall we?" Drax smirked as she energized the power pack on the pistol.

"Drax," CJ protested quietly. "You gave me your word." CJ did her best to hold her voice down.

"I'm a dead man already," Gunnar rasped. "You'll get nothing more from me."

"No? Nothing?" Drax hissed.

"QC Blair," Field Marshal Von Hoffen stepped forward. "I must protest!"

The whole room was startled by a blaster discharge that landed over Gunnar's shoulder.

"Shut up, Von Hoffen, or we'll see how you hold up under a little questioning." Drax adjusted her aim a little, taking several steps towards Gunnar. "I've been more than patient on this issue." She paused a moment, considering. "But perhaps I've been going about this all wrong." Abruptly turning around, she swung the gun right at CJ's forehead. "How loose is your tongue now?"

CJ's eyes widened, forcing a gulp down past her pounding heart. She looked around her sister's determined glare, at Gunnar. He was barely able to hold himself up. She watched him shuffle to the window and lean against the glass. The conversation with Gunnar in the *Tarzana's* maintenance bay had made their path perfectly clear. Trying to read the silent message he might be trying to send her with his tired gaze, she stiffened up.

"No," Gunnar finally said. "It won't loosen my tongue." He tried to stand up straight and defiant.

Drax's expression sharpened, her eyebrows slanting down in frustration, seeing the fortitude on Gunnar's face. She looked back at her sister, confronted with the same defiance. Shifting her eyes to the crowd breathlessly watching, she abruptly shut the blaster off and pulled it back.

"No," she repeated tight lipped. "I don't expect it would. A different approach, perhaps…" Still a slave to her theatrics, Drax moved back to the window and raised her wrist band to her lips. "Move into position and wait for my command."

Everyone in the room stepped back toward the windows, but Doctor Sann was still unable to reach Gunnar. Out in the landing bay, all of the security gun emplacements and several large gunships turned their weapons to the Starbird sitting in the bay. Outside the open doors, the two escort corvettes reappeared, with their forward cannons trained on the white alien craft.

"How about we see how much of a mess we can make of this new ship?" Drax turned back to Gunnar.

"What is the meaning of this?" Von Hoffen demanded. "You finally bring this mystery ship in and now you want to blow it up? By your own reports, if you destroy it, it could open up half of Calliope."

Drax watched Gunnar's expression lighten, a smile drifting across his face.

"Don't stop now, you're on a roll," he smirked.

Clenching her fist, she holstered her pistol and turned back to CJ, catching sight of a shadow in a corner of the room. She gave the ship below another look, it's hull glowing a light blue.

"Stand down," Drax growled into her communicator. Outside, all the guns and ships turned away and parked. She turned to the Ratronian Field Marshall. "You'll also recall from those same reports, that Ilob mind waves were used to incapacitate the ship's crew. How about we see how well that will work here?" Drax motioned for the people around her to move back a little further. "Blinda?" she beckoned the Albion Thane to the window. Blinda remained still for several moments, all eyes on her. She finally stepped slowly from the darkened corner and stood in front of the window looking down at the Starbird.

CJ developed a panicked look, turning to Gunnar, who appeared to become agitated, his respiration ramping. Blinda took several deep breaths, then closed her eyes and held her hands out, projecting her thoughts into the landing bay, reaching for the crew of the Starbird. As she began to concentrate, a sudden commotion at the far end of the observation room interrupted her and brought the four guards back to Gunnar. A ruckus ensued, sending all four guards flying. As the crowd broke to get out of the way, Doctor Sann pressed forward with the Kalinite hypo, but was knocked down and partially buried by two of the guards and several onlookers. The Doctor struggled to free his hands, only to find the hypo gone. Managing to turn over, he spotted its crushed remains at Gunnar's feet. The blue Kalinite evaporating as it spread from the broken container.

Gunnar let go a throaty howl, his face turning pomegranate. He looked past Drax and Blinda at CJ. Startled at the fierceness in his countenance and the bright colors swirling in his eyes, she stood horrified. He grimaced, groaning painfully, then flexed his arms in a twisting motion. The binders encompassing his hands suddenly cracked and he pulled a hand free. Making a winding turn, he swung the binder at the observation window. The thick clear material fractured from the blow. He swung again. This time the blow dislodged the material from its frame and fell away.

As several security teams hastily ushered the delegates from the room, Blinda and Drax watched paralyzed as Gunnar scooped up one of the guard's rifles on his way through the broken window. Jumping to one of the bay wall mounted gun positions, he moved effortlessly from one emplacement to another until he had reached the floor. Turning back to the wall, he slammed the binders against its surface, finally letting the smashed remains drop from his other hand. He looked up at the observation window, seeing CJ and Field Marshall Von Hoffen looking back at him. A quick glance was all there was time for. The madness in his head was driving him now. He could think of only one thing; getting to his ship.

Drax finally snapped back to attention, grabbing Blinda by the hair and slamming her face against the glass.

"You bring him back or don't bother coming back at all." Drax felt her eardrums popping. Her clothing constricting around her body. Waves of pressure throbbed at the sides of her skull giving her an instant headache. Instinctively, she dug her fingernails deep into the Thane's neck. "Blinda, my dear," she hissed. "Maintain control or I'll snap your neck where you stand."

Blinda tried to push away from the glass, but Drax held her pinned digging her fingernails deep into the Thane's neck muscles. She could feel blood oozing from each entry point. Angry, Blinda gritted her teeth like a wild animal, scraping her fingernails down the length of the window.

"Now, go!" Drax howled, tossing the Thane toward the other end of the room.

Putting a hand to her neck, Blinda stumbled to the window and with a wave of her hand, blew out and down to the bay floor. She turned and thrust both her hands towards Gunnar as he galloped toward the *Constellation*. In an instant, his feet were swept from beneath him and he was sent flying over a parked fighter, disappearing from sight.

*   *   *   *

"Captain Abrams," Lieutenant Nevall called.

"What is it, Lieutenant?" Dakota responded, preoccupied.

"I'm tracking a personnel transponder."

"Clear your buffers," Dakota suggested. "I saw it earlier. Probably just junk caught up in the matrix."

"Already did that, Sir. Buffers are clean. I've got a strong signal here. Coming from the far end of this hangar bay." Lieutenant Nevall gave the information a double take. "Sir, it's Colonel Conrad's signature."

"What?!" Dakota jumped to the communications station for a better look.

Lana tapped the glass display in front of her, enlarging the image.

"How in the...?" Dakota looked up at the darkened windows, then turned to Mr. Pippin. "External viewers available?"

Pippin touched his controls and several overhead monitors immediately came on, zooming in on a figure running directly at them.

"It really is the Colonel!" Nigel exclaimed.

Stunned silence prevailed on the bridge as they watched their commander sprinting toward the *Constellation*. Dakota was about to turn to communications when Colonel Conrad was suddenly upended and swept into the air by some unseen force, disappearing behind a long line of parked fighters and assault ships.

*   *   *   *

Holding her arms out from her sides, Blinda Koss levitated and drifted towards the line of parked fighters. Not immediately finding her quarry, she landed and started walking from ship to ship, looking through the undercarriage and behind the thruster ports. She was about to try reaching for his mind, when she heard the sound of revving turbines down the line of craft. It took her a moment to figure out a direction and home in on a throttling fighter lifting from the bay floor. Running toward the pivoting craft, she threw both hands out in front of her, twisting her wrists. The fighter tilted sharply, its sweeping wing slamming into the bay floor. The ship spun erratically, its damaged wingtip nearly broken off. As the cockpit came into view, Blinda's eyes widened.

There's no one at the controls! She was suddenly upended by several explosions landing at her feet. Her ears rang with the deafening noise as she felt herself flung uncontrollably through the air. Trying to regain her wits, she hit the floor and tumbled. Stunned by pain, she came to a stop next to the undercarriage of one of the fighters on the flight line. A shadow passed over her and looking up she saw Gunnar at the controls of an assault ship. Half paralyzed with pain and unable to hear anything through the ringing in her ears, she slowly rolled over and up onto an elbow. She raised a hand and bent her index finger, then twisted her wrist. As Gunnar's ship moved away from her, it suddenly careened into one of the parked fighters and stopped. Still airborne, Gunnar backed his ship up and turned around, but the bedraggled Thane was nowhere in sight. He started cautiously moving back up the row, but finding nothing, turned the ship back toward the *Constellation*.

*   *   *   *

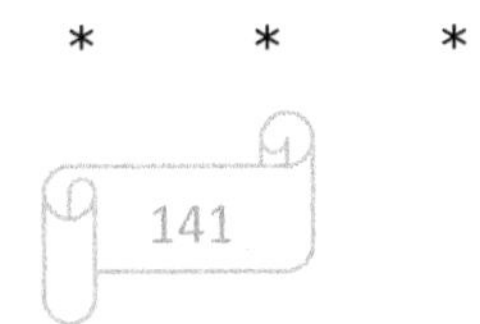

"Hail that ship!" Dakota barked glancing back at Lieutenant Nevall. He turned back to see the assault craft gliding back up the row, right at the *Constellation*.

"I've got something, Captain," Lana called out.

Dakota jumped back to the command chair and punched one of the armrest buttons.

"Colonel, can you hear me?"

"Get the ship out of here," Colonel Conrad rasped, barely audible.

"Colonel, are you ok?"

"Get out of here," came the repeated response. "I'll cover your six."

"What about you?"

"Don't worry about me, just get out of here!"

Dakota looked out at the small vessel coming right at them. He could clearly see Colonel Conrad at the controls.

"Captain," Lieutenant Hunter called from the ball turret. "There's another..."

Dakota watched anxiously as another gunship suddenly appeared, facing off with the Colonel directly in front of the *Constellation*.

"Starman, standby maneuvering thrusters. We'll drag those tugs out with us if we have to. Lieutenant Kramer, ready on your weapons," Dakota ordered.

*   *   *   *

Now facing the other gunship, Gunnar recognized Blinda at the controls. Both warriors wounded and their abilities stunted, they just stared at each other. Gunnar's mind was only partially coherent; Blinda's head was spinning and her hearing was still reeling with an unbearable ringing. Gunnar suddenly opened up on Blinda's craft and shoved his throttles forward. The Isom engines screamed, tight blue ribbons of flame spitting from its thruster ports. His assault ship jumped at the other craft, just missing it. As it passed over the top, Blinda trained her top turret at the under-belly. Bits and pieces of armor peeled off as the heavy ordnance bored large gouges into its thick metal skin. Temporarily out of control, Gunnar careened into several parked fighters and spun across the bay floor into two others. One fighter exploded, enveloping Gunnar's ship in liquid fuel and flame. Blinda circled the burning mass, examining her quarry's condition. The only thing she could see was a mass of metal in the middle of a pyramid of fire.

*   *   *   *

The landing bay's fire suppression system suddenly kicked on, followed by several automated fire control units racing toward the inferno. Before the suppression had a chance, the mass of fire suddenly erupted again, sending enormous sheets of flaming fuel in every direction. Unable to pull back soon enough, Blinda's gunship was completely enveloped in a wall of flame. Fighting the controls, she caught sight of Gunnar's assault craft rising from the inferno. There was no time to react as he opened up on her broadside, just behind the cockpit. Blinda blindly steered to get away, Gunnar pounding her. The prolonged strikes from his weapons rattled the very frame of her gunship and even as she tried to flee, she was sure vital parts were being torn away. Still taking damage, she spun around and brought her forward guns to bear. Gunnar continued advancing at her, guns blazing. Blinda's craft suddenly lurched to one side and crashed into one of the bay walls. She held the firing button down as Gunnar swung up and away, flaming fuel still running off his hull. Turning back to the *Constellation,* he opened up on the tugs holding them prisoner.

*   *   *   *

The bridge crew watched, amazed that the assault ship was still airborne, the last of the burning fuel dripping off the hull as it headed directly toward them. It suddenly opened up on the tugs holding the *Constellation* in place.
"The Colonel just popped those tugs like they were nothing," Lieutenant Hunter yelled into his com pickup. Hayden caught a glimpse of Colonel Conrad at the controls as the craft banked hard, pounding the bay security cannons. Blinda's gunship became airborne again and followed the Colonel around the hangar several times.

*   *   *   *

Feeling Blinda hitting him from behind, Gunnar passed directly over the bridge windows of the *Constellation*. They were in black out condition, but as he turned in front of the nose, he noticed the positioning lights blinking erratically and saw the turret on the aft section swing around. His assault craft continued to take hits from Blinda and the surrounding security guns in the landing bay. Gunnar throttled hard, steering beneath the fire suppression foam raining down over the burning pile of wrecked fighters.
Blinda followed, steering her gunship after Gunnar's smoldering craft as he circled inside the landing bay. With a sudden twist of the controls, he turned toward the open doors and out into the caverns of Calliope with Blinda in hot pursuit.

*   *   *   *

"Starman," Dakota exclaimed. "Lighten the windows and get this bird out of here!"

Lynette tapped the glass in front of her and yanked back on the throttles, taking hold of the control yoke. Looking down at her displays, she pulled on another lever as the Starbird sprang from the bay floor and backed smartly out the same way it came in. Hayden spun his turret around and opened up on the nose end of the corvettes waiting just outside while several other rear mounted guns on the Starbird began to fire. Astonished at the ferocity of the attack, the corvette commanders instantly pulled back. Once outside, Lynette twisted her yoke to one side while pushing her throttles forward. Answering the commands from the helmsman, the *Constellation* blasted away from the bay opening and streaked over the inner city of Calliope.

"Eyes and ears open everyone," Captain Abrams ordered, gripping his fist tightly.

"Not sure how you managed it," BachTL said. "I would have bet against this."

"Me too," Dakota was quick to admit. "No casualties, Doctor?"

"Yes there are; my wits," Fuji said, steadying herself against a console as they started taking fire from behind. "As in *scared out of*."

"What's your plan?" BachTL asked, watching the city lights fall beneath them.

"Not sure," Dakota answered, moving forward next to helm control. "Making it up as I go."

"We can't go out the same way we came in," Pippin said.

Dakota glanced back at the science officer, then forward as Lieutenant Starman turned the ship toward the opening in the asteroid.

"I agree with Mister Pippin, Sir," Lieutenant Starman said, looking at her scopes.

"As beat up as we are," Pippin warned, "I'm not sure it's a good idea trying to get through all of that."

Ahead of them, the asteroid opening was filled with the colossal form of the *Indomitable*. The massive ship couldn't fit into the opening, but its bulk worked well as a deterrent for any ship trying to get in or out of the asteroid city. Retreating further back into the asteroid caverns seemed to be the only course. Lieutenant Starman turned the ship into a tight bank and headed directly back at the pursuing corvettes.

"What can you tell us about this asteroid?" Dakota asked as the weapon's officer opened fire on the approaching corvettes.

BachTL thought a moment.

"This asteroid is comprised mainly of Corvantium. It's much like Obsidian rock; super hard. Good for cutting things to ribbons when they come in contact with it. It absorbs sensor scans really well and causes all kinds of problems with communications equipment."

"Sounds like a fun place to play," Dakota said, turning back to the science station.

"Could even the odds," BachTL suggested. "There are all kinds of tunnels running through this asteroid. Many of them exit all over Calliope's surface."

"You've been through these tunnels to these other openings?"

"When left alone with gravity bikes, boys will be boys," BachTL said from behind.

"What's it look like in here?" Dakota asked, turning back to Pippin.

"Hard to tell." Pippin tapped his glass and made several adjustments. "Our sensor scans are severely limited in here and the closer we get to anything resembling rock, the worse it gets." He glanced over at Lana. "I dare say communications are the same."

"Confirmed," Lieutenant Nevall nodded. "Channels have a lot of static in them."

"Do you still have the Colonel's transponder?"

"Yes," Pippin responded, looking at the tracking window. "Still quite strong considering the environment. He's flying low over the city, right here." Pippin pointed at a couple of returns on his displays. "I assume this one following him is Blinda Koss."

*   *   *   *

Watching the two combatants and the Starbird exit the landing bay, Drax growled into her communicator.

"Get those corvettes after that ship. Bring it back, now!"

"It won't do any good," CJ announced from behind. Drax swirled around in frustration.

"What have we got inside that will?"

"This is Calliope. There's nothing here designed to stop a warship, even a crippled one."

"Those corvettes and destroyers can take it out."

"You haven't read Dalton's report on the Warbird's operations in the Nulark over Carolon, have you?"

Drax shook her head.

"Remember the *Tarzana's* encounter with the other Starbird?"

"Where is the *Tarzana* now, and who's in command?"

"Commander SoKnack."

Drax glared at her sister.

"He's patrolling medium space around Calliope as per your orders." CJ struggled to hold a straight face. Her sister's plans were once again hemorrhaging. "The *Indomitable* is parked near Calliope's main portal, but I seriously doubt the King Commander is going to want you ordering his personal dreadnaught to mop up your mess."

Drax flared at her sister, who remained composed.

"The *Indomitable* is blocking Calliope's main portal. There's no way they can get out."

"There are all kinds of volcanic tubes in this rock," CJ pointed out. "It may be possible for them to find a way out."

Frustrated, Drax snarled, rising to a scream as she blew past CJ and Field Marshall Von Hoffen.

"Consider your security clearances revoked, Colonel! Don't plan any trips off world; I'll deal with you later."

CJ and the Field Marshall watched her sister storm out of the observation room, then looked back out into the landing bay.

"It looked like an impressive piece of hardware, Colonel," Von Hoffen said quietly. It was difficult to make out anything in the landing bay for all the smoke and fire suppression foam.

"That it is," CJ agreed, smiling slightly. "But I very much doubt anyone in Hadrian has the wisdom to have possession of it."

*   *   *   *

Dakota reached for the armrest of the command chair and touched one of the buttons.

"Tiana, how are things going back there?"

"A smoother ride would be nice, but surprisingly, everything is holding together nicely. Why?" Her last word had an air of suspicion mixed in.

"Do you have enough power for teleporter operations?"

"Boy, you never stop, do you? It was deemed a nonessential system, so no."

"Understood. Keep up the good work. We'll try to get you a smoother ride."

"It was a good idea," Fuji said, handing Dakota's pistol back to him and turning for the bridge door.

"Where you going?" Dakota asked, holstering his weapon. "Things are just getting good."

"That's code for *scary*. I don't do *scary*. I think I'd rather not know when we're going to get dashed to pieces."

As the door closed behind her, Dakota looked out at the tunnels ahead of them.

"I should be able to stay one step ahead of those corvettes behind us, but I need recommendations on a direction," Lieutenant Starman called from helm control.

Dakota looked over at BachTL.

"They all interconnect at different points back in there," BachTL shrugged.

"A real maze," Dakota surmised.

Lieutenant Starman took note of several intermittent returns on the tracking scopes in front of her and twisted her yoke control to the right and down.

"Ah, a target rich environment," Doran gleamed, taking aim at a cluster of assault ships converging on them. He noticed several other larger returns coming at them from the left. "So much to choose from," he muttered as he started firing.

"You're welcome," Lynette smirked as she steered the ship into one of the massive tunnels.

"You know me so well," Doran snickered, as the ship was swallowed by the bluish glow coming from the rock formations all around them.

"We've lost the Colonel," Pippin informed Dakota.

"Surprised you kept him as long as you did," BachTL commented.

Down together

Gunnar shook his head, trying to clear the fog clouding his mind, but the increasing pain in the back of head and neck made it difficult. The bright lights made it nearly impossible for him to see. Alternating which eye he kept open seemed to help, but the light from the structures surrounding him still made it difficult.

Atmosphere and gravity. What the heck is this place? While the assault craft was foreign to him, he recognized the characteristics of flying through an atmosphere. He pressed the throttles further forward, weaving the ship back and forth to keep Blinda from gaining any kind of a weapons lock. All around him were structures built on and into the glowing rock of the interior of the asteroid. Receiving a couple of hits from behind, he maneuvered closer to the structures. Even with his limited eyesight and concentration, he thought he was seeing civilian dwellings and people in windows. His mind in a swirl of confusion, he recognized this was not the place to be having a high speed chase with Blinda shooting at him.

Gunnar scanned his surroundings, searching for a point of reference, anything he could use to formulate a plan. He winced, pain stabbing at him behind one of his eyes and along the back of his skull. It was nearly impossible to see anything in the distance. The lights stung his eyes. Finding a dark place might mitigate his discomfort and allow him to see a little better. After lancing a straight line through what he would have described as downtown, he finally identified a possible route. Weaving further into the structures, he made a series of winding turns, then doubled back, trying to remain as close to the buildings as possible. After checking his tracking screens he pulled up and away, slamming the throttles as far forward as they would go. Steering for a dark spot he had located earlier, he checked his scope; nothing. He took several deep breaths and blinked, trying to clear his vision. He could hardly hold still. In addition to the knife points pressing on the back of his eyes and skull, the pain in his back was piercing clear to his tailbone now.

Can't let Blinda get to me… I need to keep her busy while the Constellation gets away.

Approaching the dark spot, he surmised it was an entrance to a natural tunnel shooting off from the main cavern. *Perfect!* He could

open both eyes now as there was only bare jagged glowing rock. He
made a slow climbing turn towards the tunnel walls for a closer look.
The walls were anything but smooth; jagged and unforgiving. Layer
upon layer of knifed spikes everywhere. Not a place he'd want to try
and land. Looking closer, he noticed tiny tubes leading into even
darker places.

As he pushed further into the darkness, an object appeared on his
tracking screen. *Blinda… Come on! Let's see what you've got.* He
pushed the throttles forward, maneuvering through a wide turn. As
they closed to point blank range, he was suddenly hit by a blinding
wall of searing pain piercing through his head. As if his condition
wasn't already unbearable, now every nerve ending in his body was on
fire. There was no piloting he could do. *This is the end.* Slumping
sideways, he locked his fingers around the triggers on the controls.

A dreamscape took him, seeing images of himself playing as a child
on Commenor. He longed for those carefree days. He wished he and
Audra could have retired together, like they had always talked about.
Maybe someplace like where Eldon and Rayna lived. Away from the
hustle and bustle of the mechanized and high tech world. As the
memories faded from his mind, they were replaced by his pounding
headache and the pain in his back. Bewildered, he sat up in time to
see the opening of the tunnel looming in front of him. Looking down
at his tracking screen, he saw nothing. He turned his ship around and
started back into the tunnel, noticing a brilliant flare and a column of
smoke rising from the rocks. Circling close, he surveyed the wreckage
of Blinda's gunship, but could see no sign of the Thane. As he turned
another direction, looking for a possible landing sight, his craft was
pummeled from all directions. A hail of splintered rock pounded him,
skewering clean through the hull, leaving large holes. A brilliant flash
from the rear, an unremarkable popping sound and all his consoles
went dark.

"Not good!"

Gunnar pulled his safety straps as tight as he could as he went into
a free fall. Looking down at the jagged tunnel surface rapidly coming
up at him did nothing for his already scrambled frame of mind, so he
clamped his eyes closed and grimaced, anticipating the impact.
Splintered rock continued to pummel his ship even as he struck the
ground. The craft broke apart and tumbled in several pieces into the
black crags of obsidian. Stunned, Gunnar released his safety straps,
but found himself still pinned in place by a mound of pulverized rock
and interior parts. After some pushing and prying, he pulled himself
free, but quickly realized something was wrong. He gingerly probed
his leg.

Never busted a bone before, he thought. He felt something fleshy,
wet and warm. After digging in the rubble a moment, he pulled out a

rifle and examined it. He was going to need something as he was sure
Blinda had brought him down. Applying a tourniquet and making a
bandage from the seat cushions, he pulled himself up and sat on a
piece of wreckage. He waited for the pain to subside, but it was
running all through him now. He couldn't pick a spot where he didn't
have pain to one degree or another. Looking around, he finally
spotted a crushed box still bolted to a bulkhead. Inside he found
several signaling cartridges, and a couple of portable lights. Gathering
up what he could, he started working his way out of the wreckage.

After several exhausting minutes, he pulled himself up onto a
knife's edge of rock and looked around. He could see the bright lights
of Calliope's inner city structures a considerable distance back up the
tunnel. Several hundred yards to his right was the glow of Blinda's
burning wreckage. Making his way as carefully as his condition would
allow, he started off toward the glow.

Blinda has to still be alive, he thought, carefully working his way in
the direction of the rolling column of smoke. Struggling along, he
came upon multiple holes. Shining his light into them didn't produce
much more light than what the rock was giving off. Lighting one of his
flares, he held it over the hole and dropped it in. The tube wasn't very
deep, and the air breezing through it felt cold. Water pooled between
the rocks, making the inside look as bleak as the outside. A flash
directly ahead lite a jagged horizon. *Not much further*, he thought,
continuing on. After several exhausting minutes of struggle, he finally
perched himself behind a rock overlooking the crash site. The gunship
was in only marginally better condition than Gunnar's.

She was probably still under power when she came down, he
thought, as he studied the wreckage from his hiding spot. Pulling the
rifle from his shoulder, he carefully limped out and moved along the
jagged rocks toward the burning wreck. Everything was dim and
bleak. A bonus for him, though every time the burning Isom fuel
flared, it hurt his eyes. Dark coarse sand covered the flat area where
the craft had come down. Close to one of the rock walls was another
hole similar to the many others pocking the area. Shining his light
down the hole, the bottom was barely visible. Covered in tiny strings
of running water, it was just as dismal as the others he had already
inspected.

Seeing no movement, he cautiously approached the wreck.
Something's wrong here. He studied the ground around the wreckage.
He wasn't seeing anything to suggest Blinda had walked away from
the crash site, but the light was too poor for him to know for sure.
Shining his light all over the ship, he examined the cockpit and the
ground directly around it. He found something partially buried in the
sand and picked up the spiked heel of a woman's boot. Looking back
at the wreck, he noticed a blood smear on what was left of one of the

cockpit doors. More smears of blood were found on the forward nose cone. He found several shuffle marks in the coarse sand, but they didn't go anywhere; they just stopped.

She should be standing right here. He slowly looked around the perimeter of the crash site. There was nothing but shear, black jagged rock. Adjusting his grip on the rifle, he dropped to a knee to study the ground a little closer. As he did so, he detected several small pebbles tumbling from the wall to his right. Keeping his head down, he slowly turned as if following some kind of trail on the ground. Then he saw it out of his peripheral. A pattern that didn't belong, high on the wall. Gazing in another direction, he finally stood up, cradling the gun in his arm with the barrel pointed up.

"You're going to bleed out all over that wall," he said, looking straight ahead at nothing.

"And you're going to bleed out all over the ground," came a quiet response.

"Why the chase?"

"Orders."

"You're a Thane, aren't you?"

"Yes."

"… and a Thane is a protector."

"Among other things."

"If you're here bleeding out, who are you protecting?"

"I could ask you the same thing."

"I'm not a Thane."

"Then where's your protector?"

"Rick will come when I need him."

"So he does have a name."

Gunnar's frown lengthened. The pain and brain fog had ebbed somewhat, but he hadn't meant to give Blinda anything.

"How do you feel about dying?" Gunnar raised his rifle to his hip, keeping his finger on the trigger while slowly turning and looking up at Blinda.

"It's not something I worry too much about."

"Me either."

"You know, I could blow your head apart with just a thought."

"I have no doubt, but those aren't your orders, are they?"

"No, I'm to bring you back to Drax, but it seems to me that she's gotten everything she's going to get from you, so it won't be much of a loss if you were to have... an accident." Blinda slowly raised her hands up, bringing the air in around her and floating to the ground, while Gunnar let the barrel of his rifle sink with her.

"Ah, but you're forgetting about something very important," Gunnar said.

"And what's that?"

"I have no orders and unless you have another one of Sann's Kalinite hypos on you, I'm not so sure I have any further use for you."

"I can call in medical help," Blinda said waving her wristband.

"We've both been down long enough. If you were going to do it, they'd already be here."

"There's no sense calling rescue unless I've located you."

"Well, here I am."

Gunnar tossed Blinda the heel of her boot and slowly backed up. His leg was hurting and since this pain seemed to be the only thing he could do something about, sitting was a good option. Keeping his gun trained on the Thane, he took a deep breath and blinked several times. He rubbed the back of his neck, gritting painfully. The pain was starting to ramp up again.

"You are quite the specimen," she said, just making out the heavy discoloration on his neck as she backed up to sit down as well. "The final effects of Kodiac Blu poisoning and you're still quiet and sitting upright. All the others were screaming animals at this point."

"I can assure you I'm not too far from it."

"Looking for sympathy, are we?"

"Hardly," Gunnar responded. "Certainly not from you."

Blinda carefully leaned back against the rock face and pulled her boot off. Gunnar watched with fascination as she turned the boot over and placed the heel back into position.

"Don't you need to lick it first?"

Blinda smiled, swirling her fingers around the heel, then testing the repair.

"Good as new," she said, putting the boot back on. "Explain something to me. Why are you holding that gun on me?"

"I want to make sure all your attention stays right here."

Blinda shook her head.

"Do you really think one corsair can get past an Albion battle group; a Ratronian armada and at least half the Colonian fleet?"

"Oh, you guys are suddenly good friends?"

Blinda smirked, but remained silent while she settled back a little and examined her wounds.

"That looks pretty nasty," Gunnar said.

"Spare me your benevolent platitudes."

"Why don't you just use your voodoo powers to heal yourself?"

"Because I don't know how." Blinda cringed painfully, adjusting her makeshift bandage.

"Can't a Thane manipulate matter?"

"I can, just not to the level something like this requires."

"So, you're good at messing other people up, but when it comes to yourself..."

"I grow tired of your pointless oration. Why don't we just get on with whatever you have in mind?"

Gunnar shifted his eyes from one side of the small arena to the other.

"I thought I'd already made that clear?"

Blinda stared at him.

"I need you to stay put until..."

"Yeah, yeah, yeah. Until your magnificent Starbird departs in peace. I mean, how you're gonna execute that plan."

Gunnar looked at Blinda long and hard, then let one of his fingers slide to a button on his rifle. Both sides of the business end lit up as the power generator started to whistle. He watched Blinda's eyes shift to his weapon, then back up to him. A smile developed across her lips.

"You just told me that I can manipulate matter and you're going to threaten me with a beam rifle?"

"You would rather me use something else?"

"I would rather you present more of a challenge..."

"I've stayed out of your reach this long," Gunnar snapped confidently. He winced and blinked several times, his weapon wavering.

"Quite frankly, I don't care if your ship gets away or not."

"So what do you care about?"

"Bringing you back as ordered."

"Oh, so you have principles?"

"I will see my orders fulfilled. One way or another."

"Then tell me why this weapon is holding you back? Tell me why I shouldn't just ring your scrawny little neck right here, right now? What's stopping you from using your powers on me, right now?"

"Nothing," Blinda said, getting to her feet. "But, in this instance, I think I'd rather use an old fashioned approach."

Gunnar watched her stagger a moment, teetering as she fumbled for her pistol. He hesitated, watching her bring the weapon up, but she was having trouble activating it.

"Get your thumb on the button," he instructed, trying to direct her with his finger.

Blinda fumbled to grip the gun until if finally fell. Wincing, she melted to one knee to pick it up. Gunnar noticed blood flowing over the top of her hand.

"Geez, you're in just as bad a shape as me." Gunnar struggled to his feet and limped over to help the Thane. Grabbing her wrist and the gun, he helped her back up. "Are you sure you're up for this?" he asked, holding her steady.

"I don't usually do battle in this manner," she said, keeping her head down. "I like a straightforward, head-on approach." She looked up into his eyes. "But whatever it takes," she whispered, stoic. There was a sudden muffled pop.

Gunnar's eyes widened, his grip on Blinda tightening. He looked down at the barrel of the pistol, then back at Blinda. She smiled, delighted, but the smile slowly melted as he forced the pistol back. Gale force winds instantly whipped the coarse black sand up in a deafening swirl around them as she summoned all her powers of the Thane to stop the barrel from rotating. Gunnar gritted, holding her in a vise grip and watching a wave of horror sweep across her face. Terror struck, Blinda shook her head just as the blaster discharged again. The swirl of air and sand remained active for only a few moments, then suddenly stopped, the sand instantly falling back to the ground. Gunnar's sight began to blur, sparkles dancing through his vision. He felt Blinda clutching at him as they both let go of the pistol.

"I had hoped things could have been different between us," she gasped, her eyes rolling erratically.

"It was... never possible..." As his vision faded, Gunnar felt a presence overhead as he and Blinda teetered and toppled through the hole. Feeling the rush of the fall, the nightmare in his mind suddenly became reality. Finally, he would find release from the pain and madness.

*　　　*　　　*　　　*

"I can't get any kind of lock on the Colonel, Sir," Toby Mavis announced. "Whatever's in this rock is totally obliterating his transponder signal."

Rick turned to the science officer's station and studied the readouts.

"We had him just a moment ago. Get some lights on that spot."

"I'm reading surface holes all over these formations. It's possible he's in one of them."

"Have Captain Korack and Caidin meet me in the teleporter room," Rick said heading for the bridge door. It took him only a few moments to sprint the length of the hall and into the teleporter room where Zek was waiting at the controls.

"Get me down there," Rick ordered.

"Maybe I ought to go down with you?" the Captain suggested.

"Not until I know what's there. You two, stay right here," he ordered as Caidin hurried in. "I have a feeling we're going to need you."

Zek activated the teleporter controls and Rick disappeared, materializing moments later, in a tiny clearing surrounded by jagged rocks. He instantly spotted a hole several feet away. A pistol and rifle lay next to it. He dropped to his knees, fumbling with a light. He could just make out the form of someone directly below him. Coming back to his feet, he raised his hands slightly and stepping over the hole, slowly descended into the gloom. Keeping his light pointed down, the figure of a woman gradually materialized through the darkness. He carefully found his footing as he contacted the jagged surface in the tube and knelt next to a gasping Blinda Koss. Somehow, she had survived the fall. Her eyes wide open, she sucked in shallow, chopped breaths in quick succession. Rick looked her over carefully. There was blood everywhere. The water beneath her head and back was running dark. Blinda recognized Rick as he examined the wounds under her torn, blood-soaked bodysuit.

"I know... you," she gasped. "He told... me... you'd come." She coughed hard, bringing up blood.

"Take it easy," Rick cautioned her, checking the back of her head. He grimaced and taking a deep breath, shook his head slightly.

"Your name... It's Rick... Gunnar... told me..."

"Where is he?" Rick asked, looking around. *If she survived the fall...*

"Never thought... I'd be done in... this way."

"Try not to talk," Rick said, trying to be attentive. Her injuries were extensive and he was certain he wasn't seeing the half of them. The razor-sharp rocks had done a grizzly job on her. "Blinda," Rick said, taking her hand. "Where's Gunnar?"

Blinda grasped his arm, her breathing quickening.

"He's... here." Laboring to breathe, she looked off into the darkness, then back at Rick. "That way..." Blinda gasped a couple more times. "I'm afraid..." she whispered, closing her eyes and going limp.

Rick looked around, pulling a scanner from his belt. Seeing the scans were being reflected back at him, he changed to transponder mode. Gunnar wasn't far away. Slapping the scanner closed, he stood silent; listening. He could hear drips and the gurgle of water beneath his feet, and there was something else; something soft. Summoning the air around him, he skimmed just above the surface, turning his light to one of the walls.

"Gunnar, it's Rick!"

Sweeping his light up the wall, he stopped. There was a figure huddled in a rocky recess. Rising slowly towards the form, he became aware of a soft weeping. Rick carefully touched down at the edge of the alcove and dropped to one knee. Clinging to the rock, Gunnar was still trying to climb. His condition appeared to be similar to Blinda's, with countless lacerations all over his back, arms and legs.

"Don't come any closer, please," Gunnar rasped quietly.

"Hey, Buddy. It's me," Rick said, carefully pulling something from his shirt pocket. "For crying out loud; I just can't leave you alone for one minute."

"You're too late," Gunnar groaned. "I'm done for. I told Blinda you'd be here when I needed you…"

"I'm here now, Bud." Rick reached out and put a hypo to Gunnar's shoulder.

"It's too late." Gunnar winced as the Kalinite entered his body. He was suddenly on his feet, swirling around to confront Rick. "Leave me! Let me die in peace."

"Gunnar, CJ got the antidote. You're going to be all right."

"CJ betrayed us. She went back to the Albions. She led Drax to the *Constellation*. They have it now."

"No, she went back to save you."

"No!" Gunnar suddenly exploded, grabbing his friend by the wrists. The grip was painful. Even in the dim light of the cave, Rick could see Gunnar's eyes. They were a flaming blue. His face contorting with a mixture of pain and anger, Gunnar pushed his friend back toward the edge of the recess.

"Gunnar, stop!" Rick struggled against his friend's advance, but could see he was completely out of control. Twitching his straining fingers, Rick took a big step backward and jumped. Startled, Gunnar released his hold as Rick levitated just out of reach. Anger blinding him, Gunnar backed up against the jagged wall, pressing his hands against the rock.

"Gunnar, don't…!"

"I warned you!"

"Don't," Rick lamented, watching the unbridled rage flaring through his friend's expression. Gunnar suddenly exploded from the side of the wall, leaping from the cove and diving toward Rick with his arms flung wide. Instinctively, Rick spun sideways and pushed a blast of air at his friend, sending him sailing in a different direction. Rick turned and pulled Gunnar back towards him, but as he did so, Gunnar reached out and slugged Rick across the jaw, sending him reeling. Stopping just short of the rock wall, Rick turned back around in time to see Gunnar hit the cave floor. Horrified, he swooped down and carefully turned him over. Slabs of jagged rocks had pierced his abdomen and arms, but somehow, Gunnar was still alive. He raised a hand to Rick's shoulder.

"Told you… not to call me… Bud. Geez… the Albions… really… messed me up…"

"No, I can fix this," Rick said, cradling his friend in his arms. "You're gonna be fine. I know a guy that can help."

Gunnar smiled, looking up at his friend.

"You... and your... scientific... fiddle fuddle." He coughed hard. "Rick... let me go... to her." Gunnar coughed again, his eyes bulging as he brought up blood. He tried to pull in air, but suddenly went limp.

"No," Rick choked. "I've brought her to you." Rick looked over at Blinda, then up at the outline of the hole above them. Putting Gunnar over his shoulder, he summoned the air around them and rose up through the hole. Setting him down, he looked up at the *Athena* hovering overhead.

"I've got him," he said into his communicator. "Bring him up, I'll be right back."

As Rick dropped back into the hole, he caught sight of Gunnar disappearing into the teleporter stream. Once back on the floor of the cave, he picked up Blinda and brought her to the surface.

"Ok, bring me back up."

"We're showing extra mass, General. Is everything ok?"

"Yes, bringing up an extra passenger."

Moments later, he reappeared on the teleporter pad and carried Blinda across the hall to sickbay. Moving past Caidin, he set her down on a table next to Gunnar, and moved to the desk comlink.

"Mister Mavis, let's find the *Constellation*."

"Aye, Sir."

Rick turned anxiously to Caidin who was focused on his patient. He looked up at the overhead monitor spewing information.

"CORA, are you online in here?"

"Right here, Rick. Running a complete scan now."

"Is the med assist going to do any good here?"

"Still indeterminant."

"Well, *determine* it fast."

"Caidin, what do you think? Is he still alive?"

"Richard, please," Caidin spoke softly. "You must give me time to work. CORA, the med assist?"

"Yes, Doctor Mantose."

"I think we're going to need a stasis pod."

"Internal diagnostics coming up now," CORA announced. "Med assist coming online. Rick, should I scan the woman you brought in as well?"

"Pretty sure she's already gone, but go ahead," Rick replied, glancing back at Blinda.

"Preliminary scans indicate no life signs from either patient."

Rick anxiously studied the readouts on the overhead monitor. Caidin remained focused, touching random points all around Gunnar's body.

"A full rundown of the Colonel's injuries are available now," CORA stated.

"Let's have it," Caidin said, still concentrating on what he was doing. The medical assistant maneuvered carefully around Gunnar's body, running scans and examining the more severe wounds closer.

"Multiple lacerations and contusions. Detecting a substantial entry wound below his right bottom rib. Analysis indicates a blaster discharge as the probable cause. Widespread internal damage, including his kidneys and spleen. The upper and lower hearts are in full cardiac arrest; extensive damage to the aortas and connective tissue of the lower heart. Severe bruising of the right lung and a large puncture in his left lung, on the lower quadrant. A considerable amount of blood has pooled in the Bronchioles and Alveoli of both lungs. Detecting five cracked bones along his left side in the arms, hips and legs. Compound fractures in his left femur. Multiple penetrations to the back and arms. It appears he tried to render his own brand of first aide; the bone has been set and a tourniquet applied. There is a crack in his right clavicle. Head trauma to the left side resulting in a severe concussion. Also detecting a large growth around his upper spine and neck, extending up into the cranium."

Caidin finally turned to Rick, a grim look on his bearded face.

"CORA, prepare a stasis pod."

"What are you doing?" Rick asked, worried. He looked down at his friend.

"I've never studied Dialabrons before. They have a very complex anatomy and your medical computers don't have much on him. I can't just charge into him and start pasting pieces back together without comprehensive information."

"Caidin," Rick was in a near panic. "You have to save him!"

"Richard, remember Ona's warnings to both of us about manipulating matter and causing unintended results? You of all people know how complex your friend is. I've never studied them before. Your friend is dead."

"We need Doctor Yamoto," Rick said, calming down.

"His personal Doctor?"

"Onboard the *Constellation*."

"Find that ship as soon as you can. Even in stasis..."

"You healed Dãsha."

"Yes, but she's human. I built her using Nanomech; her blueprints are in my head. Richard, you have to understand there are things that even you and I can't do."

Rick watched anxiously as Gunnar was moved into the stasis pod and the canopy closed.

"What do you want me to do about this one?" Caidin asked, examining Blinda.

Rick turned to her table as Gunnar's pod was activated.

"She's already gone," he said looking down at her.

"Well, yes and no," Caidin said, feeling her forehead and putting his head near Blinda's chest. "Yes, technically she's dead, but like Commander Atlanta, she may not be gone yet."

Rick stared at the battered face of the dead Thane. *So much trouble caused by this woman.* Anything he could recollect about her was unpleasant. His thoughts shifted to those crew members that had perished at her hand. He could only imagine what atrocities she had perpetrated on others in the name of the Albion Empire. *No, she had her chance to be a force for good. To be someone who used their great gift of the Thane for a greater cause.*

"No," Rick said, shaking his head. He turned back to Gunnar's stasis pod and examined the readouts on its side panel.

"You brought her up here. You must have had something in mind."

"I couldn't just leave her down there."

"So, just like that? We do nothing?" Caidin asked, opening one of Blinda's eyes and examining it closer.

"If you only knew what this woman has done."

"You of all people, understand life is a journey and it's how we grow through it that defines who we are and what we're to become in the end?"

Rick looked back at Caidin.

"Blinda Koss had her chance to do something good with her life."

"So, that's how it is? You're going to stand in judgement of her?"

"I think she's proven to everyone she's come in contact with, what path she wanted to take."

"Very well. I hope this isn't a lost opportunity."

Rick straightened up and watched Caidin turn away from the table. He glanced back at Blinda. Her Balkrums were still attached to her hip.

"Caidin?" Rick spoke quietly. Caidin turned back to the table. "See what you can do." Caidin nodded and motioned for the medical assistant to help him remove Blinda's torn clothing.

Rick slowly left sick bay, deep in thought. He hoped he wasn't making a mistake. Sometimes life had to be taken in order to save it. Perhaps in this, they were saving Blinda from herself.

Walking into the bridge, he didn't even look up as he sat down. He needed to find two things at once and had no idea where to look for either.

"There appears to be a great deal of interest in the deeper regions of this asteroid, Sir," Toby Mavis announced. He touched several points on his glass panels and examined his main display. Rick came out of his seat and hovered over his shoulder. "Those are corvette class ships, at least a dozen of them and those are destroyers."

"Anything bigger?" Rick asked.

"I seriously doubt they could get anything bigger than a destroyer in here. Having said that, they could easily bring in more ships if they wanted to."

"Lieutenant Habba," Rick said without turning. "Are you on the Albion channels?"

"Yes, Sir. Lots of noise out there. Still trying to separate all of it, but this rock is messing everything up. Most of it is a tangle of traffic control and ship to ship chatter; trying not to run into each other. I am hearing something about an alien runaway or something like that."

"I think we found the *Constellation*," Rick said, stepping forward to the weapons station. Pushing a small headset into place, he glanced over at Captain Dayton. "Can you figure out how to get us there?"

"I've been listening to every word," Captain Dayton replied, touching several controls.

"Where's the firing control for this thing?" Rick asked, touching several points on the glass display in front of him. Lisa looked over at him, a little worried at first but then noticed the smile on the General's lips and turned back to her controls.

"CORA, need you back up here." Rick called out.

"Already here. Handling bridge engineering functions and navigation."

"Mister Mavis, we need to pinpoint where the *Constellation* is in order to make some kind of a plan here," Rick said. "Lieutenant Habba, can you raise her on the fleet channels?"

"I've been trying, still too much noise." Laura put a finger to her ear and adjusted multiple points on her touch screens.

"Kill the running lights and dim all the windows," Rick ordered. "Let's run her dark through here." Rick pointed to a shaft next to a stream of mid-sized Albion and Ratronian ships. Captain Dayton immediately steered the ship toward the shaft as Rick glanced back at com officer.

"Nothing out there on the fleet channels, Sir." Laura tapped several points on her glass, but shook her head. "The deeper we go down this tunnel, the more static there is on all the channels."

"Should we turn around?" Captain Dayton asked, looking out at the glow of the surrounding rock.

"No," Rick said slowly.

Lisa looked over at the General. He was preoccupied with his targeting equipment.

"Find the firing button, Sir?" she asked with a grin.

"Mind your helm, Captain," Rick answered back without looking up.

"What exactly are we hoping for here?"

"Luck."

Toby suddenly turned from his instruments and stepped up behind helm and weapons control.

"General, I've been studying this rock."

"Yeah, it's your job."

"What I mean is, this asteroid is about ninety percent naturally occurring. The man made parts are where the settlements and the cities are located. These tubes appear to have been part of a volcanic planet that exploded eons ago."

Rick turned his chair around.

"Whatever is in this rock is pretty unique," Rick said, folding his arms.

"It inhibits our com transmissions," Laura said from behind.

"And it reflects most of our scans right back at us," Toby said. "If it weren't for this glowing strata, it would be pretty black in here."

"You're coming to a point?" Rick asked patiently.

"Don't you get it, Sir? We're basically inside an extinct volcano, or what's left of one."

Rick's expression suddenly shifted.

"They all lead in the same general direction..."

"Correct," Toby agreed. "Volcanic systems are derived from a central source, a magma chamber. These chambers would be full of pressurized liquid magma. That pressure would push flow out, forming tubes. We're flying in one right now. Most of these tubes will logically end up in the same place. The chamber might be gone, destroyed by the breakup of the planet, but the tubes will surely intersect somewhere. I suspect that's where we're going to find any ships that are chasing through this place."

"He makes a good point, General," Captain Dayton said. "But how are we going to find it and if we do, how are we going to see what's there? We might be flying into a hornet's nest or end up running into something."

"That's why we have Kalamar's finest at the helm," Rick said looking at Captain Dayton.

"Again, you'll have to contend with Lieutenant Starman for that title."

"You know you're the best..." Rick teased.

"Find that firing button yet, Sir?"

Rick gave Captain Dayton a quick glance. Her eyes were forward and focused on what she was doing. He looked out, trying to see anything ahead of them that would be a hint of where they were going or what they were looking for.

A Mazing

"You can't see a thing in here?" Captain Abrams complained looking down an endless glowing tunnel of rock.

"On the bright side," Doran Cartwright said from his weapons console. "Neither can those guys chasing us. I don't understand, Sir. Why don't we turn this thing around and let them have it?"

"Because this tube isn't big enough to maneuver," Dakota answered. "We need to find a place where we can't be cornered."

"I could always turn her around and run her backwards," Lieutenant Starman commented, her focus remaining out in front of them.

"You can do that?" BachTL asked, not sure if the Starbird pilot was serious.

"We could always dump that dead torpedo out the back," Doran suggested.

"Two things wrong with that idea," Pippin said. "First, we're in an atmosphere with some wicked weird gravity. That thing will fall like a fresh turd. Second, when that thing hits something, it will explode."

"Isn't that what it's supposed to do?" the navigator asked.

"Excuse me, but weren't you here for the last one that went off?" Pippin asked. "If it blows up in here, the blast will not only incinerate everything in this tube, that includes us in case you were wondering, but will likely collapse half this asteroid."

"Yeah, ok, I didn't think that one through very well."

"Lieutenant Hunter, what are you seeing behind us?" Dakota asked, turning to the command chair.

"A lot of lights moving all over the place. Probably trying to jockey for position to see who can get a crack at us first."

"All we need to do is hold this speed until we find a way out."

"What if this thing dead ends?" Lieutenant Starman asked.

"You may be showing all of us how the best Starbird pilot in the fleet does things."

"I think you've got me confused with Captain Dayton."

"Don't sell yourself short," Dakota said sitting down. "Engineering?" he called into the com system. "Tiana, how are you doing back there?"

"Hanging on by our fingernails, Captain." Tiana sounded preoccupied. "Don't give us too many bumps."

"No promises. You got everything running as it should?"

"Starting the final tests routines on the power couplings. Should be just a couple of minutes," Tiana said, still sounding preoccupied.

"Can we maneuver and fire now?" Dakota confirmed.

"Yes, just not at the same time. Ordnance is not affected. Shoot all the Pin missiles and torpedoes you want. Once the power couplings are calibrated, you can maneuver and fire pulse cannons simultaneously."

"Let me know as soon as the calibrations are complete," Dakota ordered. He glanced at the rear viewer displays over Mister Pippin's head. He could several of their pursuers gaining on them. The fighters and assault ships would come first, followed by the slower moving corvettes and destroyers.

"Mister Pippin, you said something about the gravity in here being, *wickedly weird*. Care to elaborate?"

"Certainly," Pippin said, touching the glass in front of him in several locations, then looking up. "Here's a couple of examples for comparison," he said pointing at a display. "You're familiar with this readout here."

"It's the antigrav monitor," Dakota answered.

"It's normal operating range is down here at the location marked on the scale."

"That looks very abnormal."

"You would be correct," Pippin confirmed. "Here are what normal readings look like from your average class M planet."

"Looks boring."

"Agreed. Here's the gravity where we are now. These readings are fluctuating at twenty-four Tunsils, and the further in we go, the more erratic these fluctuations become."

"Sounds right," BachTL confirmed, leaning in next to Dakota. "There's rumored to be an ionized plasma field or something like that down here."

"Rumored?" Dakota repeated.

"The exact mapping of these tunnels are classified for security reasons, so not many people get down this far. Despite not having an external atmosphere, Calliope does have magnetic fields that meet somewhere in its center. The deeper you go, the stronger the fields get. My personal opinion is that's what holds most of the atmosphere inside."

"If these variances in gravity get much larger, our antigrav system is going to have a hard time compensating."

"Something going on ahead of us," Lieutenant Starman announced, bringing everyone's attention forward. Outside, bright shafts of ionized plasma flashed passed them, fingering along the ship's hull and jumping to the glowing rock walls.

"Can you get a reading on any of this, Mister Pippin?" Dakota asked.

"Spiking above forty Tunsils now," the science officer responded. "Antigrav starting to redline."

Dakota stepped up next to the helm as an enormous cavern opened up in front of them. Entering the chamber, the ship's windows dimmed as a brilliant aberration of blue and green swept in front of them. The *Constellation* shuddered gently as flashes of charged particles blew across the hull, creating a shower of colorful sparks that spun away in its slipstream.

"This is pretty amazing," Nigel breathed.

Lieutenant Starman pulled back on the control yoke, lifting the nose toward the ceiling of the chamber. Ahead of them the colors intensified, mixing into a swirling rainbow of light and electricity. Waves of charged particles accumulated and arced from one end of the chamber to the other. Bolts of energy jumped from surface to surface, bursting randomly then dissipating and reforming again and again above an enormous lake.

"That lake is about thirty fathoms deep," Pippin announced.

"We've got company," Hayden called over the com. "A pile of fighters just came out of one of those other tubes to the right."

"Engineering? Tell me your calibrations are complete," Dakota asked nervously.

"Power coupling safety doors are closing now," Tiana answered. "She's either going to work or not."

Dakota took in a deep breath and held it, still looking at the colorful, rippling curtains of light and plasma fingers dancing all around the chamber.

"Captain?" Pip asked.

Dakota let the breath out and looked around the bridge. Everyone was looking at him.

"Pip, find a way out of here." He looked over at Doran at the weapons station. "Lieutenant Cartwright, send a couple of Pin missiles into the closest corvettes. I'm sure you and Lieutenant Hunter have a good idea what to do with those fighters." Dakota turned back to the command chair and sat down. "Helm?"

Lieutenant Starman looked back at Captain Abrams.

"I know I don't have to tell you how to do your job."

"Aye, Sir," Lynette said, smiling. She gave the weapons officer a wink and tapped her glass displays in several spots, then grasped her manual throttles and twisting the yoke, turned left and pulled up.

*   *   *   *

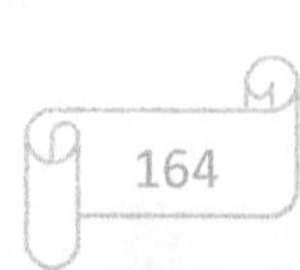

It didn't take long for the fighters to find them, but as the engagement began, it became clear the environment they were fighting in wasn't conducive to accurate marksmanship. Indeed, the missiles Doran had fired at the corvettes had a difficult time tracking to their targets and ultimately struck the cavern wall. Scratching his head several times, Doran reset his targeting systems and fired several more missiles at the corvettes. He watched carefully as the missiles streaked away from the *Constellation*, noticing tiny flashes of multicolored light flaring from the nose cap of each missile as they passed through the aurora waves. Checking his scopes, he observed each missile change its course every time it passed through a wave. He quickly pulled up the forward ordnance guns and fired several rounds at fighters banking toward the front of the ship. To his dismay, the tracer rounds veered off in erratic directions.

Dakota sat forward in the command chair, anxiously watching the engagement unfold. After several moments of observation, he noticed a general lack of damage to any of the ships, even his own.

"Bridge, this is Hunter," the turret gunner called from the com. "I'm hitting next to nothing back here with my ordnance. Are you doing any better upfront?"

Doran turned to Dakota and shook his head.

"Maybe if we were to target the walls, we might hit one of those fighters," Dakota responded.

"It's all this magnetic interference," Doran complained. "It's affecting anything we spit at them."

"Mister Pippin?" Dakota asked, turning to the science officer.

"Confirmed," Pippin replied, still looking at his scopes. "The magnetic flux in this chamber is peaking off the scales. Since our ordnance is metallic, the flux in here has an adverse effect on trajectories."

"He could have just said magnetism attracts metal," Doran mumbled. Lynette smirked quietly, giving the weapons officer a glance.

"If we're having this problem," Pippin cut in, "then those fighters out there are having the same problem. You're going to have to engage at point blank range."

"Well, that takes all the fun out of this," Doran frowned.

"Switch to pulse cannons," Dakota ordered.

"Now you're talking," Doran said, touching his glass controls.

"We're gonna see just how much strain this bird can take," Dakota mumbled to himself, glancing at the vacant engineering station.

No sooner had Dakota given the order than enemy fighters began to disintegrate before the powerful pulse guns of the Starbird. As Lieutenant Starman guided the ship through the bright dancing light waves of electrically charged particles, the colors intensified. Shades of red, yellow, green, blue and violet curtained in front of the ship as they passed close to the center of the magnetic aurora. The charged particles danced angrily all over the hull of the ship, even the forward windows, partially obscuring the outside view. Everyone felt themselves lift in their seats as the antigrav system redlined.

"Lieutenant Starman?" Captain Abrams asked, trying to see through the color field. "Your choice of course explained, please."

"We're tougher than they are," the pilot responded quickly. "We'll last a lot longer if they can't get to us."

"Good thinking, but I'm still a little nervous about our mechanical situation." Dakota looked over at the engineering consoles with a worried look. There were no alarms going off, yet, but that was usually when things started to fall apart. "How are the pulse cannons dealing with this interference?"

"They're working perfectly," Doran answered without looking back. "Sure is fun to point, shoot and watch 'em pop."

"Try to contain your enthusiasm," Lynette mumbled, leaning a little closer.

"Helm, take us back to the entrance we came through," Dakota said, sitting forward. "Lieutenant Hunter, stay sharp. Let's buy some time." Dakota noticed several other boxy shaped craft soar ahead of the ship as the weapon's officer tracked them.

"I'm good as long as they don't swarm us," the turret gunner responded.

Through the colors of the sparkling aurora, the cavern entrance became visible. There were still a few corvettes parked in the opening, even as the Starbird made its high speed approach. Noticing movement from several other openings, Dakota observed different types of ships spilling out into the cavern.

"Helm, make a flyby of the entrance and swing past those other tunnels; we'll give them a bloody nose as we go by."

Lieutenant Starman added more throttle and steered towards the wall, just skimming along its surface. A string of pursuing fighters and assault craft weaved in behind as if attached by an invisible cable. A deadly exchange of cannon fire developed between the Starbird and the fighters as they raced along the wall. The exposed corvettes took a horrific pounding as they passed by, their shields giving way and allowing secondary bursts to easily slice into their hulls. Several explosions erupted from the tunnel and the wounded corvettes spilled from the opening, plunging into the deep cavern lake.

"The bottleneck is gone," Lieutenant Hunter called from his turret. "They're pouring out of there now."

"Probably worried about scratching the paint," Doran suggested.

"Action ahead, Captain," Lieutenant Starman announced.

Dakota looked over at Pippin.

"They're all coming in from different directions," Pippin announced.

"Not sure how we're going to get out of this one," Dakota mumbled to himself. "Starman, take us back up into the core. We'll see how excited they are to follow. "Lieutenant Cartwright, you and Hunter use the ordnance at point blank if the opportunity arises. It seems to get through their shielding faster."

"Faster, Sir?" Lieutenant Cartwright repeated. "Their shields are useless against our ordnance."

Dakota smiled slightly knowing Doran was correct. While the pulse cannons were overpowering, the Starbird's ordnance weren't restricted by energy shielding.

As the ship swung away from the wall, the nucleus of the magnetically charged aurora loomed ahead of them. The ship shook as it took several prominent hits from a couple of corvettes in pursuit. Through the dense wave of colors swirling in front of them, Dakota noticed two larger destroyers emerge from a far tunnel opening. He felt the internal gravity starting to shift as the *Constellation* maneuvered closer to the nucleus. Looking at the external displays revealed a constant stream of ships maneuvering into the cavern from all directions. His attention moved to the depiction of the aft turret. Lieutenant Hunter was in constant motion now as the fighters and assault ships began to swarm, backed up by a squadron of corvettes assembling around the perimeter of the aurora discharge. As Lieutenant Starman brought the ship into the heart of the aurora, most of their outside vision was obscured by the swirling waves of color and electrical discharge spiking in every direction.

"Sir, we can't stay in here indefinitely," Pippin said, examining his readings. Dakota turned his chair to the science officer.

"Explain."

"The power being generated by the confluence of these magnetic fields is peaking our systems above their capacity. The ship can take short bursts of such energy, but nothing sustained. I'm sure Miss Mantose is getting a little nervous back there."

Dakota turned his chair around to the empty engineering station. Several displays were flashing, though there were no audible alarms.

"Any kind of estimate as to how long we can stay in here?"

"Five minutes, max."

"What if we were to go in and out; not stay in one place?" Dakota suggested.

"Yeah, that might work," Pippin said.

"Lieutenant Starman, head for the least populated corner, then swing her back here. We'll do a couple of these run and ducks and see if we can find an opening."

"Aye, Sir," Lieutenant Starman acknowledged, throttling the *Constellation* out of the middle of the magnetic aurora. Almost immediately, the Starbird was beset with an onslaught of fighters and assault craft. It didn't take long for the corvettes to try grabbing their pound of flesh. The further away from the aurora nucleus, the more hits shook the ship. As they streaked toward an empty pocket of space near a smaller tunnel, Lynette jockeyed the ship between two maneuvering corvettes. Taking the opportunity to go toe to toe with the Albion ships at point blank range, Lieutenant Cartwright let several volleys of Pin missiles go from the port and starboard missile banks. As the ordnance pierced through their shields and impacted into their hulls, he followed with a merciless burst of pulse cannon fire. One corvette wobbled sideways, then dropped into the lake. The other careened in front of the *Constellation*, then spun out of control and impacted on the cavern wall.

"Think you can do that same trick with those destroyers?" Dakota asked, pointing at the two large vessels turning in their direction.

"Certainly," Lieutenant Starman answered. "I'm assuming you remember what happened the last time we tried this maneuver?"

"Don't worry," Dakota smiled nervously. "We're not launching any Mark Vs this time."

"Sir," Lieutenant Cartwright spoke up. "Not sure how much damage we're going to do against destroyers without torpedoes. It's a good bet they'll try sending some at us."

"Hopefully, word has already spread about our little encounter over Carolon," Dakota responded. "If they don't blink, put several rounds of Pin missiles into their conning towers and get creative with your pulse cannons. Lieutenant Hunter, target their torpedo launchers."

"You people are nuts," BachTL chuckled quietly from behind Dakota.

"I think nuts is the only thing keeping us in one piece right now."

Lieutenant Starman banked the *Constellation* toward the two destroyers, aiming right between them in the same way she had over Carolon. As they came into close proximity, one ship made a sharp turn away from the engagement, but the other remained on course. Lieutenant Cartwright opened up on the larger ship with the Starbird's powerful pulse cannons and fired a volley of Pin missiles. Coursing broadside, Doran noticed the destroyer's gun emplacements disintegrating in a fiery blast as Lieutenant Hunter took careful aim and fired. Passing midship of the destroyer, it became apparent it was quite outgunned. It abruptly turned away to avoid the hail of gunfire pounding it as the smaller Starbird raked the entire side. Doran remained focused, but noticed they were veering away and worked to reset his targeting computers for the shift.

"Hey, Starman. Hold her steady, I'm working here."

"I'm trying," the pilot responded.

Doran gave her a quick glance. She was working frantically with her controls, but the Starbird continued to slip away from their target.

"What's the problem up there?" Dakota inquired, noticing their shift. At the same moment, several alarms went off at the engineering station.

"Captain Abrams," Lieutenant Hunter called out. "I've got a bit of an inconvenience here."

"Not now, Hunter."

"Someone's hit the emergency eject on my turret. I'm spinning free out here."

"You're what!?"

"Using my maneuvering thrusters to get back to the ship, but it's going to take a minute."

Dakota's thumb instinctively reached for the comlink button.

"Tiana, what's going on back there?" Silence… "Engineering? Tiana, Ensign Brandon, come in. We've lost directional control." Silence…

"Captain," Pip called out. "The outer starboard engineering hatch has been opened."

"Has it been secured?"

"No, Sir… and I'm reading someone outside."

"Outside? In all this?"

"It's Ensign Brandon."

Dakota spun his chair back to the engineering station, looking at the alarms on the displays. He abruptly popped his safety locks and bolted to the bridge door. BachTL quickly followed as Pippin and Lana looked across at the alarm displays. A pictorial of the ship's power couplings were flashing red, along with several other atmospheric readouts. Lana instinctively touched a spot on her glass displays.

"Doctor Yamoto to engineering. Emergency!"

Everyone watched helplessly as the ship continued to veer over onto her side and slow. As there was still operating control to the weapons, Doran continued to target anything that came at them as he was able, but the ship was quickly falling toward the lake below.

"The antigrav is redlining," Pippin called out. "Better hang on to something. This isn't likely to be a soft water landing. Lana, call out time to impact over ship wide coms."

*   *   *   *

Dakota and BachTL struggled to remain upright as they made their way toward the engineering vestibule. Lieutenant Nevall's voice echoed throughout the ship as the com officer counted down to impact with the lake's surface. As they passed the medical bay, the door suddenly popped open. Fuji and Alder stumbled out, following Dakota and BachTL into the vestibule and toward the darkened Mallory doors. A panel next to the doors flashed in unison with an alarm that screamed in their ears as they tried to look through. Dakota tapped several spots on the panel, silencing the alarm and turning the Mallory doors transparent.

"I don't know much about engineering," Fuji said, staring through the glass. "But this doesn't look very good at all."

Dakota studied the panel to the side of the doors, his eyes widening.

"No, it's really bad," he said, hitting an emergency air control below the main panel. Immediately, a display lit up and a hissing noise issued from somewhere behind the wall.

All four pressed their faces against the now clear glass at the scene before them. Warning lights flashed all around the engine compartment while the main console flickered erratically. On both sides, between the console and the engine bulkheads, two opposing doors in the floor stood wide open. A greenish blue mist was blasting from the floor access compartments. Dakota hit a com link button next to the panel. "Tiana! Ensign Brandon!"

"There!" Alder yelled, pointing at a motionless figure on the floor next to the engineering console.

"You've got to get us in there," Fuji insisted, looking over at the control panel. Dakota grabbed her wrist and held it away as the com officer's countdown neared its end. Glancing at the ship's attitude indicator on a separate panel, he pushed both women to one of the walls.

"Brace for impact," he yelled as BachTL slammed himself next to them. The moment the countdown reached zero, the ship pitched violently before settling.

"It can't be done from here," Dakota said grimly. He pressed his face back to the Mallory doors, looking at the open outer hatch.

"What happened?" Fuji asked, getting back to her feet.

"We've lost maneuvering power from the power couplings and crashed into the lake. With that hatch still open and no way to get to it, I'm not sure how long we're going to stay afloat."

"Isn't the ship watertight?" BachTL asked, rubbing the new bump on his head.

"We are, but with that hatch open, the engine room will flood," Dakota said looking back at the indicators on the panel. "Not too excited about getting water in those coupling compartments."

Fuji and Alder looked at him horrified.

"These readings indicate toxic plasma energy venting in there," Fuji said.

"Yes, the main control valves for the power couplings have malfunctioned, cutting off ship control. We'll have to wait until the emergency air control system can vacate the room before we can get in there to do something about it." Dakota spun back to a com panel on the wall. "Mister Pippin, I need you down here right now."

As he turned back to the scene in the engine room, everyone noticed movement next to the console. Tiana rolled over and put a hand to her abdomen.

"Tiana," Dakota called into the com system. "Are you all right?"

"That's a stupid question," she groaned looking at her blood stained hand. "Well, this can't be good." She got up on her hands and knees, trying to overcome the shock sweeping through her. "This is really harshing my buzz... I think I'm gonna puke."

"What happened?" Dakota inquired. Pippin stopped behind him, looking over his shoulder. "Status?" Dakota asked looking back at the science officer.

"Aft section is in the water, but the port side is still above the surface."

"Subject to change without notice," Dakota mumbled, looking at the open hatch, then back at Tiana as she staggered toward the Mallory doors.

"I found your spy," she labored.

"Where's Ensign Brandon?"

Tiana gave the open hatch a glance.

"He's your spy... He shot me, breached the couplings and jettisoned the turret with Hayden in it, then left the ship." Tiana looked past Dakota at BachTL. "Sorry we thought it was you."

"I thought it was me," the Castellian replied, trying to smile.

Dakota looked down at her abdomen as Fuji and Alder dropped to their knees for a better look with their hand scanners.

"This is really bad," Alder announced looking at her readouts. "You've got to get her out of there."

"Ya think?" Dakota exclaimed, turning to Pippin. "Get to the teleporter room, we'll pull her out that way."

"No, belay that order," Tiana said, blinking widely and resting against the Mallory door frame.

"You can't belay my order," Dakota said, a little surprised. "You're not in the service. I'm surprised you even know what that means."

"Someone has to stay in here and get these couplings secured."

Pippin glanced at the wall display depicting the ship's relative axis, then looked through the Mallory doors toward the far wall.

"Uhm, Sir," the science officer tapped Dakota on the shoulder.

"Mister Pippin, get to the teleporter room and lock onto Miss Mantose. That's an order. Fuji, you and Alder go with him and be ready to take her to sickbay." The tapping on Dakota's shoulder became more urgent. "What?"

Pippin's finger lanced passed his head, pointing through the Mallory glass behind Tiana. Dakota gave the far wall a double take, giving the display on the wall a glance as well.

Noticing something else wasn't right, Tiana slowly turned around and faced a wave of water gushing through the open hatch.

"Get her out of there," Dakota bellowed. "Now!"

Pippin turned to leave, but was again stopped by Tiana's insistence.

"Captain, you pull me out now and this engine room is not only going to fill with enough water to sink us, but everything in here will shut down."

"We need to abandon ship then," Fuji suggested, watching the water rise.

"In what?" Dakota growled. "We have no Interceptors and Hayden is still trying to reach us in the turret." Frustrated, Dakota turned back to Tiana, but she had stumbled away, toward the open hatch.

"We can detach the orbiter and use the maneuvering thrusters to escape," Pippin suggested.

"Without the engine pod, we wouldn't have enough power to make it back out." Dakota watched Tiana struggling against the rush of water pouring in as she reached for the hatch mechanism. "Besides, where would we go?"

"Captain," Fuji took hold of Dakota's arm. "You have to get her out of there right now. Not only has that wound ruptured vital organs, but she's received a good dose of plasma radiation poisoning and that water can't possibly be much over freezing. In her condition, it'll kill her in a matter of minutes."

Dakota's gaze remained fixed on Tiana's struggle as the water had nearly engulfed her. Remaining silent for several more moments, he mentally went through the process of activating the hatch mechanism. As if his thoughts had been heard, the door mechanism suddenly pulled closed and the hatch lock lights flashed green.

"She did it!" BachTL yelped.

They waited anxiously for Tiana to reappear on the surface of the crystal water that had already filled nearly three quarters of the engine room. Looking below the waterline, they caught sight of her, pushing toward the engineering console. Fuji and Alder noticed a ribbon of red trailing behind her as she grabbed for the edge of the console surface. They watched anxiously as she breached the waterline, gasping for air. She sucked in as much air as she could and forced herself under again, working several points on the console. They could see the display changing images, but couldn't make out what Tiana was trying to do. She came up for air again, appearing disoriented and sluggish. Even as she forced herself to submerge again, Pippin noticed the water levels starting to drop.

Tiana fought to pull herself around the console to one of the open access doors. Coming up for air again, she barely pulled in a breath and submerged again. Reaching into the access, she tugged on the manual transducer shut-off valve. Yanking as hard as she could, the greenish mist blasting from the open door suddenly stopped. Bursting to the surface again, she struggled to breath. She pulled herself back to the console and tapped its glass surface. Pausing for a moment, she looked at the other plume of plasma boiling from the other open access door. Pushing away from the console again, she forced another gulp of air into her constricting lungs and disappeared under the boiling surface. Moments later, she pulled herself back to the console, and groping for the right control, finally managed to touch the correct area. She was unable to see if she was successful, as she slumped back exhausted against the console, her face just above the receding waterline.

Dakota looked at the wall panel, checking several readings, then looked back into the engine room as the water level continued to recede.

"She's shut off the couplings and sealed the doors. But it's still going to take a couple of minutes to vent all the radiation."

"She doesn't have a couple of minutes," Fuji insisted. "Get her out of there!"

Dakota motioned to Pippin. The science officer was already heading into the teleporter room and working the controls. Moments later, Tiana lay in a puddle on one of the teleporter pads. Fuji and Alder were quick to her aide, but before they could start any kind of triage, Dakota scooped her up and rushed her into sickbay, laying her on the nearest table. As Fuji and Alder went to work, the ship's com system came alive.

"Captain Abrams to the bridge," Lana's voice rang out. Pippin and BachTL turned to leave, but paused as Dakota wasn't following.

"Fuji, is she going to be all right?" Dakota asked.

"She's still alive if that's what you're asking," Fuji responded without looking up. "Whether I can keep her that way is another story."

"Sir," Pippin said. "You're needed on the bridge."

Dakota slowly turned to the door as the Med assist dropped down into place over Tiana's limp body.

"Dakota, I'd like to stay and help out if I can," BachTL requested.

"Certainly," Dakota replied, stepping back to the table.

Alder moved in front of Dakota's view, breaking his trance. He finally left sickbay and hurried up the hallway to the bridge.

"What's our status?" he asked as the bridge door popped open.

"Dead in the water, Sir," Lieutenant Starman said, turning her chair around. "Literally. We have power, but I have zero control. The Atheons must be offline again. She's just floating here at idle."

"I can still fire, Sir," Lieutenant Doran said without looking up. "But without Hayden in the back, they've figured out they can attack from the back and, for the most part, stay out of our forward gun range. Most of our Pin missiles are gone."

"Have you got eyes on Lieutenant Hunter?" Dakota asked, looking up at the destroyers and corvettes hovering overhead.

"Yes, Sir," Lana replied. "But the aft section is under water and he's having to maneuver manually."

"What about the manual controls we created earlier?"

"Those controls were created to bypass the bridge controls," Mister Pippin said. "The problem is back in engineering with the Atheons."

"Cease fire," Dakota ordered flatly.

"Sir?" Doran asked, turning.

"I said cease fire." He turned to Lana at coms. "Signal our surrender. Tell them we have wounded onboard."

"Sir," Pippin hedged, stepping up to the Captain. "As the bridge officer, I am required to remind you that regulations state that this ship and its technology must not fall into the hands of an enemy or any hostile alien force."

"Thank you, Mister Pippin. I'm aware of the reg, but the regulations also mandate we not destroy an Indigenous culture, alien or otherwise. What will happen if we self-destruct?" Dakota looked around the bridge, making sure everyone was contemplating the ramifications of a self-destruct order. "The blast is likely to destroy most of the ships in this cavern. Now maybe, just maybe, these magnetic lines are strong enough to remain in flux and hold this asteroid together, but consider how many Mark Vs we're carrying and what just one of them can do. We'll likely blow this asteroid to pieces and everyone in it, not to mention what that kind of power would do to the fleets parked outside." Dakota paused a moment for dramatic effect. "As I am in command of this ship, I will decide if I want to die with that on my conscience."

The bridge remained silent as Dakota sat down and looked outside. Fighters and assault craft milled around overhead while more ships settled around them. Dakota turned to Lana.

"Anything?"

"Nothing but a clog of chatter. Very difficult to make any of it out."

"Sir, something's happening," Lieutenant Starman called.

Dakota turned back, looking out as the larger ships began to back away. Coming out of his seat, he looked back at Pippin and stepped forward.

"Picking up some kind of disturbance beneath us," the science officer said, looking at his scopes. Dakota looked back at the screens, then stepped up next to the helm for a better look. Directly in front of the ship, the water was moving, boiling in the center of the disturbance.

"Transponders!" Pippin called out, focusing on his displays. Lana leaned over as a list popped up.

Dakota almost climbed up onto the console to see what was happening. Two dark colored vertical fins and a double cannon barrel breached the surface. All at once, the gleaming white hull of another Starbird appeared rising from the lake. As the water fell away, the ship pivoted, facing away and hovering directly in front of them. As it turned, Dakota could clearly see her big red phoenix insignia and ident markings; *Athena SAC 09*. He watched dumbfounded as two positioning lights came on between the rear maneuvering thrusters and the main engine thruster port.

"Captain!" Lieutenant Nevall's voice pitched high as she called out. "General Niker would like a word with you."

Dakota turned, noticing the com officer wiping her eyes. He looked back out at the *Athena*, holding her position in front of them. He finally made it to the command chair and touched the comlink button.

"Captain Abrams reporting... Sir," his voice cracked.

"Hey, Captain. Haven't we seen this picture before?" Rick's transmission was garbled and heavy with static.

"My apologies, General." Dakota pushed down an emotional lump and cleared his throat. "We sort of got ourselves into a bit of a situation here."

"I saw the results of your engagement over Carolon. I'm really looking forward to hearing an exciting story."

"We've seen our fair share of action, Sir."

"The fact that your turret is just now redocking with you suggests far more. What say we get you out of here and you can tell me all about it?"

Dakota looked over at Pippin. He nodded, looking at a pictorial of the ball turret redocking.

"We're ready whenever you are."

"I'll have my weapons officer put a couple of Mark Vs into those destroyers..."

"NO!" came the collective cry from the entire bridge crew of the *Constellation*.

"Is there a problem, Captain?"

"Sir, the Mark V isn't what it was purported to be. Highly recommend you not use it at this time."

There was a brief pause.

"Very well, we'll go out the same way we came in then. Do you have any power?"

"Power we have, but we've lost our coupling transducers."

"Yikes, that doesn't help. Ok, we'll do this the hard way then."

Everyone felt the invisible tractor hawser take hold of the *Constellation's* hull and almost immediately, the ship leveled out and started moving forward.

"Just curious, Sir," Dakota asked, leaning forward. "How did you get in here?"

"There's a couple of drains at the bottom of this lake."

No sooner did General Niker finish his sentence than the *Athena* plunged back into the water, pulling the *Constellation* down with it. As the bubbles cleared the windows, the view became crystal clear. The water seemed to magnify the colors of the aurora in the cavern above. Even below, there were waves of colorful sheets of charged particles firing through the water in all directions. As they dove deeper, Dakota noticed several other ships diving after them from various parts of the cavern.

"Our scanning equipment doesn't work very well here, General," Dakota said. "You sure you can find your way back out?"

"Sure, I'm sure," Rick replied confidently. "I've got the best Starbird pilot in the fleet."

"Told you," Lieutenant Starman mumbled, smiling.

"We'll compare notes as soon as we're in the clear," Dakota smiled.

Both ships were suddenly set upon by several fighters and bombardment from a pair of corvettes diving after them. As they watched the *Athena* laying down suppressive fire, Dakota caught sight of one of the destroyers converging on their path towards the bottom of the lake. A pair of bubbled puffs exploded from the forward torpedo tubes of the large ship coming at them as the fighters swarmed around them.

"If we can't use our torpedoes, feel free to join in the fight with what you do have," General Niker suggested.

Dakota looked at Lieutenant Cartwright and nodded as he touched another button on his chair. "Lieutenant Hunter, are you back up yet?"

"Systems just coming back online. Never want to have that experience again."

"You know the drill."

"On it, Sir."

"Not sure we can hit those torpedoes, Sir," Lieutenant Cartwright announced. "Still too much interference."

"Do the best you can. Those torpedoes aren't likely to track us properly anyway."

As the torpedoes streaked towards the two Starbirds, Dakota felt a sudden burst of acceleration and a course change. As they dove toward a large opening near the bottom of the lake, the torpedoes passed harmlessly over them and turned in a wide arc, trying to reacquire their targets. The fighters and corvettes continued to press their attacks, becoming bolder and more ferocious. The water became cluttered with broken fighter and corvette parts as anything that came too close was quickly dispatched by the combined pulse cannons of both Starbirds.

As the two ships entered their target tunnel, the two torpedoes impacted on the opening and exploded. The ensuing blast shook the entire cavern area, sending cascades of rock and rushing waves of water in all directions. When the debris and bubbles had cleared, the tunnel was buried.

The Starman Maneuver

Captain Dayton held a razor focus on flying down the glowing tunnels of Calliope. It was hard enough to keep the *Athena* from striking the tunnel walls and incurring serious damage, now she was responsible for getting the *Constellation* out unscathed as well.

"Be ready to shorten the hawser stream when I make a turn," she directed General Niker at the weapons station. "It'll make it easier to navigate the curves."

"Look at you," Rick smirked. "Ordering a General around."

Lisa smiled slightly, but remained focused.

"Here comes the first turn. Shorten the stream to two hundred milli-tangs and widen the beams to thirty-six darnels. Careful not to adjust too quickly or we'll end up with the *Constellation* up our tailpipe."

"You are aware who designed most of these systems, right?" Rick asked, making the adjustments.

"Sorry, Sir," Lisa responded. "Just covering all the bases."

"I'm kidding, Captain."

"General," Laura tapped her earpiece. "Captain Abrams wants to speak with you."

"Put him up here," Rick said, pressing his headset into place. "Yes, Captain?"

"Sir, about Colonel Conrad. We have no idea where he is or what he's doing. He took one of the Interceptors over Reako and then showed up in a hangar bay here in Calliope."

"I've got him here on the *Athena*."

Static crackled over his headset.

"He's there with you? Is he all right?"

"It's a very long story, Captain..." Rick looked around the bridge, then back out at the tunnel walls passing by them.

"General," Mister Mavis said, stepping up behind him. "It will take us some time to navigate out of these tunnels. I can man the weapons console. Lieutenant Habba can transfer your conversation to your quarters."

Rick remained silent for a moment, then turned his chair around and got up.

"Thank you. Dakota, let's continue this conversation in private."
Moments later, Rick was sitting at his desk, looking at Dakota on the
viewer through intermittent distortion and static.

"I see you found the *Athena* in good working order," Dakota
commented over the noise laced transmission.

"Working? Yes. Good? Hardly," Rick responded. "Thankfully, she
was no worse for wear when I caught up to her. CORA 500 had
somehow managed to track Jayda's pod and still keep from hitting
anything."

"Then you found Jayda?"

"*Found* her? I'm not sure *found* is the right word. It was certainly
an adventure, but yeah, she's fine." Rick paused a moment. "You
said Gunner left for Reako in one of your Interceptors?"

"Yes, Sir. Said he had to check something out and that he'd meet
back up with us when we finished the mission on Carolon."

"Tell me a little about your engagement."

"Intense would be a good descriptor. We sent our extraction team
down to retrieve the highest concentration of Asium we could easily
locate, the old Carolon base. They went in after it while we drew off
the orbiting ships to the other side of the planet. We engaged the
corvettes first, but they turned out to be more of a nuisance than a
threat. They sent in a couple of destroyers when we came back for
the extraction team."

"What about fighters?"

"There were a lot of them, approximately three squadrons, but
Captain Dalley kept them busy in a lower orbit."

"All of them? By himself?"

"Yes, Sir. We kept tabs on him until we dropped into Carolon's
atmosphere to perform a hot pickup of the extraction team, but lost
track of him."

"You've heard nothing from him?"

"There was no way to make contact with him as our systems were
down. My only hope is that he bugged out. The recall instructions
were to perform a three tier lightspeed jump back to Aster, so no one
could follow."

"We can only hope he made it and is with Aster Command. Did
your team get the Asium?"

"Yes, I have it here onboard. We have enough for both ships and
then some."

"What happened to the destroyers?"

"I had Lieutenant Starman drive between a pair and we put a Mark
V into one of them. General, I've never seen anything like it. One
torpedo blew that ship in two and severely damaged the other one.
The shockwave threw us clear and scrambled all of our systems."

"Maybe you hit a weapons magazine or main power source," Rick suggested.

"If we did, there was no way to know. Before we were able to get everything back up to full power, the Albions brought in what they call a Torag."

"Sounds ominous."

"It has the ability to overload and scramble another ship's command and control systems. They had to chase us down and hit us several times before they took us. We lost our engineer, Billy Moon during those engagements. He basically saved the ship, delivering a pin missile to the Torag after we lost the ability to fire our weapons."

Rick scratched his head.

"Delivered a Pin missile? Care to elaborate a little?"

"When we lost the ability to fire, Billy was outside the ship working on the exchanger vents. He went forward and with the help of Tiana, pulled a Pin missile from a launch tube and pushed off from the ship as the Torag came at us. We never did see the explosion, but did see the damage it did when they towed us aboard their destroyer."

Rick remained unmoved, thinking of the loss of more crew members.

"General, about the Colonel? You said you have him? Can I speak with him?"

Rick remained silent, thinking. After a long pause, he looked back at Dakota.

"Apparently, Janox convinced him and Alex to take her to Reako. They found her nanny and then went to an Albion moon called Dither to retrieve some documents and tokens?"

"Documents and tokens?"

"Yes, Janox... is a Princess of Hadrian; Jana Oxlind. The documents and tokens are proof of her identity and right to the throne. There was a firefight as they were trying to leave..."

"When they brought us into the Calliope hangar bays," Dakota cut in, "we picked up the Colonel's transponder. He and Blinda Koss had a pretty good running battle in the hangar bay, then took off. We got out right after them, but didn't see which way they went. We took off down this maze of tunnels and ended up here."

"I must have just missed you guys. We picked up Gunnar's transponder shortly after coming out of one of these tunnels. But by the time I got to him, he and Blinda were both dead. I'm not sure what happened. Blinda shot him, he shot her, they fell through a hole... No idea. I have them both here in sickbay. I need Doctor Yamoto over here as soon as we clear this asteroid. She's the only one that can help save Gunnar."

Dakota stared at Rick for a moment, then cocked his head slightly.

"I'm not sure I heard you correctly, General. What good is Fuji if he's already dead?"

"Well, as I have learned during my time in the Spartus quadrant, dead may not necessarily mean, dead."

"I don't follow."

Rick took a deep breath.

"There is a lot to take in here and not enough time to understand it all." Rick checked a smaller screen next to the viewer. "This explanation will require an open mind."

Dakota sat back and folded his arms.

"As you are aware, a Thane is a protector of persons and or things. I'm not at liberty to disclose who I'm assigned to protect, but I think everyone has it figured out. The powers of the Thane allow me or anyone else given the capability, to manipulate matter at the subatomic level. This person can come from any walk of life or profession. As a scientist and military man, my talents fall along those lines. A medical doctor would have abilities consistent with his or her field."

"Are you suggesting that Fuji is a Thane?" Dakota asked.

"Unless she's told you something she hasn't told anyone else, no. But she is a gifted doctor with a knowledge of the Colonel's physiology."

"I'm trying to follow, but..."

"I'm getting there," Rick said smiling. He hesitated for a moment, trying to find the right words. "You've heard of Caidin Mantose?"

"Who hasn't?"

"While tracking the *Athena* in the Spartus, I happened upon a derelict submerged in a water planet called, Acuity. Caidin Mantose and his wife Dãsha were marooned onboard. It's a long complicated story of how they got there, but like me, he was given these Thane powers. His medical knowledge spans multiple galaxies and he's learned to blend his medical expertise with that of the Thane. He can make repairs at the subatomic level. Under certain conditions, he can bring back the dead."

"General, I think you're grasping..."

"Dak, he brought his wife back."

"Uhm..."

"While I was helping repair their ship, Dãsha was caught in a flooded tunnel and received a traumatic head injury. Not only did she drown, but half her skull was crushed. I was able to retrieve her body, but Caidin was able to repair her injuries and brought her back."

"So you think he can do the same for Gunnar?"

"It's possible, but in Gunnar's case, he can't do it alone," Rick replied. "He has very limited knowledge of Gunnar's physiology; he needs Fuji to guide him."

"So what are you proposing?" Dakota asked, trying to digest everything.

"Teleporters won't work in these tunnels; the interference is too concentrated in here, but once we get out I'd like to have Fuji transferred over here to help Caidin."

"Well, that's a bit of a problem, General," Dakota objected carefully. "Fuji already has her hands full with our wounded."

"How bad is it?"

"Apparently we picked up a saboteur on Aster. Right before you found us, the engineer's mate, Ensign Brandon shot Tiana, opened the control vents on our power couplings, ejected the turret and then left the outer engineering hatch open. Tiana was able to close the hatch before the engine room was completely flooded and shut off the coupling transducers, but was exposed to some significant plasma radiation. She suffered extreme hypothermia from exposure to the cold water when the engine room flooded, not to mention her gunshot wounds and the Plasma burns."

"Sounds pretty bad. How's she doing now?"

"Still being assessed by Fuji."

"What about Ensign Brandon? What happened to him?"

"Don't know, don't care. He's not on this ship."

"It sounds like Fuji could use Caidin's help," Rick suggested.

"General, if it's all the same to you, I'd like to transfer them over to the *Athena* where I know they'll be safe. I have no idea what else Mister Brandon did."

"Sounds reasonable," Rick said. Their transmission was interrupted by Lieutenant Habba.

"Excuse me General, but we've reached the surface. Captain Dayton is holding us just inside the entrance."

"I'll be right up," Rick said motioning to Dakota. Rick was out the door and to the bridge in a matter of moments.

"Anything on sensors?" Rick asked entering the bridge.

"At the moment, we're still under the influence of Calliope, so we're only slightly improved," Toby reported.

"Anything in the immediate vicinity?"

"Not that we can detect." Toby made several adjustments while looking at his displays.

"Any idea which side of Calliope we came out on?"

"If we could see where the Hadrian ships were positioned, maybe. But there's nothing out this hole to reference to."

"Lieutenant Habba, anything?"

The coms officer slowly shook her head as she studied her displays and listened to her headset.

"Gibberish. I can make out a word or two every now and then, but..."

"Anything on our battle frequencies?"

"All quiet."

Rick looked out at the moltey array of stars beyond the tunnel entrance and considered for a moment, then stepped up to the weapons station and sat down.

"Captain Dayton," he said, pulling his headset down into place. "Nose her out, slowly. Lieutenant Habba, as soon as you can, raise Mister Dacey and tell him to standby to open the hangar doors."

The moment the *Athena* and *Constellation* cleared Calliope, Rick noticed his scanners start to clear.

"How are those channels now, Lieutenant?"

"Most of the static and warble has cleared, but I swear, every ship with a communications device is out there using it."

"Hold your position right here. Be ready to move on my order." Rick shifted his attention to Toby at the science station.

"All our scanning equipment has cleared somewhat," Toby announced acknowledging the General. A myriad of information and images began flashing across his displays. "Rush hour out there," Toby said, adjusting his scopes. "*Calypso* appears to have become the center of attention. That Albion dreadnaught and its battle group have taken up positions on her port side. The *Maxell* is sitting off her starboard. The Colonians look like they're trying to squeeze Diord's ships out from in front of *Calypso*. Fighters and assault craft flying formations everywhere." Toby adjusted his scopes, then turned to Rick.

"A pretty good mix of Hadrian vessels."

"Has anyone scanned us?" Rick asked.

"Not that I can tell."

"Helm, how long will it take us to get to *Calypso*?"

"Two minutes," Lisa replied. "More if we have to engage any hostiles."

"Back us up a little," Rick ordered.

"Sir?"

"We need to stay hidden for a while."

"Captain Abrams on the line," Lieutenant Habba announced.

"I'll take it up here. See if you can raise the *Intrepid* and *Montego* on the battle channels. Captain Abrams, getting a little nervous, are we?" Rick asked, working the targeting computers in front of him.

"We're seeing the same thing you're seeing, General. I'd love to be a part of the planning stages here."

"I'd love that too," Rick replied. "I'm entertaining any good ideas?"

"Only the good ones? Still want me to send Fuji over there?"

Rick looked back at Toby who shook his head.

"We're not clear enough of Calliope yet. No one can see us at the moment, so we've got a little time to think."

"I wouldn't take too much time," Dakota cautioned. "It looks like a powder keg out there. We're going to have one heck of a time getting either of us through all that ship clutter."

"It's a wonder they aren't shooting already," Captain Dayton remarked quietly.

"I wonder what tipped them off?" Toby asked.

"They can't scan *Calypso's* interior," Rick said, thinking.

"General," Lieutenant Habba broke into the conversation. "*Intrepid* and *Montego* on your display."

Rick looked at one of the quadrants on his glass as two familiar faces appeared through a cloud of distortion.

"What's your status over there ladies?"

"Completely surrounded," Jayda answered. "We've gotten several demands to be boarded."

"General," Dakota cut in. "Who are you talking to? Sounds familiar."

"That would be Jayda."

"Where is she?"

"*Calypso*." Rick turned his attention back to Jayda. "Any details on their demands?"

"They're spewing all sorts of accusations about how we're harboring Diord's super weapons. I have no idea how to respond to any of this."

"How did they get an idea like that? Their scanning equipment can't penetrate *Calypso*."

"I'm afraid this is my fault," Audra cut in quickly. "I went outside."

Rick closed his eyes and rubbed his forehead.

"It was a rescue op," Audra added.

"A rescue? Whose rescue?"

"General," Dakota broke in again. "Permission to tie into your transmission?"

"We detected Kalamarion transponders," Audra continued. "From a transport that dropped out of light speed shortly after you went in after the *Constellation*."

"General," Dakota interrupted again. "That doesn't sound like Jayda."

"Captain Abrams," Rick held his hand up, frustrated. "Standby on this channel. I'll be with you shortly." Rick motioned to Laura to mute the channel. "Sorry," Rick said, taking a deep breath. "Kalamarion transponders?"

"Yes," Audra replied. "The *Constellation's* number two Interceptor. Its light speed engines were on a buildup to overload. Two other transponders ejected from the transport right before the Interceptor exploded. Captain Dalley and Engineer Moon."

Captain Dayton's eyes widened and looking over at the General, slowly shook her head.

"We should have just picked a random spot in space and waited for everyone to come to us," Toby commented, shaking his head as well.

"Are they all right?" Rick asked, trying to think the scenario through.

"A little bewildered about who they found in command," Audra answered smiling.

"I can relate," Rick replied. "Ok, we've got to get to *Calypso*. I'm taking recommendations."

"We assume you were successful?" Jayda asked.

"Can you see us?"

"Barely, we have you on the surface of Calliope as a primary target only," Jayda said looking at her scopes. "You keep dropping in and out."

"I have the *Constellation* in tow. Captain Abrams reports they have power, but their coupling inducers are offline. They can shoot, but have no thruster control."

"Wounded?" Jayda asked.

"Yes, both of us."

"Who?" Audra inquired quickly.

"Let's stay focused on getting aboard *Calypso* first, then we'll worry about everything else." Rick looked away from the display to hide his expression. It was difficult enough keeping information hidden for a more appropriate time. Now he needed everyone focused solely on the business at hand.

"The Hadrian ships have us completely surrounded and demanding we surrender and grant boarding access," Jayda repeated. "Diord's command group is negotiating, but as you can see, none of the factions are taking no for an answer."

"Where's Diord?"

"On Calliope. The Commander of the *Merc* reports he's scheduled to check-in about an hour or so from now."

Rick considered for a moment, then turned to the tactical displays over Toby's head.

"Where's the *Maxell*, currently?" he finally asked.

"Holding its position on *Calypso's* starboard side. A couple of heavy cruisers and a frigate are with her," Toby responded, looking at his displays.

"Other side, where the hangar doors are," Rick concluded, thinking carefully. "Captain Dayton, do you think you can swing us close enough to *Calypso* for CORA to teleport our people directly into their medical bay?"

"I can get us in there close enough..." Lisa said, turning to Rick.

"…and I can get them in there if there aren't too many bumps," CORA 500 announced.

"Put Captain Abrams back on," Rick ordered. "Ok, Captain, new plan…"

"Uhm… here it comes."

"Safest place for our wounded right now is onboard *Calypso*."

"Gonna be a tough maneuver even getting close to it."

"Oh, I'm just getting started," Rick came back quickly. "I'm going to lengthen the hawser beams and we're going to do a flyby of *Calypso*."

"A flyby…"

"Stay with me, it gets better," Rick said. "As we go by we'll teleport our wounded over to *Calypso's* medical bay and make a wide loop back out. Any change to your weapons status?"

"Still the same. We have a full load of Mark Vs, minus one, and only about twenty Pin missiles. Pulsar cannons are fully operational."

"Perfect," Rick said quietly.

"So what happens once we've got our people over to *Calypso*?" Dakota asked.

Rick leaned over to Lisa's displays and pointed at several targets on her scanner.

"Right there is where we'll run the gauntlet. What's the maximum distance the hawser can maintain operational integrity?"

"Wouldn't the systems designer already know that?" Lisa teased.

"Touché, Captain." Rick smiled, shaking his head. "Touché."

"Three hundred fifty two Kaldants. Glad I'm not making all this up."

"Why? You could be one of the authors of a great adventure."

"This has been one never ending adventure and I swear I'm never going to complain about having nothing to do, ever again."

Rick chuckled lightly, but the smile faded when he thought of their passengers.

"Captain Abrams, once we've got our people safely aboard *Calypso*, we're going to make a wide swing around Calliope and penetrate the blockade where the *Maxell* is sitting.

"Uhm…" Dakota hedged. "You're aware of our current operational status?"

"Yeah, you told me that earlier."

"Then you know there's no way we can get in there by ourselves."

"Correct."

"I don't like where this is going…" Dakota said, remaining uneasy.

"Do you remember when we first launched over Kalamar, and the *Athena* wasn't quite ready? We got tangled up in the yard wreckage during Croft Heckla's bomber attack and Gunnar kept us from splatting into the surface of the planet? We're going to do something similar here. We're going to bring you around to the far side, then slingshot you past the *Maxell* and straight into *Calypso's* landing bay. Once we release you, we'll run cover for you."

A soft hiss echoed back from his headset.

"Captain?" Rick looked back at Laura, who shrugged.

"You're serious?" Dakota finally replied.

"Yeah, why not?"

"You want to sling shot us through a couple of heavy cruisers and a battleship at an unfamiliar landing bay with only partial maneuvering thrusters?"

"It's a big landing bay."

"You understand that even under optimal circumstances, a maneuver like that gives us almost no time to make course corrections and get slowed down for a proper entry and touchdown? Not to mention everyone out there is going to be shooting at us."

"You said you still have maneuvering thrusters and shields, didn't you?"

"Yes."

"And you have braking thrusters, don't you?"

"Yes."

"Then you're good."

"General, there has to be a better way. Anything…"

"I am open to suggestions."

There was a long silence again as time quickly slid away.

"I can't think of anything," Dakota finally conceded.

"Don't worry, Captain. This will prove once and for all who has the best Starbird pilot." Rick touched a button on his coms panel. "CORA, status of Caidin and his patients?"

"It appears Doctor Mantose has completed his work on the female you brought in. Colonel Conrad's condition remains unchanged."

"The female's condition?"

"Alive, but heavily sedated."

"Have him get his patients to the teleporter room and standby for transport to *Calypso*." Rick looked around the bridge and took in a deep breath. "Ok, Captain. Let's go," he said looking at Lisa.

Lisa turned and gave the *Athena* throttle. Pulling away from Calliope, the two tethered ships accelerated as the full vision of what lay ahead of them tilted into view. *Calypso* held her position not far from Calliope. Surrounding her were a myriad of Colonian, Albion and Ratronian capital ships, corvettes, fighters and assault craft. While there seemed to be no engagements happening, the flurry of maneuvering going on was apparent. As the *Athena* and *Constellation* approached, there was a marked change in the focus of Hadrian's attention. Entire squadrons of fighters and corvettes shifted their flight patterns, moving to surround the two ships as they approached.

"I understand what the goal is here," Dakota called. "But what about these guys that are in our way? We gonna treat them the same way we did down in the core of Calliope?"

"You let us worry about clearing a path," Rick responded, configuring his pulsar canons. "Your job is to keep anything from getting behind us. Make sure you let them fire first."

As the two Starbirds approached the outer perimeter of the blockade, Rick glanced over at a display in front of Lisa. Various readouts flashed information in quick succession around a pictorial of *Calypso*. A thin blue line encompassed the image, indicating the maximum distance they could safely teleport their wounded. Even as they skimmed along the boundary of Hadrian ships, he could see they still had to penetrate much deeper in order to be within the circle. Checking his tracking screens revealed a string of ships lining up to give chase.

"Helm, we need to get a lot closer."

"Permission to punch through, Sir?" Lisa asked.

"Have you got a weak spot in mind?" Rick responded.

"It's got to be right here, or we string these guys along around to the other side where the *Maxell* is."

Rick checked the tactical displays behind him, then gave his tracking screens a glance. Trying to get to the other side with this many ships chasing them could prove problematic. If they weren't towing the *Constellation*, there would be no contest, but until they were able to get their wounded and ship safely onboard *Calypso*, they were going to have to take the closest route.

"Do it," Rick abruptly ordered.

Rick hadn't even finished giving the order before Lisa was turning the ship in a tight curve through a small cluster of medium ships. Because of the large group of corvettes and fighters tailing them, the larger Hadrian ships held their fire to a minimum. The *Athena's* turn was tight enough that by the time they had come full circle, they were overrunning the line of ships chasing them. Several pursuing ships collided in the confusion, as both Starbirds dodged the mayhem and streaked through another cluster of larger ships.

Checking his displays, Rick gave helm control's displays another glance. They had to penetrate deeper. The ship was constantly buffeted by a hail of gunfire from their pursuers. Now, even the capital ships were intensifying their fire.

"Are we supposed to be firing back at these guys," Captain Abrams called in Rick's ear.

"Leave the big ships alone, but if those corvettes and fighters get too close, remind them what happened over Carolon," Rick said, watching a flurry of targets converging on them. He examined his arsenal inventory. The Starbird held a large supply of Pin missiles, in the forward and aft magazines, including a large complement of Mark V torpedoes. He considered for a moment, pulling one up into a status of ready, but then Captain Abram's warning about its destructive power came to mind.

Rick targeted a cluster of smaller ships converging on them from different directions and fired. He didn't even take the time to track the outcome as they snaked between three frigate class vessels that had opened up on them. The missile disbursement hit their marks with pinpoint precision, a few continuing on to several ships bringing up the rear of the attack pattern. Secondary damage was created by primary explosions, which in turn broke up any remaining formations trying to follow the lead attackers.

"Hard to believe there are this many ships in one place," Lisa commented.

"Hard to believe our tractor hawsers are holding up with all this maneuvering," Rick said, making adjustments to the beams holding the *Constellation* in tow.

"How am I doing?" she asked, making another turn directly under a bulky, boxlike cruiser.

"Better you not know," he said after firing more Pin missiles.

"Shield strength dropping below seventy percent," Toby called from behind.

"I'm seeing a hole, Sir," Lisa said a moment later.

"I have the utmost confidence in you, Captain." Rick targeted a pair of corvettes trying to sneak around several larger ships to intercept them.

"General," Captain Abrams called in his ear. "My turret gunner reports things are starting to get a little intense behind us."

Rick glanced at Lisa's display of *Calypso*. They were just skimming along the thin blue line indicating the teleportation safety zone. They had to get closer. Before he could make a suggestion to the pilot, Lisa was twisting the *Athena* into a hard turn. Rick let several more Pin missiles go and adjusted the tractor hawser beams to compensate for the inertial forces affecting the beam's ability to hold onto the *Constellation*. As they came out of the turn, *Calypso's* light grey hull took up most of their view. Lisa pushed the *Athena* harder, even as a myriad of fighters and corvettes increased their fire while converging on them.

"We've gotta get closer," Rick grunted amid the constant buffeting of missile and energy strikes. "CORA, are you ready?"

"Caidin is standing on the teleporter pads with his two patients."

"Captain Abrams, are you ready?"

"Ready over here."

The buffeting from the surrounding firefight intensified. Rick looked down at the hawser controls. They were beginning to redline as the ship pitched violently through energy an ordnance strikes.

"No guarantee we'll get back out to smoother space," Lisa informed Rick, holding her focus forward.

Rick glanced at the scanners. They were well within the limits for teleportation, but trying to send multiple people at the same time under these conditions was risky at best.

"We have to get closer," he mumbled.

Lisa held her controls firm as she added more power. Even without the scopes in front of her she could see clusters of ships coming from all directions. Warning alarms flashed in front of them as the *Athena* and *Constellation* took several direct hits, rocking the ship nearly out of control. A message popped up to prominence in front of Rick. Horrified, he reached for his comlink control, but Captain Abrams' voice was already blaring in his ears.

"We've lost the hawser beam!" he yelled as both ships were again raked with gunfire.

"Veer off!" Rick ordered. "CORA, Dakota, initiate teleporters!" Rick checked his scopes, then glanced at Lisa's. "Bring her around and cover the *Constellation*."

Seconds groaned by as Lisa went to work while Rick concentrated on the targets surrounding them. Without the *Constellation* in tow, they were free to maneuver unfettered at greater speed and agility.

"Teleportation successful," CORA announced.

"Fuji and Tiana are on *Calypso* as well," Dakota said. "Unless you have a better idea, we're going to try and make it to the hangar bay."

"We'll cover your six. Mister Dacey suggests using the middle bay. It's bigger and it's empty. That way you don't hit anything valuable if you can't get her stopped in time," Rick suggested.

"Wished those doors were on this side," Dakota responded.

"You and me both. Tell Lieutenant Starman... Lisa says... good luck."

*   *   *   *

"Nearly ran up her tailpipe," Dakota said watching the *Athena* take a hard hit from a corvette as they banked across its path. "Just so I know when to scream in terror, what's your plan, Lieutenant?"

Using the maneuvering thrusters, Lynette rolled the *Constellation* on her back relative to *Calypso*. Several corvettes and a squad of fighters swarmed after the Starbird, but a formation of Flightstreaks from the *Trax* swept through the attacking fighters, breaking them up and sending them scattering in different directions.

"We need all the speed we can hang on to," the Starbird pilot said confidently. "I'll make a half loop over *Calypso's* starboard side, drive right in and park."

"Just like that..." Dakota stated. Being a fighter pilot, he knew the maneuver very well. But doing it in a fighter and doing it in an assault corsair that had only maneuvering thrusters were two very different scenarios.

"The *Trax* and *Realistic* have ordered their fighter squadrons to sweep ahead of us," Lana reported, turning from her consoles.

Dakota looked up through the overhead windows as *Calypso's* grey hull engulfed his view. Ahead of them, clusters of ships turned in their direction as scores of fighters and assault craft buzzed angrily toward them. Several waves of Flightstreaks from the *Trax* and *Realistic* swooped into the middle of them, breaking their ranks and leaving them in disarray.

As *Calypso* disappeared from view, Lynette held her course, straight out and away from the ship. Using her scopes in tandem with her visual scanning, Lynette had to time her maneuver just right. Get the arc wrong and the *Constellation* would hit the hangar bay floor when it came in or worse, strike the side of *Calypso*. The Starbird shuddered with every strike as the fighters pressed their attacks. Lynette touched several points on her glass displays, firing the maneuvering thrusters. Outside, the stars began to shift under them as the ship began its arc back to the bay doors.

As *Calypso* came into view, Dakota gave the Starbird pilot a glance. Lynette's focus was as intense as he had ever seen.

"You gonna make it in, Lieutenant?" he asked, becoming nervous at the speed and angle they were approaching.

"Piece of cake," she replied confidently. Lynette held her attention on the controls and the doors lining up ahead of her. Her fingers danced around the glass controls as if the surface were too hot to touch. As her displays flashed information, she grabbed a short lever to her right and pulled it back. Everyone felt the braking thrusters fire as the threshold of the landing bay took up their field of vision. Lynette made several quick course corrections, then held her fingers in place. Doran gave her a quick look. Her expression was steely as the *Constellation* entered the landing bay. Dakota tensed in his chair, his knuckles turning white and his feet pressing hard against the floor, trying to stop the ship with his will. As the inner wall loomed in front of them, Lynette touched another set of controls and the Starbird's landing gear extended. The nose tipped up slightly, holding in place for a moment, until the ship gently settled onto its lifters. Lynette held the thrusters at station keeping for a moment while she checked her other controls, then shut everything down. Moments later, everyone noticed the *Athena* settle to the floor next to them.

"Oh, that was smooth," Nigel said smiling as the helmsman slowly turned her chair around. Her rear console disappeared under the other consoles.

"Excuse me while I go change my underpants," Doran said standing up on wobbly legs.

"Congratulations, Lieutenant Starman," Dakota said, rubbing his forehead. "I think we've settled the issue of who the best Starbird pilot is once and for all."

"I have no doubt that Captain Dayton could have successfully performed the same maneuver."

"I think the correct response is, thank you," Pippin said, still safety locked in his chair.

"Thank you, Sir." Lynette smiled, stretching.

"General Niker sends his compliments to the best Starbird pilot in the fleet," Lana announced. "Captain, he'd like to see you outside immediately."

"Everyone stay put until you hear from me," Dakota said heading for the door.

As the Captain disappeared behind the bridge door, everyone sprang from their seats and rushed Lynette with hardy congratulations and a recap from the Starbird pilot.

Reclamation

Amid continuous jolts of exterior weapons fire thumping at the hull of *Calypso*, Captain Abrams stared at the four new Starbirds sitting in the hangar bay adjacent to the bay the *Constellation* and *Athena* had just landed in. General Niker stood a little behind him, letting Dakota digest what he was seeing.

"These were in the holds of those freighters?"

"They were almost ready to go when we found them," Rick replied.

"We're all pretty anxious to get away from this place as soon as possible. How long before we can get under way?" Dakota looked around at the interior of the bay.

"I'd love to see things cool down a little first. I'm hoping to avoid any more entanglements," Rick said. "But if push comes to shove, this ship is tuff enough we could just push clear of everything and jump to lightspeed. It might be nice to have these Starbirds out there clearing a path, but since we're way undermanned and none of them have loaded crystal assemblies, that's probably not a good idea."

"Pretty sure we recovered enough Asium from Carolon to refit the *Athena*, *Constellation*, probably these four and still have some left over." Dakota started down a set of stairs for the main hangar deck.

"Captain," Rick called. "We need to teleport to *Calypso's* infirmary first. I want to see how the wounded are doing."

Dakota started back up the stairs, but stopped and looked back. Several officers were milling around one of the Starbirds. He recognized Jayda Niker standing next to a shorter brunette officer. He gave them a double take, thinking he recognized the other one, but shook his head and started off after a fast paced General Niker.

Using the *Athena*, Rick and Dakota teleported directly into *Calypso's* infirmary, or what looked more like personnel quarters attached to a makeshift laboratory. As they finished materializing, Caidin Mantose turned from a table and recognizing Rick, grinned broadly.

"Richard, it's good to see you've made it aboard," he said, throwing his arms around Rick.

"It was a little tricky," Rick admitted. "How are your patients?" The ship's hull shook gently with more weapons strikes, giving cause for everyone in the room to look up, as if they could see their assailants buzzing around outside.

"In very capable hands," Caidin said, delighted to see Tiana.

"What about Blinda?" Rick asked, looking around the room. Dakota stepped over to Gunnar's stasis pod next to one of the examination tables.

"She's in the next room," Caidin gestured to a door on the other side of the table Doctor Yamoto was working at. "She's still sedated. I've fitted her with a neuro inhibitor band, just in case. You didn't tell me she was a Thane."

"Must have slipped my mind in all the excitement," Rick responded quietly. He stepped over next to Fuji.

"What are the odds that my daughter would be serving aboard one of your ships?" Caidin asked, standing next to BachTL and looking at Fuji on the other side.

"Well, she's not exactly serving," Dakota spoke up, looking at the controls on Gunnar's stasis pod. "As she's not in the military, she isn't considered a part of the crew."

Caidin looked back at Dakota and BachTL.

"Independent contractor," BachTL added.

"She hounded the Colonel until he let her help Billy with the ship's systems," Dakota said. "Then when we lost the engineer, she stepped in and did what she could."

Rick smiled, but never got a chance to speak.

"Gentlemen, medicine is not supposed to be a spectator sport," Fuji complained, trying to focus.

"How is she?" Caidin asked.

"I'm just closing up. It'll take some time, but I think she'll make a full recovery."

"Why didn't you just let Caidin have a go at it?" Rick asked, stepping around for a better look. "She is his daughter after all."

"BachTL and I were pretty rushed to get her to the teleporter room," Fuji said, still focused. "And, this is my patient. I'm perfectly capable of taking care of her."

"It's ok, Doctor," Caidin said. "It would have been inappropriate for me to intrude on another surgeon's procedure without their expressed invitation. Besides, it would be unwise for someone with an emotional attachment to be involved."

"Well, all I know is we wouldn't have gotten as far as we did without her," Dakota said.

"Don't forget the others," BachTL suggested calmly.

"Yes, the crew have all performed above and beyond."

"I think we've all performed pretty extraordinarily under difficult conditions," Rick said as his communicator went off.

"General," Laura interrupted. "I'm getting a priority call from the Albion Admiral commanding the dreadnaught, *Indominable*..."

"Not their King Commander?"

"No, Sir. He's attending the Rite. This is an Admiral Hilton. He's demanding boarding access to *Calypso*."

"Isn't everyone?" Rick replied.

"He says if we don't comply, they'll start firing torpedoes."

"How big of him to let us know," Dakota grumbled.

Rick considered for a moment. Hadrian scans were unable to penetrate into *Calypso* and her hull was tough enough to withstand most bombardment, but torpedoes could present a significant threat."

"Keep them guessing; no reply," Rick ordered. He turned back to the medical table just as Fuji and BachTL started tucking warming blankets around Tiana.

Fuji moved the medical bed to a darkened corner and after a moment of fussing with the monitors, she stepped back to the others.

"There is some nice equipment on this ship. Where did all this come from?"

"I take it you two haven't had much time to get acquainted?" Rick asked.

"We had time to get to the teleporter pads and transport directly here," Fuji said, stepping toward the stasis pod on the other side of the room. "Besides the woman you brought with you, who's this?"

"Doctor Yamoto," Rick spoke up, bringing Fuji to a stop. "I'd like to introduce you to Doctor Caidin Mantose."

Fuji turned around, a puzzled look forming. She looked at the older gentleman for several passing moments, then back at Rick.

"Thee, Caidin Mantose?"

"Am I really that famous back home?" Caidin asked.

"Apparently legendary," Rick said.

"Caidin Mantose?" Fuji finally blurted out. "Inventor of..."

"Yes, I invented all that stuff," Caidin finished waving his hands in front of him. "Why couldn't the Spartus quadrant have been this excited to see me?"

"Must be your looks and fiery demeanor," Rick smirked.

"But how? You were presumed lost long ago." Fuji leaned against the stasis pod and folded her arms.

"At least in your version, I'm not dead," Caidin grinned. "Lost is a good descriptor, so we'll leave it at that for now as it's a very long story; one meant to be told with time. Right now, Rick and I are in need of your help."

"Ok, not sure what I can do that you can't."

Caidin glanced at Rick, who stepped toward Fuji and touched a control on a panel located on the side of the stasis pod. An interior light brightened, revealing the occupant. Fuji turned around and peered inside.

"This is Colonel Con... rad..." She faded off, studying the readouts on the panel, then suddenly turned back to Rick and Caidin as Dakota and BachTL stepped to the pod for a closer look.

"What happened?" Fuji asked, feeling the pangs of shock starting to develop.

"He's dead," Dakota said looking back at the other three.

"The Kodiac Blu caught up to him," Rick answered.

"Kodiac Blu didn't cause all this," BachTL observed.

"He's had the daylights beat out of him," Fuji said, looking closer at the readouts on the pod, then gazing at Gunnar.

"Based on Dakota's preliminary report, Gunnar was running interference for the *Constellation* while trying to escape custody. He and Blinda Koss got into a fire fight. As near as we can tell, he led her off into the tunnels of Calliope. Somehow they managed to shoot each other down. From there, it appears they got into a hand to hand fight and somehow ended up shooting each other. They weren't dead when I found them, but by the time I got them back to the ship we couldn't detect any life signs."

"Gunnar's physiology is so complex," Caidin cut in, "there is no way I could do anything for him without someone who has knowledge of his alien makeup. The best thing I could do was put him in stasis until we were able to catch up with you."

"What good am I going to be? You said he's dead," Fuji asked, glancing back at the pod.

"It's possible, I can still revive him," Caidin said carefully. "But I need your help to do it."

Fuji blinked. She looked at Rick, hoping he would add something that would explain Caidin's statement.

"My help? To do what? He's dead!"

"Maybe you better have her sit down for this?" Dakota suggested.

Rick gestured to a sitting area and as they got comfortable, Rick activated his communicator.

"Yes, General?" Laura Habba answered.

"Have Dãsha and our two Lieutenant Commanders ready to teleport to my location on my signal?"

"I'll have them standing by."

"As you are aware," Rick began, "I am what's called in this galaxy, a Thane. You understand what a Thane's responsibilities are and the special abilities they possess?"

"Yes, General," Fuji answered. "I would love to see a scan of your brain. If what I understand is true, it could be the door to opening up a wide variety of human advances."

"I have no doubt," Rick smiled.

"But I'm still a little fuzzy on how this happened," Fuji said. "Someone changed your brain?"

"Yes, but I was never given any information on the mechanics of it. The reality is, I have abilities and the responsibilities that go along with them. As I am considered a warrior and scientist, my abilities follow along those lines. Caidin has been designated a Thane as well with his abilities following his profession. He can do things medically that even our latest tech can't duplicate."

"Does that include raising the dead?"

"Under certain conditions," Caidin interjected.

"And what are these conditions?" Fuji asked skeptical.

"Well, there has to be a physical body to repair, but more importantly, the body must still be in flux."

"In flux?"

"Yes," Caidin continued. "The energy that drives the body must still be present. For example, you wouldn't be able to bring back one of your long dead ancestors to answer questions about family matters, but imagine if you could repair critically damaged or terminally ill patients. Those mortally wounded by gunshot, brain injuries, drowning or being crushed; whatever the circumstance may be. Individuals injured beyond conventional medicine's ability to overcome. As long as the energy that drives the body remains in flux, their injuries could be repaired and they could be revived."

"You're suggesting you can somehow cheat death?" Fuji asked.

"Cheat death?" Caidin thought a moment. "No, death is inevitable; it's a part of life. I'm talking about giving life a chance to continue. If a patient can be reached in time and put into stasis, the damage to their body can then be repaired and they can be brought back."

"Impossible," Fuji said, shaking her head. "You can put a patient into stasis, but there's no way to work on them while they're in that condition. You can't just open the pod and perform surgery. It has to remain closed in order for the stasis environment to remain active."

"Exactly," Caidin agreed.

"Well, then I don't see how you think you can even attempt this." Fuji replied, seeming to end the conversation.

Rick activated his communicator.

"Send them over, Lieutenant."

"Who's them?" Fuji asked.

"Well, you haven't had time to ask, but yes, I did find the *Athena*," Rick said as three forms began to materialize in the room behind them. "...I also found Jayda."

Fuji turned in her chair to see three women standing behind her.

"Doctor Yamoto," Jayda spoke up with a grin. "It's so good to see you!"

A tall slender woman moved closer.

"I am Dãsha Mantose," Dãsha said, taking Fuji's hand. "I am pleased to make your acquaintance. Commander Atlanta has told me so much about you. From her description, there can be no finer doctor... well, with the exception of my husband, Caidin."

Fuji turned back to Rick and Caidin, trying to catch up. She looked back up at the tall woman and came to her feet as tilting her head back that far hurt.

"She's your wife?" Fuji asked in disbelief. "How is she your wife?"

"She's married to him," Rick chuckled.

"I've been getting this reaction a lot lately," Caidin said smiling.

"Ok..." Fuji accepted, working through the initial shock. "I've heard of stranger things, I guess." Fuji looked back and forth a couple of times. "Wait, you said Commander Atlanta told you..." Fuji twisted around again, this time stumbling backward into Dakota, who was equally surprised as Audra stepped out from behind Dãsha's tall frame.

"Hello, Fuji," Audra smiled. "Surprise..."

"This some kind of sick joke?" Dakota asked, holding Fuji up.

"No, not a joke," Rick said. "And before you ask, there is no Hallavertor technology on this vessel."

Regaining her composure, Fuji steadied herself against Dakota as Audra stepped over to them.

"It's really me," Audra said with her infectious grin. She grasped Fuji's hand and reached for Dakota's.

"The Hallavertors can reproduce this sensation," Fuji gasped.

"Didn't you hear me?" Rick said. "There are no Hallavertors on this ship. She's the real thing."

"You can't be. You're... dead. I watched Gunnar teleport your pod into space..."

"Rick found my pod," Audra said, looking over her shoulder at the General. "Caidin and Dãsha brought me aboard *Calypso* and repaired my injuries."

"How?"

"Well, I wasn't awake for that part of it," Audra said. "But it really is me. He brought me back."

"As he has done for me as well," Dãsha chimed in.

"You were dead too?" Fuji asked, looking up at Dãsha as she stepped forward.

"You are one tall woman," Dakota said as he and BachTL admired her figure.

"I have so many questions," Fuji said overwhelmed. "I hardly even know where to begin."

"Your questions, while well founded, will have to wait for now," Rick said. "According to Caidin, we're on a time constraint."

"You're talking about..." Fuji pointed to the pod.

"Yes," Caidin said. "The sooner we start the procedure, the better his chances of survival are. Right now, I'm afraid the odds are very much against him."

"Who are we talking about?" Audra asked bewildered. She turned to Rick, who motioned to the pod. He stepped next to her as she peered inside.

"I found him in Calliope," Rick said quietly.

Audra put her hands to the glass and gazed at the battered face of her husband.

"His last words were of you."

Audra remained still, emotionless, just staring through the glass. After several moments of silence, Dãsha and Jayda looked into the pod with Audra.

"This is your Gunnar Conrad?" Dãsha asked, examining the readouts on the pod. "Husband, these readings have no life indications; he is dead."

"I think that condition has been sufficiently established," Jayda commented, irritated with Dãsha's apparent lack of sensitivity. She pulled her back to allow Audra some privacy.

"I think what Dãsha is trying to say is, under normal conditions, even a deceased person will continue to produce some kind of residual energy signature," Caidin said as the group came back together. "A biofeedback if you will. We see no such indications."

"Then I'll ask the question again." Fuji lowered her voice so as to not disturb Audra. "If he's dead and his energy is no longer in flux, then what do you hope to accomplish?"

"Because he has such an unusual body chemistry and physical makeup, it's possible that the pod's sensors are unable to detect or interpret his biosignature."

"That's assuming he's even putting one out," Fuji suggested.

"Are you willing to bet his life on that question?" Rick asked.

Fuji looked around the circle, then beyond at Audra still gazing into the pod. She took a couple of deep breaths, then looked back at Caidin.

"Ok, what do you need me to do?"

"We need someplace that will give us the maximum privacy with the most information about him."

"The *Constellation's* medical bay will have everything we need."

"Do whatever you have to," Rick said motioning to Dakota as he brought his communicator to his mouth. "Lieutenant Habba, standby to bring personnel back to the *Athena*." Rick looked at Dakota. "Have everyone transported back to the *Constellation* and make sure Caidin and Fuji have whatever they need. I'll work on diplomacy with our spectators outside."

"General, wouldn't it be better to just get *Calypso* away from all this?" Dakota asked. "Light speed is a great place to be uninterrupted."

"Can't argue with that," Rick said motioning to Jayda as his communicator went off again. "Right now, we're sort of surrounded."

"Sir, a large cluster of ships dropped out of light speed a couple of minutes ago."

"And you're just telling me?" Rick responded curtly.

"We wanted to make a positive identification," Toby interrupted. "Sensors indicate a small armada of Tomplie and Colonian ships."

"What do you make of this?" Rick asked, looking at Dakota.

"It's possible this is Diord's friend, General Aurora."

"Sir, I'm getting a signal from Administrator Vandmire," Lieutenant Habba said.

"Never rains but it pours," Rick said. "Put him through."

"General Niker," came the worried call from the Cross administrator. "You gave us quite the scare. Were you successful?"

"Somewhat," Rick hedged cautiously. "We have the *Constellation* and Gunnar."

"That's a relief," Diord responded.

"It's pretty tense out here. *Calypso* is completely surrounded and I get the distinct impression trigger fingers are a little itchy. To add to the mix, I think your friend Lou just showed up. A Tomplie flotilla with some Colonian ships."

"It's about time," Diord griped. "You should be able to identify his flagship, the *Executioner*. Sounds like he has the Aster contingent with him. If you can get a hold of him and let him know what's happening, I'm sure he can bring his ships in close with mine and provide you more protection."

"What about you? Were you able to get the Princess into the Rite?"

"The Rite starts in a couple of hours, but I've just been informed that all delegations from Cross have been barred; we've been shut out of the proceedings."

"I thought you had a ticket?"

"I did, but the Albion Secret Service is refusing our entrance. I've lodged a formal protest with Calliope security, but I suspect this is coming from the highest levels. Is there anything you can do to help us?"

"We just got aboard *Calypso* by the skin of our teeth," Rick replied. "Every ship in Hadrian is out there banging on our doors now. I'm not sure how we'd ever get back to Calliope in one piece. Even if we did, short of starting a firefight during the proceedings, I can't imagine what we could do."

"There's got to be something," Diord said. "I can't believe we've come this far only to be tripped up at the last second."

Rick paused, sensing something. After a moment, he looked toward a darkened doorway. Caidin noticed it too and stepped up beside him. He gave Rick a glance, whose gaze remained focused on the open door to an adjacent room.

"Rick, are you all right?" Jayda asked from behind. "What is it?"

"I'm fine," Rick said quietly. "I want you to help Fuji with Gunnar and Tiana. Caidin and I will be along in a couple of minutes."

"What's in that other room?" she asked, moving around them. She was stopped by a firm arm across her chest.

"Commander Niker," Rick said quietly, but forcefully. "I'm ordering you to leave immediately."

Jayda hesitated a moment, but reading the urgency in Rick's voice, turned without another word to help the others. Rick looked at Caidin, then back at the doorway. He considered for several more moments, then brought his communicator back up.

"Diord, if we teleported over to the *Merc*, could you have one of your transports take us to your location on Calliope?"

"That should be no problem," Diord said. "But how does that get us into the proceedings?"

"I think I may have a solution. I'll get back to you."

Rick and Caidin approached the darkened doorway and peered inside. A dim light shone down on the form of Blinda Koss lying on a medical bed, her eyes closed.

"The sedative is wearing off," Caidin whispered from behind. "I didn't know the extent of her abilities, so I set the inhibitor to full power."

Rick nodded and stepped quietly into the room. For a moment he felt pressure on his ears drums, but it was quickly replaced by a ringing sound that slowly faded. Remaining cautious, he stepped to the foot of the bed.

"Is this my punishment?" Blinda asked in a whisper, her eyes remaining closed.

"Punishment?" Rick answered quietly.

"Since I'm dead, am I to be endlessly tormented by the ghosts of those I've fought in life?"

"Would those be people you've killed?"

"Yes."

"...and do I look dead to you?"

"You do in my mind..."

"Blinda, open your eyes," Rick said leaning forward.

The Albion Thane cracked her eyes, then blinked several times. She looked at Rick, then around the darkened room.

"This is not what I pictured being dead would be. I am... dead?"

"No," Rick responded. "I mean, yes, you died. But you're not dead."

"If I died, but I'm not dead, then what am I?"

"You did die... you were dead, but now... you're not."

"Now, I'm not..." Blinda repeated in a whisper. "Still feeling a sense of punishment here." She tried to sit up, but was unable to move. "No, I'm dead."

"I found you in one of Calliope's tunnels and pulled you out. This is Caidin Mantose. He healed you."

"How do you heal someone who's dead?" Blinda closed her eyes again, trying to grapple with what she was experiencing.

"It isn't easy," Caidin answered from behind.

"Why did you bring me back?" she asked, opening her eyes again as Caidin materialized from the shadows.

"To give you a second chance," the doctor answered.

Blinda developed a puzzled expression and looked at Rick.

"I can't move."

"You're wearing a neuro inhibitor," Caidin said. "The mind does all sorts of funny things when coming out of anesthesia. I didn't want you hurting yourself."

"Myself, or someone else?" Blinda replied.

"Both," Rick responded, hardening.

"You're still afraid of me..."

"Afraid?" Rick repeated, straightening up and folding his arms. "You killed half my crew and my friend Gunnar."

"I will offer no apologies," she responded defiant.

"I don't expect any. What I do expect is for you to consider the opportunity being offered to you."

"I feel so blessed," Blinda replied snidely.

"You've used your Thane abilities for your own self-serving agendas," Rick said.

"I serve the Albion Empire..."

"Bull!" Rick retorted. "You used the Empire as an excuse to serve your own interests. Not any sort of greater good."

"I tried serving the greater good once. It didn't work out so well."

"Who did you serve, Blinda? Who was your appointment? Who were you supposed to guard at all costs? Drax Blair? Thoene Dismon? Stephanie Benatar?"

"A Thane does not reveal their charge."

"You don't have a charge anymore, remember? It didn't work out so well for you."

Blinda glared at Rick. He could feel her trying to reach for him, but the inhibitor held her back.

"Why did you bring me back?" Blinda hissed. "To take pleasure in tormenting me?"

"As my friend here said before," Rick said gesturing to Caidin. "To give you a second chance."

"There are no second chances."

"There are if you recognize them and act." Rick held eye contact with Blinda until she looked away into the darkness. He turned and motioned for Caidin to join the others in prepping Tiana and Gunnar's pod. After the doctor had left, Rick looked back at Blinda.

"What do you want from me?" she asked in a lower, subdued tone.

"Your help."

"Doing what?"

"I need to get into the Rite of Pintar."

"Have a good time with that."

"Your Secret Service is blocking our entry."

"It's an exclusive club. Why do you need to get into the Rite?"

"To present the rightful heir to the throne of Hadrian."

Blinda went to speak, but hesitated. Her demeanor suddenly shifted even further, almost emotional.

"There are no Oxlind heirs. They're all dead."

"How do you know?"

Blinda gave Rick an uneasy stare, then slowly shifted her eyes back into the darkness.

"It's common knowledge," she said quietly.

"That's what Hadrian has been led to believe," Rick said, pausing. "The youngest daughter... has survived."

Blinda's eyes instantly shifted back to Rick, her expression filling with a mix of skeptical hope.

"The youngest died in a drowning accident..."

Rick slowly shook his head.

"No, she didn't. Her Au Pair saved her. Gunnar and CJ found her in a biosphere inside the planet Reako and took her to Cross."

Blinda closed her eyes, allowing her mind to flash back to the encounter with Gunnar on Dither. The station; the two women with him. The older one looked somewhat familiar. The younger one was rather wild looking, yet there was something about her... *The young lady in Diord's apartment... Was she...? No...* Blinda opened her eyes again, but Rick was gone.

*   *   *   *

"What do you want done with Blinda?" Dakota asked as Rick reemerged from the room.

"She stays right where she is for now," Rick said looking back into the darkness. "Have a couple of armed crewmen keep an eye on her." Rick looked across the room as Fuji and Tiana dematerialized. Audra was still standing at Gunnar's stasis pod gazing inside.

"I'm trying to figure out why I'm not bawling my eyes out," Audra said quietly as Rick stood next to her.

"We all respond differently," Rick said quietly. "I cried when I found him."

Audra looked up at nothing, despondent.

"Before all this started, I always knew what or how I was supposed to act and feel. It just came naturally. Now, I'm having a hard time figuring out why I'm not responding the way I think I should. Is something still wrong with me?" She looked at Rick.

"I don't think I've cried since I was a teenager," Rick said looking into the pod. "When we found out Jayda couldn't have children, I was more angry than sad. I might have been angry enough to cry, but us guys manifest emotion differently than you girls."

"I guess I'm still feeling the effects of being in stasis." Audra looked back down at Gunnar. "He looks like he's just asleep."

"I know a bunch of people who would be very upset if that were the case."

"We once talked about what we wanted if one of us should die. It was one of those conversations you have where you never really think anything will ever happen, but you're curious what the other person thinks, ya know?"

"Jayda and I have had conversations like that. I think all couples do at one time or another."

"I told him I wanted him to find someone and be happy. We even talked about potential candidates."

"I'm sure that was a fun list."

"It's hard to be serious about something like that when you're never faced with it. I think I have a better sense for what Gunnar was going through all this time. Now that we've both had to face it, I have to admit, I don't think I'm doing very well with the prospects of trying to be happy in a life without him."

"I admit I'm not doing very well with it either."

"I feel so lost," Audra tranced. "How can I possibly function in any capacity without him?"

"You remember what he would have you do and focus on that."

Audra gazed back at Gunnar.

"Do you really think Caidin can bring him back?"

"I don't know..."

"What would you like me to do?" Audra asked, taking a step back and folding her arms.

"What do you want to do?" Rick asked.

"I can't just sit around feeling sorry for myself. I need to be focused on something else."

"How about you assist Alex 7001 in getting as much of that Asium processed and fitted as quickly as possible. Billy Moon will help. Right now, you guys are our best hope for getting these ships up to full operating strength. I've got a hunch we're going to need them before we figure out a way to get home."

"Then you've decided to stay and help Administrator Vandmire?"

"I'd like to think that I'm still open to doing the right thing," Rick smiled.

"Will *Calypso* protect us well enough from what's outside?"

"Unless they decide to get really serious about getting in," Rick responded. "All the more reason to have those crystals processed and installed."

"I'm on it," Audra nodded. She brought her communicator up, but hesitated, looking back at the stasis pod as it, Dakota and BachTL dematerialized. An intense sense of loss pierced deep. *Gunnar is gone...* Taking several deep breaths, she recovered quickly.

"Hey, you ok?" Rick asked, noticing her posture.

"Yeah," she said, bringing her communicator back up. "I'm fine. *Athena*, I'm ready to teleport over."

After Audra was gone, Rick turned around to find Blinda standing in the doorway holding the neuro inhibitor. Neither warrior had any weapons, only what their abilities would allow. Straightening his startled demeanor, Rick motioned to one of the seats in the room and sat down. He watched Blinda step tentatively to an opposing chair and take a seat. She continued to finger the inhibitor even after she had gotten comfortable. Rick wasn't sensing any hostility, but something different. She appeared emotional.

"Princess Jana Oxlind," she muttered awkwardly. Her eyes remained on the device in her hand.

"What about her?" Rick finally asked.

Blinda remained silent for several moments. She opened her mouth to speak, but brought a hand to her lips instead. Rick waited patiently, noticing her lower lip quivering and a glint developing in her eyes as she worked up the courage to speak.

"You asked who I was assigned to protect." Blinda took a deep breath and looked away. "It was the Princess; Princess Jana."

"What happened?" Rick watched her carefully.

"Ona Tusk instructed me," she started slowly, her voice shaky and uneven. "She told me that forces within the factions of Hadrian would seek to destroy the traditions of the royal line. I watched as the

Duchess eliminated the Oxlind family one by one, starting with the Queen, Mila Oxlind. But I couldn't produce the proof needed to take action. Seeing that she would eventually come after Jana, I arranged to have her and her Au Pair moved to what I felt was a safer environment on Eno; a small moon near Albia. But when the King Commander discovered what I had done, he ordered his secret service agents to take them into custody. Shortly after that, they were transferred to a secure facility in a remote location. I tried to find them using every resource I had at my disposal. I was later told there had been an accident. They were both drowned in a flood." Blinda's eyes pooled, tears overflowing down her cheeks. "I wasn't there to protect her from them."

"Who is them?" Rick asked quietly.

"Them," Blinda choked. "All of them. The King Commander, the Duchess, the Secret Service, several high ranking faction delegates and military officers."

"But why? What did they have to gain?"

"Greed, a lust for power. If an Oxlind were to return, the Duchess, the King Commander and all who've been conspiring with them, could be thrown out and likely executed. If an Oxlind doesn't return, then the Colonians will forfeit because of the perceived might of the Albion Empire. The King Commander and the Duchess would join the two Empires and rule together with an iron fist."

"How do we fit into that?"

"Gunnar and your Starbirds present a heightened technological advantage to anyone who has control of them. That's what Drax has been chasing."

"What exactly did she want with Gunnar?"

"Cloning him to produce super-soldiers."

"Does she expect to take the throne of Hadrian by controlling our Starbirds and producing these super-soldiers?"

"I don't know," Blinda blurted. "She's unstable... erratic; it's hard to know her mind."

Rick ran his fingers through his hair and let out a deep sigh.

"I've lied to you," Blinda whispered through a sniffle.

Rick looked up at the Thane. Wiping her eyes, she stared off into a dark corner.

"Doctor San Sann and I ran our memory probe on Gunnar. Not only can it display memories from the mind, but the person actually feels everything associated with those memories. As if it were actually happening. I saw his memories. I watched him experience all of them. I'm ashamed to admit that I even replayed them several times, just to watch him suffer. I am sorry for what I have done."

"Why?" Rick asked, somewhat appalled.

Blinda drew up, gaining a little confidence and resolve.

"Gunnar represents everything that I wasn't... but... wanted to be... could have been. I guess I was angry with... myself because things didn't turn out the way I had planned. I was attracted to him for that very reason. I even made several passes at him, which he promptly rebuffed. I realize now why. I don't know Audra, but what I observed from his memories of her, they were an amazing couple together."

"You have no idea," Rick agreed.

"I have little comprehension of how a man and a woman can be so close and share so much love."

"Audra and Gunnar are a unique couple that come around once in a lifetime. It's hard to explain."

"I'm sorry for what he has had to endure without Audra. And I am sorry for what Audra now faces. The fault is mine and I accept responsibility for it; your crew and so many others. I was blinded by the hate I harbored for myself and my failure to protect the Princess."

Rick leaned forward, his hands clasped together.

"It's not too late to make some things right."

Blinda slowly looked back at Rick.

"I cannot bring your friend back, and neither can you."

"No, you can't, but you can consider doing the right thing."

"You're still asking for my help?"

"You are a Thane," Rick affirmed. "Entrusted with the care and protection of a certain individual, just as I am. The Princess needs her protector, now more than ever. I need you to help us get her into the Rite and provide protection as she claims her rightful place on the throne of Hadrian."

"But with all that I've done, how can you possibly trust me?" Blinda's expression was one of disbelief.

Rick had to think a moment. *Is it my nature to be so trusting?* No, he was aware of many of the atrocities Blinda had committed, but he had to believe that the knowledge of the existence of the Princess and the pursuit of her rightful place on the throne of Hadrian had helped to change her. Being dead certainly should factor into how someone views life and the living. He took a deep breath and let it out.

"You trusted me, shouldn't I do the same for you?"

"I've never trusted anyone, least of all you."

"You've taken me at my word about second chances. More importantly, you've disclosed your charge."

"Doesn't that make me even more untrustworthy? I was instructed never to reveal my assignment."

"I think in this instance, Ona would approve." Rick came to his feet, stepped over to Blinda and held out his hand. She looked at it for a moment, then carefully took it and stood up. "I don't think we're going to need this anymore," he said, taking the inhibitor from her and tossing it into one of the chairs.

"So what happens now?" Blinda asked.

"Now, we go see if we can hitch a ride back over to Calliope."

"Just you and me?"

"We'll need a little help."

"No weapons?"

"Yeah, about that," Rick cringed. "I'll have them sent over to the *Merc* after we get there."

"And how are we going to do that? This ship is likely surrounded and the moment you try to get out, the factions will be all over you."

Rick brought his communicator up.

"Magic... Lieutenant Habba. Have BachTL teleport to the *Merc* with mine and Blinda's things. I'm ready to teleport directly there; along with a guest," he said, noticing Blinda giving him a skeptical look.

"Magic?" she repeated.

Rick motioned for her to stand next to him, then stood still, waiting. He felt the teleportation process begin and glanced over at Blinda who was looking at her hands with wide eyes.

"Magic," he repeated as they dematerialized. Moments later, they rematerialized in another room.

"Welcome aboard the *Merc*, General Niker," a gentleman said a little apprehensively. "I'm Temari, Administrator Vandmire's aide."

"First time seeing teleportation?" Rick asked, smiling at the gentleman's expression.

"Yes, Sir. The administrator said you'd be coming in this manner, but I have to say, I wasn't sure what to expect."

"Your first time is a little unnerving," Rick admitted, giving Blinda a quick look. She was still examining her hands. "You ok?"

"You could have warned me," she finally said.

"Where's the fun in that?"

The aide looked around Rick at Blinda. "Aren't you Blinda Koss?"

Blinda gave Rick a nervous glance, then stepped forward.

"Yes."

"The administrator didn't say anything about transporting you."

"What did the administrator say?" Rick asked stepping between Blinda and several guards that brought their weapons up.

"Just that you would be coming with a group to be taken over to the delegation landing platform on Calliope."

"Is there a problem?" Rick asked firmly. He felt something affecting his ears. "Blinda, please... I'll personally vouch for her."

The aide fidgeted nervously, looking back at Blinda. An awkward silence stretched until Blinda shifted, resting a hand on her hip.

"Very well, General," Temari motioned to a figure stepping through the door behind them. "This is General Aurora. He arrived a few minutes ago."

Lou stepped forward sizing Rick up and looking over at Blinda.

"This is quite the plot line," Lou said hesitantly. "General Niker, you're aware who this woman is?"

"General Aurora, meet Blinda Koss. Blinda, General Aurora."

"Yes, we're well acquainted," Blinda grinned, reading the worried look on Lou's face.

"Would someone like to explain what's happening outside?" Lou asked. "That hotbed out there looks a little too intense for my liking. Usually I keep my ships as far away from situations like this to avoid any unnecessary entanglements."

"Are you here to help or complain?" Blinda sniped.

"A little of both," Lou fired back.

"Diord said you were delayed for a rescue operation?" Rick cut in.

"Hope it was worth it," Blinda mumbled beneath her breath.

"Were you successful?" Rick firmed up, giving Blinda a look.

"Yes," Lou said, digging into one of his pockets. "We kept the Drake from getting their hands on an Albion munitions transport in distress."

"You diverted for a munitions transport?" Blinda was incensed.

"Turns out the pilot had something she was trying to deliver to Diord."

"I hardly think the Administrator is in need of a shuttle full of munitions," Rick said.

Lou opened his hand. A large, ostentatious ring sparkled in his palm. Blinda's demeanor shifted, glancing up at the Tomplie General.

"Who was the pilot?" she asked, stepping closer.

Lou closed his fingers around the ring and smiled.

"Malina Cass. Said she knew a Colonel Conrad? They recovered the ring from the *Tarzana*."

Rick looked at Blinda.

"Do you know anything about this Malina Cass?"

"Very little, other than she hooked up with your friend. We caught up to him, but she escaped using a munitions shuttle before we could take her. She must have been trying to get it to Cross."

"Where is she now?" Rick asked.

"I have her safely tucked away in one of my suites aboard the *Executioner*."

"So what are your intentions, General Aurora?"

"I need to get this ring to Princess Jana. She's going to need it in order to claim her right to the throne of Hadrian."

"Your timing is impeccable," Blinda smirked.

Lou gave the Albion Thane a hard look, then turned back to Rick.

"Why is she here?"

"Diord and the Princess are in a bit of a jam on Calliope," Rick answered. "She's here to help get the Princess into the Pintar hall."

"Blinda Koss?" Lou asked suspiciously, pointing at the Albion Thane. "This Blinda Koss... The one right behind you? She works for KC Dismon..."

"That's right," Rick responded, nodding.

"You remember that I'm standing right here?" Blinda shifted. "You know, scariest woman in the galaxy?"

Lou looked at Rick.

"You seem to be missing any sense of logic. You're not actually going to trust anything about this woman, are you?"

Rick glanced at Blinda who remained silent, smiling.

"Would you like to come along?"

Lou considered for a moment then shook his head.

"I suspect things are probably going to get a lot more dicey out here before they settle down." The Tomplie General handed Rick the ring. "Please see that the Princess gets this. I'll stay out here to help make sure these gun happy factions don't get too frisky."

Rick held the ring up and tipped his head toward Lou.

"And General," Lou said, stepping away. "Never turn your back on Blinda. She will betray you at the first opportunity that presents itself."

Blinda opened her mouth to deliver a scathing rebuke to the pirate turned patriot, but Rick held the ring up. She looked at it, then at Rick. He motioned for her to take it.

"When the time comes, this should come from you. Besides, I'm going to have my hands full with other things."

Blinda remained silent, looking at the ring, then at Rick. She touched it carefully, still looking at Rick. Rick smiled slightly, then held Blinda's hand open and placed the ring in her palm. She stared at it for a moment then looked back at Rick. Her demeanor seemed to soften as she put it in one of her pockets. Rick turned back to Diord's aide.

"I think we're ready."

"If you'll step this way," Temari motioned to a doorway. "We have a shuttle waiting in our executive hangar bay."

"Just a minute," Rick said, stopping him. "We've got some items on the way."

"Very well," Temari said, moving back a little as a figure materialized next to them. Everyone looked at BachTL holding two bundles of equipment. Rick stepped over and took his utility belt, then handed Blinda hers.

"Welcome aboard." Rick smiled at the look on the Castellian's face.

"You could have warned me," he responded tentatively.

"Where's the fun in that?" Blinda snickered.

"How long does it take to get used to this?" BachTL asked as Blinda pulled her blaster partially from its holster and examined it. Shifting

her eyes to Rick, then at Temari and back at BachTL, she pushed it back in and strapped the belt around her hips.

"Captain Abrams told me you know exactly what you're doing," BachTL stated, giving Blinda a long look, then looking back at Rick.

"These clothes aren't what I would call comfortable," Blinda complained, making sure her Balkrum was secure.

"At least you've got clothes on," Rick teased, referring to their first engagement.

Blinda gave Rick a quick look.

"You know, you do have the ability to make them whatever you want."

"Oh, yea..." Blinda responded, running her hands down the Kalamarion flight suit past her hips. The clothes instantly changed to a slick black bodysuit, all the way down to where her shoes changed into her black knee high spiked heels.

"Yes, so much better." Blinda looked up at Rick and smiled.

"You could have chosen white."

"I like black."

"So a couple of questions," BachTL said after waiting for them to finish their exchange. "What am I doing here, and what is she doing here?"

"Nice to see you too, BachTL," Blinda smiled.

"She's here because she's the Princess's protector."

"Princess? What Princess?"

"Princess Jana Oxlind."

"There's no such person," BachTL refuted. "All the Oxlinds were killed off by the Duchess."

"I beg to differ," Rick said. "She's on Calliope with Diord trying to get into the Rite of Pintar. That's why you're here. I need you to help get us into the Rite."

BachTL looked at Blinda.

"Come on," Blinda finally said. "We're on a schedule here."

"All right," BachTL relented. "But I would keep my eye on her if I were you. She can't be trusted."

"Ouch, that really hurt," Blinda chided sarcastically.

"Don't make me sit between you two," Rick said, turning to Diord's aide. "Lead the way."

Temari hesitated, watching Blinda pull her hair back into a ponytail and make sure her outfit was straight. He finally motioned for them to follow. Once they reached a main hallway, they boarded a ground transport that quickly moved them down to an awaiting shuttle. As they got comfortable and waited for departure, Blinda turned to Rick.

"How far are you expecting me to go with trying to accomplish what you're asking?"

"Only you will know that," Rick replied.

Familiarity

Jana looked out her apartment window at the flurry of ships moving back and forth across her view. The main cavern of Calliope was ablaze with lights of every kind from the Castellian city lining the cavern walls. Most of the military ships were moving quickly for the main exit. A bottleneck had formed behind the Albion dreadnaught, *Indominable,* still stationed in the mouth of the great cavern. It was difficult to see anything beyond the *Indominable*. She wondered if all this activity was a direct result of Rick Niker and his search for the *Constellation*. As she watched, the last images of Gunnar came to mind, bringing pangs of guilt and thoughts of *if only...* repeating themselves in her mind.

"Nervous?" Tonnie asked, stepping up behind the young Princess. The nanny fussed with her outfit.

"I can barely stand," Jana answered quietly.

"Good, it means you have your wits about you."

Jana turned to her caretaker and smiled.

"You look pretty good... for an android," she said admiring the gown Tonnie was wearing. The Au Pair looked down at what she was wearing and held it out.

"I do look nice in this," she said smiling. The smile was short-lived. "With my luck, some alien is going to hit on me."

"Oh, you never know," Jana teased. "You might attract another human male."

"As long as it's not that worm, Lou Aurora," the android snickered.

Jana turned back to the window, slipping back into a melancholy mood. She remained silent for some time until there came a summons at the apartment door. Tonnie was quick to answer and led Diord into the room.

"Your highness," Diord bowed, out of breath. "We need to leave immediately."

"What's the urgency?" Tonnie asked. "The Rite isn't scheduled to start..."

"The Albion Secret Service is on their way here to arrest you. We must get you to safety right now."

"And where is safety located?" Tonnie asked, ramping up.

"Back on the *Merc*," Diord replied instantly. "But we have to leave now," he beckoned, becoming somewhat frantic.

"What happened to making the presentation?" Tonnie asked. "We were assigned to the first of the Rite."

"I don't know what happened," Diord fumbled trying to figure it all out. "Somehow the Duchess and King Commander have found out we're up to something. My guess is Blinda Koss had something to do with this."

"I knew this was going to happen." Tonnie grabbed Jana's book and started past Diord, but Jana didn't follow.

"Your Highness?"

"Administrator, how will we get back to the *Merc* undetected?"

"I have a shuttle craft on its way to the delegate docking bay as we speak. They should already be there waiting for us."

"How did you know we would need a shuttle?"

"When I received word that we had been barred from the proceedings, I called over to General Niker for help. He and several of his men will be waiting for us at the shuttle. Please, your Highness, we need to leave before the secret service gets here."

Jana finally nodded and all three hurried out of the apartment and down one of the side hallways. Constantly looking over the balcony walls at the floors below, Diord caught sight of several Albion officers and a garrison of troopers making their way to the upper levels. Weaving and alternating their modes of descent from stairways to elevators, they somehow avoided the troopers and dissolved into the masses of people working their way toward the Grand Pintar hall.

Spotting a cluster of troopers and more officers at various places in the main concourse leading to the grand hall entrance, Diord ushered his charges into the back of a small café. He made sure he and Tonnie had their backs to the bustling outer doors.

"This isn't good," he puffed. "They're crawling all over the place."

"How are we supposed to get back to the docking bay?" Tonnie asked.

"I think the best thing to do is split up," Jana suggested.

"Absolutely out of the question," Tonnie refused flatly.

"The Administrator is easily recognized," Jana countered. "Two lavishly dressed women will stick out."

"In this crowd?" Tonnie argued. "Everyone out there is dressed in their best. No one is going to give us a second look."

"But what happens when we're out of the crowd? They're looking for two women, like us," Jana responded.

"We can't just stay here," Tonnie exclaimed in a whisper.

"The Princess may well be correct, but in a different way," Diord said looking over his shoulder at the crowds moving past the café. "I am the most recognizable. Most won't recognize Tonnie either, but there is a chance, especially if they've had your image sent to them from your encounter on Dither."

"Yuck, hope no one associates me with what that picture must look like," Tonnie said.

"Princess, no one really knows what you look like," Diord said. "You should be able to move anywhere freely."

"Until she reaches a checkpoint," Tonnie pointed out.

"She has her entrance ID," Diord fired back. "She could easily walk right into the Rite and take her seat in my box."

"You don't think the entrance security wouldn't have her ID on file and flag her? She'd be caught before she could make it down the concourse, if she even got that far."

"Yeah, all right," Diord admitted. "Just trying to figure out if we can salvage this somehow. We're so close."

"I admit that she is the least recognizable," Tonnie surmised.

"Exiting a space in times of heightened security is generally easier than trying to get in," Diord said.

"Splitting up is the only chance we have," Jana maintained.

"They'll be looking for a man and two women," Tonnie confirmed.

"Your Highness, do you know how to get back to the delegate docking bay?" Diord asked tentatively.

"Yes."

"There is an executive lounge just off the bay area; we'll split up and meet there in twenty minutes." Diord went to get up, but Tonnie stopped him.

"Let her go first, we may draw attention, she won't."

Diord and Tonnie looked back at the apprehensive look on Jana's face.

"Be confident and move with purpose," Tonnie coached.

Jana nodded and got up, moving straight for the crowded hallway. As she melted into the masses, Tonnie patted Diord's clasped hands and got up. She too got lost in the moving crowds. Waiting and watching the crowds move, Diord observed several officers in dark uniforms moving in the same direction Tonnie had turned. Quickly out into the hall, he kept a visual of the dark uniforms; *Secret Service*. They looked Albion, but it was difficult to make a positive identification in the sea of bobbing heads. Pushing hard through the crowds, he followed as close as he dared, all the while trying to see ahead of them for any signs of Tonnie. As they moved off into a branch hallway, he caught a glimpse of the android weaving through the thinning crowds. Diord watched the agents speaking into communicators as they worked to close the gap. He picked up the pace, moving faster as he picked his way through the lines of people.

He had almost caught up to the agents when he detected several more file in behind him. The crowds had thinned, some of the people making way for the chase. Catching a glimpse of Tonnie as she stopped at an elevator, he broke out into a sprint. As the elevator opened, Tonnie turned around to the sound of heavy footsteps and charging blasters. She had only a moment to step back as Diord flung himself into the back of the three agents, all of them toppling through the open door of the elevator. He looked up from the crumpled mass of tangled arms and legs as Tonnie frantically pushed on the door control, but several agents kept it from closing. A moment later, Diord and Tonnie were pulled up together and pressed against a wall. Diord looked over at the android as they were both searched for weapons and fitted with wrist binders. Her expression was one of confusion mixed with anger.

"Let's go, Administrator," one of the agents said, taking Diord's arm. He and Tonnie were pulled down a side hallway. "Someone would like a word."

"Who?" he asked.

He got no reply, only a faster pace and another tight hand around the opposite arm. He looked back at Tonnie, who was being held in a similar manner. At length, he recognized the executive lounge doors directly in front of them. His heart sank as they came to a halt just inside. Jana was sitting in one of the chairs in front of the large bay windows looking into the docking bay. There were a couple of small shuttles and a few larger transports parked inside the closed hangar. Several agents held pistols pointed at the young Princess's head. Leaning against a darkened corner, Drax Blair frowned and took a couple of steps toward the administrator.

"Captain Hawkins, have your agents wait outside, but you are to remain" she ordered, giving the lead commander a glance. She kept her attention on the other three in the room. "I have every right to shoot you for treason, right here, right now," she growled.

"Then let's get on with it," Diord complained defiantly.

"All in good time," she said, settling a little. "I have a few questions first, and I will have straight answers." She pointed at Jana. "Who is this girl?"

Diord looked back at Tonnie, then over at Jana.

"And if you try to call her Kara Americ, I will shoot right now and have these other two interrogated Ratronian style."

He took a deep breath, reading the fiery determination in Drax's continence.

"Princess Jana Oxlind," he said, sounding defeated. He then straightened up, standing bold. "True heir to the throne of Hadrian."

Drax stared at the administrator for several seconds, then burst into laughter.

"True heir to the throne?" she repeated through her gasps of laughter. "How many times did you rehearse that line?" she asked, quelling her grin. "Ok, ok, please tell me how you arrived at that conclusion?"

Diord motioned to Jana's book that had been left on a small table in front of her. Drax straightened her face and took the book, examined it for a moment, then turned it around.

"Nice try," she said, tapping a fingernail on the locking mechanism. "I've seen stacks of these. You got a key or do I just take your word for it?"

"I have the key," Jana said, still sitting.

"Well then little girl, bring it up here and let's see what this is all about."

Jana slowly came to her feet and glided across the room in front of the Queen Captain. Pulling her necklace from under her neckline, she pressed the small orb to the indentation on the book mechanism and turned it in several directions, pulling her finger away after pushing on the ball. The mechanism suddenly lit up, beeping several times. As the orb went dark, she took a step back and watched as the latching mechanism began to expand while several mechanical fingers released the front binding of the book.

"Now that's impressive," Drax smiled. "I should believe you just on this alone."

Jana stepped back to her seat as Drax opened the book and examined its contents. Several sheets of parchment lay smooth across a flat screen. Drax picked up the facing page and held it up.

"Very fancy," she smirked, gazing at the lavish signature at the bottom above a curious seal.

Diord glanced at Jana. She held a tense expression, watching Drax read through the documents. For several anxious minutes, everyone waited as Drax cruised through the material. Occasionally she looked up at Jana, then at Diord and Tonnie. At length, she slowly closed the book and handed it to Captain Hawkins.

"Captain, see that this is secured in my personal box in the Rite hall. I'll be along shortly."

"You can't take it," Tonnie objected, stepping forward. "It belongs to the Princess."

Drax developed a patronizing look.

"This blaster says I can." She motioned for the Captain to proceed. As he left the room with the book, Drax stepped back against a dark corner, deep in profound thought.

"Do you have any of the royal tokens?" she asked looking back at Jana.

The young Princess nodded quickly, a spark of hope flickering into her face. She held her hand up, revealing her mother's ring.

"Bring it closer," Drax commanded.

The young woman came forward again and held her hand up.

"May I?" Drax asked.

Jana hesitated.

"It's all right, I'll give it back."

Slowly pulling it from her finger, she held it up to the Queen Captain who took it and examined it closely. After several tense moments, she gave it back and motioned her back to her seat.

"Well, this is certainly a turn of events. Who else knows about this?"

"Only a few others," Diord said.

"My sister?"

"What difference does it make?" Diord replied cautiously.

Drax grinned broadly.

"You're right, it makes no difference." Drax shifted. "Does it make any difference if I told you, I believe you?"

"You do?" Diord and Tonnie responded together. Optimism filled Jana's expression.

"Then you'll see us into the Pintar hall to present her at the Rite," Tonnie spoke up.

Drax looked at her with a blank expression, finally shaking her head.

"No."

"No?" Diord repeated. "What do you mean, no? You just said you believe her. She must be presented at the Rite."

"No," Drax responded stone faced.

"Why not?"

"Hadrian is at a tipping point. Without her father's ring, presenting the Princess would throw the factions into a fit of chaos leading to a bloody civil war. For the good of Hadrian, I can't allow that to happen."

"That's not right!" Tonnie exclaimed.

"But if you don't," Diord cut back, "the Duchess will be made Queen of Hadrian. The other factions won't allow it and war will break out anyway."

"Yes, and the strongest will survive."

"The Albion Empire," Diord sneered.

"Yes, the Empire is poised to take its position as the new leadership of Hadrian. Once it's clear the transition of power has stabilized the government, I can take my place in a position more worthy of my abilities."

"This isn't about what's best for Hadrian," Diord snarled. "This is about the mighty Drax Blair taking as much power as she can get."

"Why shouldn't I be in command? In this time of transition, isn't proven leadership what Hadrian needs?"

"I'll make sure more than three of us know about this," Diord flared.

"My dear, Administrator. Surely you understand that given the current state of affairs, I can't have anyone trying to tip this delicate balance?"

"It doesn't matter where you banish us to, Hadrian will know of this."

"Banishment? Who said anything about banishment?" Drax asked, giving him a surprised look. "Your Highness, if you'll join your two friends over there." Drax waved them to the wall with her pistol. Diord reluctantly stepped to the wall as Jana rose and walked toward the royal nanny. As Drax watched her go, she was somewhat surprised at her poise. Perhaps she didn't understand what was about to happen? As she stood next to her companions, Diord and Tonnie moved in front of the Princess.

"Ok, so this is impressive," Drax gloated. "I admire courage and loyalty as much as the next person, but what good is this going to do? Come on, don't make this any harder than it has to be," she said motioning them back into place. Tonnie remained defiant, standing resolute in front of the Princess.

"I doubt this will be hard for you," Diord retorted.

"I have a schedule to keep. The Rite ceremony is starting and I have a box to get to." She looked at the time while powering up her pistol.

"Tonnie, it's ok," Jana said, trying to pull the Au Pair back into place. "I am not afraid."

"My prime directives are in play, your Highness," Tonnie replied, standing up straight, looking directly at the barrel of Drax's pistol.

"Fine," Drax sighed. There was a loud pop accompanied by a sudden flash. Jana flinched, feeling Tonnie jolt, then stagger slightly. Jana looked around the nanny at Drax who held her blaster still pointed at the Au Pair. Tonnie's head suddenly cocked hard to one side at the same time she made a half turn. Trying to remain in front of the Princess, she put her hands to her chest. As she turned and faced Jana, her expression was one of surprise. There came another loud pop and a flash. Tonnie jolted again, this time leaning into the young Princess's arms. Diord quickly helped settle the android to the floor.

"Tonnie!" Jana exclaimed, kneeling next to her. The nanny looked at a greenish yellow fluid coating her hands. Her chest now bore a gaping hole, revealing her internal workings. Inside, tiny fiber-optic strands glowed from frayed and melted ends. Sophisticated electronics flashed from her chest injury causing her hands and fingers to twitch erratically. Her body stiffened as she looked up at the Princess.

"I think I've been damaged," she gurgled as fluid overflowed from her mouth. Her voice had turned high pitched and full of digital artifacting.

"Stupid android," Diord muttered, trying to stem the flow of fluid coming from her chest and back.

"I don't feel so well," the android stuttered.

"Stay with me," Jana said, looking anxiously into the eyes of the nanny. "We'll get help."

"My... child," Tonnie smiled trying to steady her jerking eyes at the Princess. "I have followed... my directives to... the, the, the... best of my ability. I... am sorry... I wasn't with you... when you needed... me most; growing up... alone."

"Quiet, you silly ol' woman," Jana gasped through flowing tears.

"I, I, I suppose this... time I really, really, really will sleep... though I'm at... a loss to comprehend what, what, what... that is. When, when, when... I slept in the well... I was con... con... conscious of my, my, my... surroundings. This feels a... little more permanent."

"What would an android know about such things?" Jana wept. "You can't die, you're not human."

"Yes... I can. I don't know, know, know what... it will be like. I, I, I... only know... that of all the things... I will leave in this, this, this life, I will miss... you the most. I have nothing, nothing, nothing to compare... having a, a, a daughter to. I can only... feel, feel, feel like I was a mother, mother, mother to... you."

"I could have asked for none better." Jana cradled the nanny's head in her arms, gently rocking her.

"How touching," Drax drooled. "Feelings for a robot."

Jana looked into Tonnie's lifeless eyes. She gently set her head down and came to her feet, facing the Queen Captain.

"If you're hoping your little emotional display will change my mind..." Drax suddenly froze, her eyes widening. She took a couple of steps forward.

The tension in the room shifted as CJ Barker stepped out from behind the Queen Captain with a blaster pistol pressed into her side. CJ carefully took the pistol from her sister and tossed it across the room at Diord's feet. He quickly scooped it up.

"It would be nice to get these binders off," he said pointing the pistol at Drax.

CJ pulled a tiny boxlike device from her pocket and tossed it to him.

"Tonnie, I presume," CJ stated looking at the lifeless android.

"She shot her in cold blood," Diord said, removing the binders from Jana's wrists.

"How do you shoot a robot in cold blood?" Drax asked.

"I saw what happened," CJ responded angrily. "Sorry I didn't get here sooner, your Highness."

"What's your plan here, Sis?" Drax inquired.

"Shut up," CJ said, motioning her to a chair. "Your Highness, why don't you see how well your binders fit the Queen Captain?"

Drax took a seat and holding her wrists out, the Princess activated the binders around them. CJ backed up against one of the windows, still holding her gun up.

"Now, as for my plan. We're all going to go to the Rite... And you're going to host us in your box."

"I don't think so," Drax said calmly. "Castellian Security operates independently of me. With or without me, they'll have orders to arrest or shoot these two on sight."

"Who barred my delegation from the Rite?" Diord asked.

"Can you really be that thick minded?" Drax chuckled. Something caught her eye outside in the docking bay. Recognizing a figure walking across her vision, an arrogant smile quickly formed. "The King Commander and the Duchess have everything under control."

CJ became aware of a clicking noise. It was steady and deliberate; getting louder. She looked at her sister. An annoying smile had formed across her lips, indicating she had more cards to play. As the clicking grew louder, CJ recognized it as footsteps. The spiked heel of a woman's boot. Reality shifted into slow motion as Blinda Koss slowly strode into the room. Before Diord or CJ could get off a shot, their blasters flew from their hands. Blinda caught both pistols, but instead of pointing them back at them, she carefully set them on a nearby table and turned to Jana as two more figures entered the room.

"BachTL!" CJ exclaimed. "What are you doing here?"

"Tour guide," he brightened, starting for CJ and Drax. He stopped short, noticing the binders around Drax's wrists. "CJ... What's going on?"

"Right now, it's very complicated," CJ responded with an uncertain look.

"Drax?" BachTL inquired, hoping to get more from the Queen Captain.

"Your sweet mother is correct, Son," Drax smiled. "It's quite complicated at the moment."

"Oh, that's smooth," CJ gritted, pressing her sister back into her seat.

"I think we all need to take a deep breath here," Rick suggested eyeing everyone in the room.

"We don't have time for that, General," Diord exclaimed pointing at Blinda as she stepped toward Jana and took a knee. "What is she doing here?"

"Your Highness, I am Blinda Koss. Albion Thane and protector...
Your protector."

The room fell into stunned silence. Jana looked down at the
kneeling warrior. Fumbling with what to say, the young Princess
finally cleared her throat.

"Thank you for coming to our aide. Please rise." Jana offered her
hand to the Thane.

Blinda looked up and smiling, came to her feet.

"You visited me earlier," Jana said, passing Diord a glance. "In the
apartment. You said you were an acquaintance of Administrator
Vandmire."

"An acquaintance?" Diord blurted. "What are you talking about?"
He turned to Rick and pointed at Blinda and BachTL. "General Niker,
this isn't what I had in mind when you said you had an idea."

"What? BachTL knows the layout of Calliope and... Let's just say
that the realization of the existence of the Princess has changed Miss
Koss's perspective," Rick said looking back at Diord. "She has agreed
to help see that the Princess is presented at the Rite."

"Well, aren't family reunions just the nicest things," Drax smirked.
She went to get up, but CJ pushed her back down. "I knew my sister
and the administrator have been plotting something, but you want
everyone to believe the mighty Blinda Koss, who has butchered
countless people in the service of the Empire, has suddenly joined the
side of benevolence?"

"As if you would know what true benevolence really looks like," CJ
commented.

"I know Blinda well enough to see that this is an elaborate ruse to
claim your capture and present you to the factions. She'll even
execute you in front of everyone herself just to prove the point. Tell
them I'm wrong, Blinda."

All eyes turned to the Albion Thane. Blinda looked back at Rick,
then around the room.

"With the exception of the Princess, everyone in this room knows
who and what I am. CJ and Drax have been witness to the horrific
things I have done and can do in the service of the Empire. Part of the
General's crew perished at my hands. Looking back, I can recall many
things I regret, but ultimately, can do nothing to rectify." She looked
directly at the Princess. "But I can look ahead at what I can do now.
That is to see to your safety, as I had once pledged." Blinda reached
into her pocket and taking a knee again, held up a ring.

Shocked, Jana carefully took the ring and examined it. Diord
stepped over for a closer look as Blinda rose and stepped back next to
Rick. Beaming, Jana put her parent's rings together, admiring them.
Diord suddenly turned to Rick and Blinda.

"She took the Princess's heritage book," Diord said, pointing at Drax. "Without it, there is no point proceeding, unless we can get it back."

"You let me worry about that," CJ said. She tugged on Drax's cloak, but BachTL stopped her.

"No. If what you say is true, Calliope security won't let you get very far. It has to be someone that's not under suspicion. As the top aide to the Queen Captain, I know right where everything is and how to get to it."

"Good enough for me," Rick said, picking up the blasters and handing them back to Diord and CJ. "Can you direct us to the fastest way to the Pintar hall?"

"Are you kidding?" Diord exclaimed. "After all this, putting aside the fact that we're talking about trusting Blinda Koss…" Diord pointed out into the shuttle bay. "There's virtually no chance of getting out of this room other than that way. The Secret Service is right outside those doors," Diord said looking back at the closed hallway doors. "One little peep out of the dragon lady here and they'll lay us flat, not to mention all the Castellian security on high alert with the proceedings under way. They wouldn't let us in even with Drax standing there telling them to."

"Then we won't ask," Rick said.

"What about the Secret Service outside?" CJ asked, holding a gun on her sister.

"The General and I will take care of that," Blinda said, feeling the distrust in the room.

"Big mistake," Drax grumbled.

"Shut up," CJ said, slapping Drax across the back of her head. "Come on, let's go," she said, grabbing her by the collar.

"Why not just shoot me now and get it over with?"

"I'm not letting you off that easy," CJ said, pushing her toward the closed doors. "You're going to witness history, one way or another."

"Your Highness," Diord said, taking Jana's arm and pulling her away from the doors. "Are you sure you still want to do this? There's a shuttle right there. We can still make it back to the *Merc* and get out of here."

Jana looked down at Tonnie, then nodded.

"Yes, we have to try."

"Ok, please stay close."

Rick and Blinda stopped at the double doors and pulled their blasters.

"I'm thinking we use these only if necessary," Rick said looking over at Blinda.

"You do things your way, and I'll do them mine," she said.

"You go out those doors and there's no going back," Drax hissed from the side.

"Gun in one hand, rings in the other," Blinda replied, pulling her Balkrum glove on. She gave Rick a glance. "This is what I do best."

"There are better ways," he replied in a hush as they activated the doors.

The moment the door cracked open, Blinda's blaster and Balkrum were powered up and pointed at... nothing. Secret Service agents lay motionless all around the floor of the hallway. Surveying the scene, Blinda lowered her weapons. BachTL checked several as Rick checked either direction.

"They're unconscious," he said, surprised.

"Unconscious?" Blinda repeated. She looked around trying to figure it out. "How?" The Albion Thane turned to Rick.

"I told you there are better ways," Rick said tipping his head. "Which direction?"

BachTL pointed down a short hall. Blinda looked after Rick leading the Princess and Diord.

"Looks like there are always alternatives," CJ said, directing Drax at gunpoint.

Blinda looked around at the agents on the floor as she lowered her weapons. *Alternatives? Manipulating matter requires a command of mind and element, not soft whisperings. Doesn't it?* Noticing the group turning a corner ahead of her, she quickly trotted off after them.

The Rite of Pintar

Weaving their way through the Castellian complex, BachTL lead the group past several contingents of agents undetected, as well as regular security personnel patrolling the areas directly around the Great Pintar hall. Finally reaching a delegate entry portal, the group huddled in a darkened corridor across from a small foyer. Two sets of guards stood at the closed entry with blaster rifles cradled in their arms.

"This is it," BachTL said in a hush. "This entry leads directly into the main hallway near the front. There will be personal Albion and Colonian guards stationed in small alcoves all around the Pintar Hall, but Castellian security controls the entire hall. The personal guards are there only to escort the Duchess and King Commander."

"Even if we reach the platform, the Duchess and the King Commander probably have this whole thing rigged," Diord grumbled.

"Can't we appeal to the Grande Wamar and the rest of the council?" Jana suggested.

"While the Wamar is Castellian and in command of the ceremony, he's likely just a puppet of the King Commander," Diord replied.

"Don't expect to get too far," BachTL advised. "Security will be on you before you reach the platform."

"We have an appointment," Diord said, trying to sound optimistic.

BachTL looked back at the entrance.

"This is where I leave you. I'll bring the book down when it's safe to do so, but if you guys get into trouble, which I know you will, I'll make sure it stays safe."

As BachTL started down a different hall, Blinda sized up the guards standing watch at the entrance.

"You gonna put these guys to sleep too?" Blinda whispered looking over at Rick.

"You could always give it a try," he suggested passing her a glance.

Blinda looked at the General for a moment, then noticed Jana staring at her. The Thane considered for a moment, then raised her hands and closed her eyes. Concentrating on what had to be done, she pressed her eyes tight. Exasperated, she finally twisted her wrists and grimaced. A blast of air suddenly flung the guards into opposing corners. Several shots exploded against the wall near Blinda and Rick's position as one guard came back up firing. Before Blinda could bring her own weapon up, a glowing disc buzzed past her and into the guard, instantly dropping him. As the Balkrum flew back, Rick caught it in his gloved hand and shut it off.

A clamoring from different directions quickly developed, as the small group hurried to the entry portal, only to find it locked. Blinda pointed her blaster at the control, but was stopped by Diord.

"That's not going to do any good," he said looking at it closer. "This is a reinforced double sided blast door with a dual redundant locking mechanism. You blow this panel and the panel on the other side still has command of the door."

"Then we'll just pull the whole thing out," Blinda said, raising a hand.

Rick stopped her with a careful hand to her wrist.

"You must consider the unintended consequences of any action you take. I take it you don't have an access code for it?"

"Why do you think I was going to blast it?" Blinda replied.

"We've got company," CJ said, pushing Drax to the control pad. "She's got access. Open it."

"Sure, I have access," Drax smiled confidently. "But I'm not going to open it for you or anyone else."

CJ pressed the barrel of her pistol into Drax's side a little harder.

"Shoot me," Drax resisted. "Wasn't I going to do the same to you and Gunnar?"

"How does the door work?" Rick asked, looking at the security pad.

"Pneumatic piston axis with a redundant hydraulic pressure arm controlled by a Class A alpha numeric code," Diord said looking down one of the halls. Several figures were running in their direction. "I know mine's been disabled. CJ, you have it, don't you?"

"I did, but I've been locked out of everything too as I'm considered a security risk now," she said looking at her sister.

"Wonder why that happened?" Drax winced as CJ jabbed the pistol barrel into her side a little harder.

Before anyone could say anything else, Blinda threw her Balkrum down one of the darkened halls and waved her hand. A blast of air took her airborne as she followed after her buzzing weapon. Moments later, flashes of light and blaster discharges echoed up the hall. Rick spun in the other direction as several bolts of energy streaked past their position and impacted on the walls. He fired off a couple of shots, then spun a hand above his head and launched himself in a barrel roll into another corridor. Blaster flashes followed his departure, along with a sharp buzzing sound as his Balkrum was sent sailing.

Diord checked down the third hallway and finding it empty, relaxed a little.

"Well, I guess it's up to us to get this door open."

"You know, you could accidently end up as collateral damage just as easily as any of us," CJ jabbed at her sister again.

"Wouldn't that just be the worst?" Drax replied sarcastically.

"Come on, Sis. Do you really want to do this?"

"I have a better chance of getting through this by simply doing nothing. The chances of me getting hit are pretty slim. If you manage to get in and mess everything up, I can still just sit back and watch it all happen. At the end of the day, I'm still Queen Captain and if you guys get captured, which is inevitable, I take the credit. Have I left out anything?"

"Yeah, a couple things," CJ replied. "If we get in and we're heard and the factions accept Jana as the heir apparent, all your plans dry up."

Drax looked over at the Princess.

"What's the other one?"

CJ put the gun to her sister's head.

"I do the same thing you threatened to do to me and Gunnar."

"Pu-lease," Drax drawled. "You don't have it in you to shoot your own sister."

"We can't just stand here and wait for somebody to come pick us up," Diord said, interrupting the family exchange. Blinda suddenly reappeared from the darkened hallway.

"We've got a couple of minutes now," the Thane puffed, stopping in front of the group. "Figure this out yet?"

"No," Diord responded. "CJ's been locked out of the system and Drax won't give her code up."

Blinda looked at the Queen Captain for a moment, then took a step back.

"How about if I convince her," she said, closing her eyes and bringing one of her hands up.

Drax suddenly pressed herself against the wall. CJ held onto her, keeping her pistol in her side.

"Blinda," Drax gasped, her eyes popping. "Don't..." She suddenly brought her hands to her head, grimacing. "No..." The Queen Captain began to convulse in agony.

"The code," Blinda demanded stoic, her eyes still closed.

"No, I won't..." Drax struggled.

The others grimaced, watching the terrible scene unfold in front of them. Drax tried to scream, but nothing came, only a horrific hissing sound from an open mouth. Blood appeared in her nostrils and ears.

"Blinda, stop!" Jana pleaded. "You're killing her!"

"The code!" Blinda insisted.

Drax moaned in agony, jerking her head back and forth. She looked up into her sister's eyes, her face contorting, experiencing a

taste of what so many others had endured at the hands of Blinda Koss. Every fiber of her being felt as though it was ready to explode. Invisible knives slowly worked through her flesh, all over her body, trying to pierce through her skin. Hot pokers bored their way from the center of her skull outward, trying to make an exit.

"The... code," Blinda demanded.

Drax grasped at CJ, clutching her uniform as if it could somehow provide some kind of relief from her exploding nervous system. CJ turned away, closing her eyes. Drax lost her grip and slowly sank to the floor.

"Nine... three... seven... four... two... Alpha, Zulu, Tango..."

Diord quickly keyed in the security code as the Queen Captain gasped out the code.

"Blinda," Jana cried. "Enough!"

The Albion Thane relaxed and opened her eyes, looking down at the sobbing form of Drax Blair. As the door opened, several guards on the inside turned bringing their weapons up. Before they could get a shot off, their weapons promptly tumbled to the floor next to CJ as they fell unconscious. Diord looked back at Blinda; she only shrugged. Behind her, Rick Niker held his Balkrum and pistol up. He looked down at Drax and CJ.

"What's this all about?" he asked.

"The Queen Captain was just providing us with the security code," Blinda replied.

Rick pushed through the group. As Diord passed, Blinda noticed Jana looking right at her. Her expression was a mix of fear and disapproval. Blinda glanced down at CJ and Drax, then looked after the Princess. The realization of what she had become stung her to the core. Perhaps for the first time, the Thane understood what she had become through the eyes of others.

Blinda helped CJ get Drax to her feet. As the group made their way through the door and up the corridor, they could hear the voice of the Grande Wamar echoing from the Pintar hall. Everyone came to a stop at the entrance, while Diord peered into the immense hall, trying to listen to the oration of the Wamar.

Shiny black floors reflected the enormous clusters of chandeliered lighting high above in the glass ceilings. The wall at the head of the hall was entirely transparent, revealing the starfield and all the faction ships moored outside. From their location, they could see the mammoth shape of *Calypso*, surrounded by a myriad of other craft. Tiered seating stretched to the vaulted glass ceilings on the other three sides of the enormous hall. Open executive boxes were scattered about the rest of the tightly configured rows.

The entire hall was packed with those who came to participate and those that came to spectate. Seated in open cubicles on either side of the podium were dignitaries from many of the larger factions. In a place of prominence, the Duchess Benetar sat a couple of seats down from the Albion King Commander, Thoene Dismon. Directly in front and below the podium, sat a large lavishly decorated chair. It was white with raised ornate patterns of silver and gold woven through its fabric. Standing at an elevated podium, the Grande Wamar was speaking of the heritage of Hadrian.

"Are you ready, your Highness?" Diord whispered, turning back to the group.

"Scared out of my wits," she said, trying not to hyperventilate. The young Princess straightened herself, trying to put on a dignified look.

"You're all going to die," Drax growled defiantly, still holding onto CJ and Blinda.

"Shut up," both women whispered in unison.

"I'll go out first," Diord instructed. "Then the Princess, then General Niker, then Blinda. No doubt we'll get stopped by security."

"We can take care of them," Blinda said, letting go of Drax and reaching for her weapons.

"No," Diord warned. "We don't have to make it to the podium, only into the room. Yes, the closer we can get, the better, but under no circumstances are you to engage any security forces. Protect the Princess, yes, but no shooting or flying body parts."

Blinda frowned while securing her blaster and Balkrum. As Rick was doing the same, he leaned a little closer to the Albion Thane.

"Sometimes doing nothing makes a greater impact. Besides, do you really want to make a big scene in front of all these people?"

"They'll know we mean business," Blinda responded quietly.

"I suspect we're going to get plenty of attention without violence."

"CJ," Diord leaned back looking at the two sisters. "Are you two going to be ok here?"

"I don't think she's too excited about going anywhere right now," CJ said, helping Drax to a seat just inside the portal doors.

Diord took a deep breath and looked at Jana.

"Here we go," he said, standing up straight and walking with conviction out into the hall with the others right behind him. The Grande Wamar continued to speak, even as the small group started up the aisle toward the front of the grand hall. Diord's attention was momentarily drawn to several individuals on the stand who had taken notice of them. As they came closer, people in the floor seating turned as several bands of security guards rushed in from all directions. Diord came to a stop as his small group was surrounded, a myriad of blaster pistols and rifles pointed at them. The Grande Wamar finally stopped speaking and turned to his aides. After some consultation, he turned back to the podium and looked down at Diord.

"Administrator Vandmire, we were informed that you had abdicated your time of historical instruction at this presentation. Why are you here now?"

"If I may be allowed to approach the platform, I still have a presentation to make."

"This is highly irregular, Administrator. There are long established protocols and procedures in place for any presentations or announcements given at the Rite. You know this better than anyone else in this room."

"Yes, your eminence, but my approved request for presentation time was unduly revoked without hearing or reason."

"I was not made aware of this," the Grande Wamar said looking over his shoulder. "Who informed you of this revocation?"

"The Director of the Albion Secret Service."

The Grande Wamar turned to the Albion King Commander.

"That's not possible," the Grande Wamar insisted. "They have no jurisdiction on Calliope and especially not at the Rite of Pintar."

"Nevertheless," Diord maintained, "at the last minute, I was informed that my presentation would be barred and no entry given to either myself or my guests."

A commotion rolled through the hall as the Rite guests began to talk amongst themselves. The Grande Wamar's aides huddled behind him conferring for several long moments, then one of them spoke quietly to the Wamar, who finally turned back to Diord.

"Do you still wish to make your presentation and if so, are all these people behind you necessary?"

"They were all approved as part of the presentation," Diord responded confidently.

"I see Blinda Koss behind you. How is she connected to your presentation?"

"She will be reciting the Hadrian creed... from memory."

More murmuring rolled through the crowd. Rick detected Blinda repositioning herself next to him.

"Bet you didn't know you would be performing in public when you got up this morning."

"I think we've hit that threshold we talked about earlier," Blinda mumbled back.

The Wamar glanced over at the Albion King Commander, then in the other direction at the Duchess. Neither responded.

"Since we're still early in these proceedings, I'm going to allow your presentation. I apologize we have nowhere for your guests to sit as we were informed that you would not be attending."

"I understand," Diord agreed. "It will work out better if we all stand. May we approach?"

The Wamar nodded and stepped back, taking his seat as the small band was escorted the rest of the way to the front. The guards then turned and exited the hall. Diord gave quiet instructions for Blinda to stand to the right of the podium and Jana and Rick to stand on the left side. Taking hold of the podium, he looked out at the vast audience on the floor. Following the angled tiered seating that stretched up into the recesses of the hall, he could barely make out the faces of the audience in the top rows. Nervous, he took a drink, swallowed, cleared his throat and addressed the hall.

"Delegates, Heads of State, Military Leaders and most importantly, my fellow Hadrian patriots. Only a handful of you have ever witnessed a Rite of Pintar. My colleagues and I are here today to help you understand its significance along with the whys and the hows. Although I can't see everyone in this great hall, I want to ask those who were present for the Rite of Pintar of his royal majesty, King Tiev Oxlind, to stand up." Diord carefully scanned the room. Catching sight of several individuals scattered on the main floor, he searched the tiered seating. A number more were spotted. He turned around, seeing the Grande Wamar, the King Commander and the Duchess come to their feet.

"Thank you, please be seated. I was but a lad when King Tiev was made ruler of Hadrian, but my parents brought me to this same grand hall to witness history in the making. Now, here we are, about to witness history being made again. Another Hadrian Monarch must be appointed at this Rite. The laws and regulations governing this process are very clear and have been upheld for a millennia. When King Tiev was crowned, it was by unanimous accord of all the factions of Hadrian." Diord paused, feeling a little more comfortable, then continued.

"Eons ago, several traveling factions of humans, Colonia, Albia, Drakonia, and the Drake, colonized this galaxy, calling it Hadrian. To guarantee order and peace, a set of simple laws were established to help govern these factions. These wide reaching laws are the glue that has helped to hold this galaxy together."

"The purpose of the Rite of Pintar is to ensure that all of Hadrian have the opportunity to participate in confirming their devotion to the monarchy of our great galaxy. Hadrian law clearly dictates that only a blood heir of the royal line be made monarch. If one cannot be found, then a new blood line must be formed by the strongest faction of Hadrian. Until now, Colonia has been the oldest and largest of all the factions... Albia now considers itself to be the strongest in this galaxy. How does that happen? I'm not here to expose how so many smaller factions have been absorbed by Albia. I'm not here to lay out how Colonia has been weakened and systematically dismantled from within. Those subjects are best left for special investigations and political committees dedicated to finding, verifying and revealing the truth to all of Hadrian."

Diord looked around, searching the audience, hoping to see BachTL. But there seemed to be no one moving in the sea of heads. Somehow, he needed to give the Castellian more time.

*   *   *   *

CJ leaned back on the bench just inside the portal, out of sight from the audience in the Pintar Hall listening nervously to Diord's orations. She glanced back up the corridor for any sign of BachTL. *What's taking him so long? Is he having trouble accessing Drax's box?* She looked back at her sister. Dark circles ringed Drax's closed eyes, giving her a bedraggled and somewhat hollow look. She appeared to be breathing normally now.

"You gonna be all right?" CJ asked.

"I think I would have been better off if you had just shot me," Drax replied quietly.

"Blinda's talents can be very effective," CJ agreed looking back out the portal, trying to listen.

"Have you ever had the pleasure of her assaulting your mind?" Drax grumbled.

"Not the mind, but I've seen it at work as often as you have. I did get a dose of the whole ear pressure thing she does. That gave me a migraine for a couple of hours."

"Multiply that by a factor of one hundred," Drax countered. She cracked an eye, looking toward her sister who was trying to listen.

"Well, serves you right," CJ finally said.

"You've sent your friends out there to die," Drax muttered.

"Maybe," CJ responded, helping her sister get a little more comfortable. "But it's a chance we had to take."

"Depends on your point of view I guess," Drax rasped. She glanced at the still open security door at the other end of the hall.

"Getting in was most of the battle," CJ continued, distracted by the dialog out in the hall.

"I hope you enjoy seeing executions."

"Jana only has to show her parent's rings and book to the factions."

"I keep telling you that's not going to happen."

"And you still haven't explained to anyone otherwise..." CJ's vision suddenly spun as a terrific blow to the back of her head burst stars through her vision. Blackness quickly enveloped her mind as she slumped to the floor, unconscious.

Drax scooped up the locking control for her binders and quickly deactivated them. Dropping them next to CJ, she carefully reached for the blaster pistol in her sister's limp hand. Resting for a moment, she checked CJ's pulse, then slowly struggled to her feet. She looked down at her sister as she steadied herself against the wall.

"I'll say it again, Sis. I'll deal with you later." Glancing back at the Pintar Hall entrance, she turned the other way and made her way toward the open security door and out into the main corridor. Looking both ways, the Castellian security guards were nowhere to be seen. But it seemed reasonable to think that after the ruckus the two Thanes had caused, it wouldn't be much longer before their comrades came to investigate. Making her way as quickly as her condition would allow, she arrived at a small elevator and headed for the upper levels.

The mid-levels of the Pintar Hall were the best for seeing everything that went on in the hall. The rich and powerful paid a high premium to reserve viewing boxes on these levels. Most of the boxes were much like little apartments. Lavishly decorated to meet the tastes and requirements of the individual client, many of these living spaces were designed with sitting rooms away from the viewing windows and open balconies. Food replicators and consumption areas were commonplace next to lavish sleeping accommodations.

Drax moved through the outer halls unhindered by the myriad of security personnel guarding the entryways to the box corridors. Stopping at one of the inner doors, she worked the code for the lock and stepped cautiously inside. Other than the indeterminate conversations coming from the Great Hall beyond the open balcony, there were no other sounds in the room. She jammed a finger at the door control, then turned with her blaster drawn. Moving slowly passed the eating area, she looked into the sleeping quarters. The box appeared to be vacant. Turning to a small desk, she rifled through several storage compartments.

"Where is it?" she mumbled frustrated. *It should just be here on the desk.* Drax looked around perplexed. Stepping back out into the middle of the room, she carefully scanned her surroundings. *There's nothing here!* As she finished her turn, she came face to face with BachTL standing in the open balcony doorway.

"Drax...?" BachTL blurted quietly. His vision twitched down to the pistol in her hand. He reaffirmed his grip on the large book tucked under his arm. The two stood frozen, staring at each other. "You look a little tired," he finally said, smiling.

"Having Blinda rape your mind will do that to you," Drax responded.

"People usually aren't standing after getting a taste of her in their head."

"Yeah, well, it took me a little time to recover."

"Where's CJ?" he asked looking beyond her.

"She wasn't feeling up to any reunions."

"You didn't shoot your own sister, did you?"

"Do you really think that little of me?"

"I've been your aide for how long and you're going to ask that question?"

Drax took an exasperated breath.

"Give me the book," she said, holding a hand out.

"Or what?" BachTL asked quietly, looking down at the blaster. "You'll shoot me?"

"I'll do what I have to do to keep Hadrian from descending into chaos," Drax firmed up. "Now please..." she beckoned. "Give it to me."

BachTL looked down at the book, then behind him to the open balcony. He took a short step back, stopping at the threshold of the open door.

"You talk about wanting to keep Hadrian from descending into chaos. But it has already descended and conscious of it or not, you've been the one leading the charge to take it there. This book..." BachTL held it to his chest. "I believe this book and that young woman out there fighting for her life and what she believes in, are Hadrian's only hope to climb back out of the mess you, the Duchess and the King Commander have created."

"BachTL... finally all grown up and making his own decisions, skewed as they may be," Drax smiled coolly.

"Depends on your perspective."

"That's deep."

"I've seen the view from several angles. How many angles have you looked from?"

"The only ones that count."

BachTL leaned his head back a little, trying to hear the proceedings out in the Great Hall.

"Come away from the door," Drax requested politely.

"It must be exhausting, exerting so much energy maintaining control of that temper of yours?"

"Son, please come away from the door and give me the book," Drax repeated even softer.

BachTL looked back at the Queen Captain, an odd expression developing.

"You know, Mother, I don't know that I've ever heard you call me that before. At least not like this." His eyes shifted momentarily to the state of the small blaster in her hand; her other hand was still extended.

"Come away from the balcony and close the door, please."

BachTL glanced back, then nodded and stepped forward, touching the sliding door mechanism. The glass door instantly slid closed.

"Thank you," Drax sighed, relieved.

BachTL took a couple of steps forward and held the Oxlind book out. He looked Drax directly in the eyes as she stepped up and took hold of the book with both hands.

"You're doing the right thing," Drax reassured the Castellian.

"I know, but not because you told me so."

"I'm proud of you, BachTL," Drax smiled. She fumbled momentarily, trying to take possession of the large book and hold the blaster as well. As the full weight of the book became apparent, Drax felt the blaster coming out of her hand. She glanced at BachTL. He grimaced, his focus on the gun. Drax let go of the book as they tussled with the weapon. Twisting and tugging, the pair strained against each other, trying to wrest control of the pistol. BachTL pushed back as hard as he could trying to maneuver the Queen Captain away from the viewing windows. In doing so, they stumbled over a small table in the living room, ending up in a tangle on the floor. Coming back upright, neither was holding the pistol.

"Fool," Drax growled, punching him as hard as she could. The blow laid BachTL flat. Tasting blood as he rolled over and came to his hands and knees. He felt something pop as a blow to his ribs upended him again. Moving to get up, a boot struck him squarely across his jaw, knocking him into a wall. Dazed, BachTL slumped a moment. Hearing a shuffling sound and soft but deliberate footsteps, he brought his hand to the stabbing pain in his side. He felt something against his hand and cracked an eye. It was the book. He pulled it to him, cradling it in one of his arms. A moment later, he heard the power generators on a blaster begin to wind up. He looked up, his left eye starting to swell. Drax loomed above of him, the blaster in her hand pointed at his head.

"I can see I've misjudged you," Drax puffed. "A mistake I won't make again. Give me the book."

BachTL looked back down, reaffirming his hold on the book.

"I can't." He looked back up at Drax. "I won't."

"Such a disappointment." Drax hissed, firming up to squeeze the trigger. There came a medium pop accompanied by a brilliant flash. BachTL jolted. Grimacing, he waited for his battered body to register the shot. When nothing happened, he opened his eyes. Drax crumbled to her knees in front of him. She made a half turn, revealing her assailant leaning against the open outer door. She let the blaster fall and turning back to BachTL, dropped to her hands. Gazing at him a moment, pain rippled through her eyes. She turned over onto her back as CJ shuffled over to her, holding a blaster rifle. Blood matted her red curls on one side and her facial expression quivered as her eyes filled with tears. Drax looked up at her sister as CJ sank to her knees next to her. BachTL crawled over next to CJ and took her hand. No words were spoken as life quickly drained from Drax. She looked at BachTL a moment, then at CJ. Opening her mouth to speak, her eyes glazed toward the ceiling.

Hearing the muffled sounds of conversation in the Great hall, BachTL turned and picked up the book.

"I've got to get this to the Princess," he rasped painfully."

CJ looked up at BachTL, her cheeks wet.

"Go..." She pulled the security key from Drax's wrist and handed it to BachTL. "I'll stay here."

"Are you going to be all right?" BachTL asked, checking the injury on the side of CJ's head.

"No," she rasped, grief stricken. "Hurry..."

BachTL carefully forced himself to his feet and stumbled to the door as several guards appeared. After waving the security key in their faces, they allowed him to pass and pushed into the box to survey the scene. Recognizing Colonel Barker, they knelt beside the Queen Captain to assess her condition.

"Don't bother, Lieutenant," CJ said, coming to her feet. "She's dead."

"What happened here, Colonel?" the Lieutenant asked, looking at the blaster rifle and pistol on the floor next to the body.

"She took her own life before I could stop her," CJ responded, wiping her eyes and sterning up. "Speak of this to no one. Inform the Wamar that nothing was found in the Queen Captain's box."

"But the Queen Captain's aide?" the Lieutenant gestured to the door.

"You have your orders," CJ ordered gruffly.

The young officer and his men quickly exited the box, leaving Colonel Barker alone. She opened the sliding glass door to the balcony and shuffled out, sitting down in a far corner where she could watch the proceedings without being noticed. Looking back into the box at Drax, her grief overpowered her and she buried her face in her hands.

*   *   *   *

Diord scanned the crowd before him, feeling confident that he had everyone's attention.

"Shortly, the Grande Wamar will again address you, present a new candidate for monarch, and ask if there be anyone who would contest and give just cause for not confirming that candidate. Here you will have to make your will known. But remember, you must make sure you have all the information so you can make an informed decision."

"As part of the Rite, all of the faction representatives will have the opportunity to participate in the confirmation of a new ruler. Once the candidate has been presented and no objections are raised or unresolved, the Wamar will instruct all the faction delegates to come forward and confirm their allegiance by kneeling before the candidate and repeating the following; Representing the house of, state your name and the faction house you represent, pledge their devotion and fidelity to the House of Pintar and his or her Royal Majesty, the candidate's name. The delegate then bows and returns to their seats. After everyone has made their pledge, a chosen patriot recites the Hadrian creed and then all must bow to the new monarch. Grande Wamar, Delegates and patriots, thank you for this opportunity to instruct and declare my devotion to Hadrian."

Diord took a couple of steps to one side, standing directly behind Jana. The entire hall applauded with approval as the Wamar again took to the podium.

"Thank you so much for your insights into the Rite and the ceremony that is attached to it. Without further interruption, I'd like to present to you the monarch candidate to lead Hadrian into the next generations. Will the Duchess Benetar please rise," he said, turning to Stephanie. "The Duchess has been acting Queen of Colonia since the untimely death of her late husband, his highness, Kind Tiev Oxlind. If there be anyone in this great hall who can provide evidence as to why she should not be confirmed as Queen of Hadrian, please come forward and present your reasons and proof."

The audience fell silent, some people looking over their shoulders to see if anyone would come forward. The Wamar smiled and opened his mouth to speak...

"She is not a blood heir," a young voice spoke up.

The Wamar fumbled for a moment, looking for the source of the objection.   Rattled, he quickly recovered.

"As the honorable Administrator from Cross has already explained, if a suitable blood heir cannot be found, then the strongest of the factions must create a new bloodline and assume rule."

"She is not a blood heir and therefore is not eligible to rule," the young voice reiterated.

"Who makes this objection?" the Wamar asked, visibly aggravated. "Step forward and present your case!"

"I do," Jana said, turning around and facing the Wamar.

The older gentleman gave Jana an odd look, then looked at Diord.

"Administrator, who is this young woman?"

"You asked the question, your Eminence," Diord smirked. "Let her answer."

All eyes fell on the young woman as she approached the podium and the Wamar stepped aside. She looked out at the vast audience before her. Her throat was trying to swell as her heart pounded uncontrollably. Terrified, she fought the urge to run for the nearest exit. Grasping the podium, she forced a dry gulp down.

"I have proof that a blood heir to the royal Oxlind line still lives." She looked down at Blinda. After several agonizing moments, Jana pulled her mother's ring from her pocket and held it up for the recorders to broadcast to the entire population of the hall. An audible commotion rolled through the crowd, giving Blinda and Rick cause to move a little closer to Jana's position. Blinda gave the Duchess a quick glance, then looked toward the King Commander. A worried look was whisping across his face.

"This is my mother's ring. I am Jana Tilee Oxlind, Colonian Princess of Hadrian. Fifth daughter and seventh child of their royal majesties, King Tiev and Queen Mila Oxlind of Colonia and the House of Pintar." By the time she had finished, she stood confident and poised.

"This is not possible," the Grande Wamar shouted, rushing to the young Princess. Before he could reach her, Blinda was standing in his way and anyone who would try to lay their hands on her. Rick took up a position directly in front of the podium, to keep anyone in the audience from rushing the platform.

"That ring must be examined and verified," the Wamar demanded, trying to get around Blinda. The Albion Thane brought her Balkrum up in her gloved hand and released the safety. Tiny colored activator lights instantly started blinking, giving anyone trying to get any closer, cause to think better of it. The audience quickly fell silent, those that had rushed toward the stage had stopped and backed up. Rick noticed the King Commander come to his feet and move toward the podium. Blinda shifted as Thoene held his hands up to calm the crowd.

"Ladies and gentlemen," the King Commander shouted to the audience. "Please, take your seats. We should be able to sort this out in short order. There are a couple of things that can provide preliminary verifications of this young woman's claim. As you know, there are certain criteria that must be met in order to make a verification of a blood heir. We should be able to ascertain if those basic criteria are verifiable right now. May I approach?" he asked looking at Blinda.

Diord stepped in front of the Thane.

"No, I think you're close enough. The Grande Wamar can make any visual verifications required."

"Very well," Thoene agreed carefully, stepping back. "If the factions of Hadrian will be patient, we can proceed. First off, is it possible for the Wamar to verify the authenticity of this ring? Are there any supporting documents that can substantiate this claim?"

Jana motioned for the Wamar to approach the podium. After several moments examining the ring, he motioned to one of his aides who stepped forward with a hand scanner and held it over Jana's hand. The Wamar studied a myriad of information appearing on the small screen of the scanner. After he had finished reading and comparing the images on the screen with the ring on Jana's finger, he looked back at the King Commander.

"Yes, it is the Queen's ring."

A murmur rolled through the crowd, and it wasn't until the King Commander held his hands up, that they fell silent again.

"And the documents?" Thoene asked.

Jana paused, then looked over at the King Commander.

"I don't have the documents with me, but they are being retrieved as we speak."

Thoene looked at Diord and slowly shook his head.

"Young lady, why are you wasting this council's time, by coming in here and making claims that cannot be verified? The Grande Wamar will see to your claims after this session is over."

"By that time, it will be too late," Jana countered.

Thoene smiled and looked out at the vast audience.

"I'm sorry, but unless you can provide this body with further proof, you will need to be escorted from this great hall."

Jana looked at Diord and fumbled with her sleeve, finally pulling something from under the fabric. She held her father's ring up so the recorders could once again broadcast the image to the entire assembly.

"This is my father's ring. His royal majesty, King Tiev Oxlind. Colonian ruler of Hadrian."

The crowd erupted with jubilation mixed with disbelief as the Wamar's aide scanned the other ring. Several noisy moments passed as the verifications were made. The look on the Wamar's face changed as both rings were positively identified. Diord looked back at the Duchess. Her face had turned white as she shuffled nervously back toward her seat. The Administrator smiled confidently as he looked back at the King Commander, who was holding up a hand to silence the crowd.

"This is all very compelling," Thoene admitted quietly. "But again, Hadrian law requires that you provide verifiable proof that you are who you're claiming to be. I'm sure you can appreciate this body's concern about someone finding these rings tucked away in a box somewhere and making a claim to the highest position of power in this galaxy. The laws governing any claims are quite clear and not subject to interpretation. You must have signatory documents signed by King Tiev Oxlind and witnessed by a third party. Do you have such documents?"

Diord's heart stopped. His greatest fear had now manifested itself in front of all the factions of Hadrian. Without her father's book, Jana could not lay claim to her blood heritage. He looked back at Jana's blank look. Blinda Koss kept her Balkrum held up in front of her as she looked directly at Thoene.

"The book was confiscated by Queen Captain Drax Blair and taken to her personal box, here in the hall," the Thane reported. "Her personal aide is retrieving the book as we speak."

"Very well," Thoene said, remaining uncharacteristically agreeable. "Where is this aide? Let's have Calliope security search the Queen Captain's box immediately." He turned to the audience and held his hands out motioning to the crowd so they would be pacified.

Jana looked frantically around the great hall for BachTL, but could only see the bobbing heads of the crowd as they conversed in muffled whispers. Blinda remained silent as she had nothing to offer. After a number of minutes, several security guards approached the platform. The Wamar leaned closer to one of them as they made their report. After a quiet exchange, Jana stepped aside as the Wamar moved to the podium.

"Calliope security has just come from the Queen Captain's box and found nothing."

Thoene looked at Diord, then at Jana and shrugged. Turning again, he held his hands out to the crowds.

"Without further evidence, these proceedings have been interrupted long enough and I move the council table this matter until such time as further evidence can be forthcoming." Thoene turned to the Wamar and bowed. "Grande Wamar, please proceed."

The room erupted in an outrage of jeers for the apparent power grab happening before their eyes. Diord looked helplessly back at Jana. Now thoughts of how they were to escape before the King Commander's Secret Service arrested them flashed through his mind. Surely they would be taken out and executed. The Grande Wamar raised his hands and motioned for quiet as the King Commander confidently took his seat. It took several minutes for the noise to subside enough for the Wamar to be heard. He finally settled the room.

"Does anyone have anything else to bring into evidence on this matter?" The Wamar searched the crowd as the air in the hall turned heavy. "Very well," he said reluctantly, looking at Jana. "We shall proceed." He motioned to a group of security forces waiting at the sides of the platform. "Please escort these people out and hold them until after these proceedings have been completed."

As Calliope security moved to surround the small group, Jana remained unmoved, defiant to the will of the council. One of the guards nudged her, but she refused to move. The Wamar noticed her defiance and leaned in her direction.

"I'm sorry, Miss. Will you please follow these gentlemen?"

Jana glared at the Wamar, giving him cause to recoil in surprise.

"You would betray Hadrian when you know the truth?"

"Without verifiable documentation, there is no truth here, whether it's actually fact or fiction. I can do nothing further."

"What is to become of us?"

"You will be held and heard after these proceedings are over."

"It'll be too late then."

"I'm sorry, there is nothing more I can do."

Jana turned to Diord. He was staring out into the crowd with a distant look. She noticed Rick searching the crowd; he seemed to be following someone. Looking hard, she too noticed a short form making their way behind the row of guards. BachTL abruptly stopped and faced Rick, cradling something in his arms. There came a sudden blur streaking toward the platform and in an instant Rick was holding Jana's book. The guards brought their weapons up, training them on him. He froze, holding the large book out in front of him. Thoene slowly came to his feet to get a better look. Rick grinned, watching the blood drain from the King Commander's face. Moving methodically, Rick stepped over to Jana and presented her father's book to her. She nodded and took it, then turned to the Grande Wamar.

"Proof," she announced confidently.

The Wamar took the book and held it flat while Jana worked the lock with the orb. After several fascinating moments, the book's security mechanism released the binders and the Wamar opened the book.

"This Orb was given to me by my Royal Au Pair," she said. "It is coded to my DNA and I am the only one that can work it. Inside these pages, you will find the required documents you seek."

Astonished, the Wamar picked up several pages from the book's interior and examined them closely. A couple of his aides joined him at the podium, making their assessments. The audience talked amongst themselves while several agonizing minutes of discussion ticked away. Finally, the Grande Wamar closed the cover and after giving Thoene a glance, looked at Jana.

"I think we're ready to proceed," he said smiling. "Delegates, Heads of State, Dignitaries, Military Leaders, Ladies and Gentlemen, honored guests, and as Administrator Vandmire so eloquently put it, my fellow Hadrian patriots. After careful review of all the evidence presented to this council, the authenticity of these documents have been verified and is undisputable. They contain the required signatures and witnesses, making them legal and binding. Therefore,

I am pleased to present to you, her Royal Majesty, Jana Tilee Oxlind, Princess of Hadrian."

Silence remained entrenched for only a moment, then as if on cue, the hall erupted into a thunderous display of jubilation and applause. The Wamar glanced again at the Albion King Commander, then bowed respectfully to Jana and handed her book back to her. He then presented her to the factions of Hadrian and stepped away from the podium. The hall shook with the deafening roar of pounding feet, clapping and yelling. Overcome with emotion, Jana could only stand and look out across the crowds of people applauding her, many bowing with respect.

Blinda and Rick remained vigilant, watching the Calliope guards in front of them. The sentries appeared bewildered, not knowing what they should be doing at this point. The Grande Wamar finally motioned them back. As they filed out, Blinda again moved up next to the Princess while the multitude continued to applaud. Diord finally leaned closer to Jana.

"It appears you've become famous."

"Now what happens?"

"The Wamar should get control of the crowd and start the Allegiance Rite."

"Explain that again."

"Each faction representative will be required to approach you and declare their allegiance to the Oxlind heir."

"How long is that going to take?"

"Until everyone has had a chance to declare their allegiance."

Jana looked back at the Wamar, who continued to clap. After several more moments of thunderous applause, the Wamar came back to the podium and brought his hands up, calming the crowd.

"Ladies and Gentlemen, if you'll take your seats, we'll come back to order. I'm sure everyone is anxious to continue into the next phase of these proceedings." The Wamar motioned to Jana. "Your Highness, if you will take a seat," he said, motioning to the chair in front of the podium. Jana hesitated, then carefully sat down as Blinda took up her position next to her. Rick and Diord stood back with arms folded, listening as the Wamar began to address the hall once again.

The Right Thing

Commander Dalton SoKnack sat in a corner of the *Tarzana's* bridge, away from the activity bustling at every station. From here, he could look out either the front or the side bay windows of the massive battlecruiser. Listening to the chatter of his escort ships and the other faction craft, he occasionally gave the main scanning screens a glance. His orders were to remain clear of any of the faction ships. Perhaps he had been conditioned by Drax to remain stubbornly firm in his resolve to follow his orders. *Even if things go wrong, I follow orders and it's someone else's fault. Remain on patrol clear of all faction traffic.* CJ had echoed those same orders, but at the same time, she had indicated that at some point he would have to assess the situation and do the right thing. Whatever the right thing was.

He was well aware of the focus of all the commotion. Based on the reports he had received before and after he had been put in command of the *Tarzana*, the enormous ship sitting among the Cross and Tomplie contingents, was carrying the alien Starbird ships that they had been looking for. While he was concerned that the ship he had helped to capture had escaped and was hiding in the mammoth ship, he wasn't surprised. Indeed, he was relieved that he was not to blame and didn't have to deal with any of the repercussions.

As more detailed reports came in on skirmishes taking place, his aides frequented him with information. Some even pressed him to join the action. Each time a recommendation was made, he held up a hand to silence them. Growing tired of the constant barrage of suggestions coming at him, he became agitated as yet another aide approached.

"Admiral Hilton of the *Indominable*, Sir." The Officer of the Deck handed him a communications display.

Dalton snatched the display from the officer, glaring at him. Once alone, he turned it on.

"Yes, Admiral," he said, trying to relax.

"Commander, I am aware of your standing orders from QC Blair, but I am under orders from the King Commander. We will be firing Thug torpedoes at the alien ship shortly. I want you to come in closer to the other side and provide backup."

"And why are you going to put torpedoes into that ship, Admiral? Has there been any hostile activity from it?"

"Two of the smaller super weapon ships came through our defensive perimeter a short time ago."

"Yes, we tracked them as well."

"Then you observed them engage our sentries, destroying a number of our fighters and assault vessels."

"Yes, our analysis indicates they were laying down suppression fire, not going after any specific ship. Admiral, with all due respect, I think we should rethink what's going on here and how we're responding to it."

The Admiral paused a moment, looked over his shoulder and lowered his voice.

"Exactly what are you suggesting we rethink, Commander?"

"What are we doing here, Sir? Why are we treating this like an offensive? Why are half our forces deployed around an old ship that by all our scans, has zero defensive weaponry?"

"We're here by order of the King Commander, to protect what's going on inside Calliope," the Admiral replied, sounding irritated. "Those super weapons attacked and destroyed an entire squadron of corvettes and a couple of destroyers over Carolon."

"Yes, Admiral," Dalton cut flatly. "I was there. I commanded the operation."

"Then, that same ship and its companion lured a squadron of ships and fighters into the interior of Calliope where they were ambushed and savagely attacked."

"Savagely attacked..." Dalton repeated slowly. "Admiral, did you even look at the configuration of those two ships? One was towing the other. Clearly it was a rescue operation... and we were in the way."

Admiral Hilton gave Dalton a funny look, then looked down at a display for several moments. The dreadnaught commander studied the information in front of him, finally looking back up at Dalton.

"Their configuration is of no consequence, Commander. We've been given a direct order by the King Commander to open up that alien ship and retake the Alvadorian super weapons in the name of the Empire." The Admiral paused a moment and lowered his voice further. "Dalton, I need the *Tarzana* on the back side between the alien ship and the *Maxell*. Shipley is a trigger happy idiot that would destroy those ships rather than let anyone else get their hands on them."

"Perhaps no one but the rightful owners should have their hands on them... Sir."

Admiral Hilton stared at Dalton for a moment, considering. He finally drew up in decision.

"I'm giving you a direct order, Commander. Bring the *Tarzana* in on the opposite side of the alien ship and await further orders."

Dalton looked away from the viewer for a moment, then back at Admiral Hilton and nodded.

"Yes, Sir. I'll bring her in close. Request you hold your fire until we are in position. Certainly that will keep the Ratronians guessing."

"Excellent tactic, Commander. We'll be watching for you."

Dalton shut the viewer off and leaned back again. He looked out at the Shampoli Nebula that arched through the heavens over their position. For a moment, he wondered if the alien ship's home world was numbered among the trillions of stars within the massive expanse of dust and gases. *Were they just trying to get back home and got caught up in the internal squabblings of Hadrian?* Presently, he came to his feet and clasped his hands behind his back. Looking around the bridge, he spotted the deck officer standing next to one of the control pits.

"Captain Gulch, new course. Take her in close to the alien ship. The *Indominable* will be firing torpedoes shortly. We need to maneuver in between the alien ship and the *Maxell*."

"Are we to fire on the alien ship?"

"Not unless ordered to do so," Dalton answered.

"What if Admiral Shipley objects to our position?" Captain Gulch inquired standing beside his commander.

"He'll know better than to object when he gets a face full of the *Tarzana*," Dalton replied, watching the bridge crew go to work to make the course change. As the enormous bow of the Albion battlecruiser began to swing, Calliope drifted across their view, then the *Indominable* and the rest of the faction ships surrounding the gigantic alien ship.

*   *   *   *

The Grande Wamar's speech stretched out for quite some time. Certainly longer than Rick and Diord were prepared for. As they listened, Rick's communicator went off.

"Sort of in the middle of something here," Rick mumbled quietly with his wrist raised to his lips.

"Sorry, Sir," Laura Habba answered. "Commander Niker needs to talk to you."

Rick gave Diord a glance.

"Put her through."

"Rick," came the anxious call from Jayda. "You need to get back here, now."

"Sort of saving the galaxy at the moment."

"It's really getting hot out here. We've intercepted communications from the *Indominable* and the *Maxell*. They are preparing to fire torpedoes."

Rick gave King Commander Dismon a glance, watching him working with a small device in his hand, holding it near his mouth. Rick turned and looked out the windowed wall behind him. The massive bulk of the *Indominable* had moved away from Calliope, closer to *Calypso*.

"I was hoping they'd hold long enough for the Princess to take the throne. She could stop all this with just a word. Have Alex and Commander Atlanta been able to refine that Asium?"

"Yes, Engineer Moon is fitting our ships with them right now. But you said it yourself, *Calypso* probably can't withstand multiple salvos of torpedoes coming at us at the same time and Dakota is the only one here qualified to take a Starbird into battle."

"Ok, calm down and continue prepping for launch. We'll get back as soon as we can."

Rick looked at the Princess sitting in front the podium, then over at Diord.

"You go," Diord said. "I'll make sure she gives a great speech."

"Can you get me out of here?"

"I can," BachTL said, leaning close.

"Thank you, General," Diord said, looking back at the Princess.

"You going to be ok with Blinda?"

"A lifetime of distrust will be a hard thing to overcome," Diord admitted. "But yeah, I think the Princess is in good hands."

"Agreed," Rick said. He inconspicuously turned and followed BachTL passed a line of Calliope security guards and to the nearest exit.

"I still don't fully understand you people," BachTL puffed as the two worked their way through the empty corridors toward the shuttle bay.

"What's not to understand?" Rick asked as they stopped at a checkpoint for clearance.

"You could have easily just jumped to lightspeed after you got your ship back, but you stayed, why?"

"Because it was the right thing to do," Rick said as they entered the bay and started toward one of the empty shuttles.

As they took off, they abruptly stopped, hovering over the city. Their way out was completely blocked by a mass of Castellian sentry ships charged with keeping the fight outside.

"Got any good ideas?" Rick asked, thinking of the maze they had to travel in order to get out with the *Constellation*.

"Sure," BachTL reassured him. "Stupid Albions and Colonians think they know everything." BachTL turned the ship toward the back side of the city, steering toward a dark impression in the rock wall.

"Please tell me we're not going chasing around inside this thing again."

"Not exactly. We went that way because those were the only tunnels big enough to handle your Starbirds. This way will be much shorter." They were suddenly engulfed by the blackness of the wall. BachTL activated the exterior lights on the shuttle exposing a narrow shaft winding ahead of them.

"And small," Rick observed. "Not much rock glowing in here. You sure this leads outside?"

"General, Please. I grew up here... off and on," he added. "No glow because the Corvantium is really thin here." Moments later they were passing through an opening and outside the asteroid. "Now, getting back to *Calypso* will be a different matter altogether."

"You let me deal with that," Rick said, taking the controls. "You work your scanning gear and I'll fly the ship."

"A shuttle in a battle zone is typically not considered a threat, but everyone out here is going to be trigger happy, so we're going to have to be really careful."

"Just find me a way through, preferably away from any Albion or Colonian action. I'd just a soon avoid any Ratronian traffic too."

"Well, that's the trick, isn't it? We're not going to be able to get back to the *Merc*, but I think we can board the *Executioner* if General Aurora is feeling generous. He seemed a little antsy."

"He did, didn't he?" Rick responded quietly as he steered in the direction of the Tomplie ship. As the shuttle closed on the *Executioner*, Rick noticed several specks maneuvering in their direction.

"We've got some kind of activity going on out there," Rick said, trying to see the scanner in front of BachTL.

"Yeah, I've got them. A set of Black Tigers and Flightstreaks. Probably just coming in for a closer look. Kind of weird seeing them flying formation and not shooting at each other."

"Does this thing have shields?" Rick searched the console in front of him.

"Not the kind you're hoping for," BachTL replied. "Just really thick armor..."

"Alvadorian shuttle," a voice from the console speaker boomed out. "We have you on our scanners. This space is a restricted battle zone. You are ordered to turn back to Calliope or risk destruction."

BachTL gave Rick a glance.

"I think we better rethink this," he said after a moment of waiting.

Rick looked out at *Calypso* and the myriad of ships surrounding it. The *Merc* and her sister ship were positioned between the mammoth shape of *Calypso* and the *Indominable*. The *Realistic* and *Trax* trolled slowly between the *Maxell* and *Calypso* while on the other side, the *Executioner* and her flotilla sat beyond the Alvadorian ships. Swarms of tiny specs buzzed all around the open spaces between the leviathans. Moments later, two Black Tigers and two Flightstreaks darted past them, firing warning shots.

"We're not going to do anyone any good if we get blown to bits," BachTL suggested as they continued on.

"We're not going to do anyone any good if we turn around either," Rick countered. "We've got to give Diord and the Princess more time." He brought his communicator up. "*Athena*, this is General Niker, come in."

"*Athena* here," the com officer responded instantly.

"We're on one of the Alvadorian shuttles inbound, but are being intercepted by hostiles."

"We have you on our scopes, Sir," Dakota cut in. "You're too far out for us to get a lock on you from inside *Calypso*. Would you like us to launch; come get you?"

"Negative," Rick stated sternly. "Under no circumstances are you to open those doors. What's the status outside."

"Mostly small skirmishes. The Administrator and General Aurora's escort ships are giving the other factions a good reason to keep their distance. The *Trax* and *Realistic* are keeping the *Maxell* at bay for now. *Calypso* has taken several hits from light torpedoes, but the damage has been minimal."

"What about the Starbirds?" Rick asked as the shuttle was buffeted by several near misses.

"All five ships are fully operational," Captain Abrams responded. "The *Constellation* is under repair as we speak. I've assumed temporary command of the *Athena*. We're sitting at idle just inside *Calypso's* doors as ordered. General, I think we need to launch."

"You do and all hell is going to break loose."

"It looks like that's inevitable," BachTL warned looking at the shuttle's scopes.

"We've got to give the Administrator and the Princess more time," Rick maintained.

"The *Indominable* has just locked a Thug torpedo on *Calypso*," BachTL exclaimed. "It's one of their hull busters."

*   *   *   *

Dalton gazed out at *Calypso*, considering what was happening around him.

"Commander, should we launch our escort fighters?" Captain Gulch inquired amid the bustle of the bridge activity.

"Standard approach procedures, Captain." Dalton turned from his position on the command deck and dropped down into the navigation pit, stopping at the main guidance terminal. "What location are you heading to, Lieutenant?"

"Here, Sir," the navigation officer responded, pointing at the screen in front of them.

Dalton looked closer, then stood back for a moment, thinking. He studied an overview screen above them, then glanced back at the navigation instruments.

"Adjust your destination to here," he said pointing at the screen.

The navigation officer developed a puzzled look and glanced up.

"But, Sir, that will put us…"

Dalton turned to the helm officer.

"Alter your course on my command only," he insisted, ascending the stairs back to the command deck. "Flank speed," Dalton called out.

"Is the Admiral really going to torpedo an unarmed ship?" Captain Gulch inquired standing just behind his commander.

Dalton glanced over his shoulder, then back out at the cluster of ships ahead of them.

"Send Tiger squadron four ahead to intercept anything the *Maxell* might launch," he said quietly.

"Sir?"

"The Ratronians tend to act rashly when excited," Dalton said. "Have bomber squadrons seven and four follow the Tigers."

"Understood, Sir."

"Captain," Dalton stopped him as he turned to carry out his orders. "As commander of this vessel, I am responsible for its actions. You have only to follow my orders."

The Captain paused a moment, then nodded and continued with his assignments. As the alien ship loomed closer, Dalton observed a myriad of fighters and assault craft buzzing around the other faction ships as they surrounded the enormous vessel.

"Commander," the Deck officer called out. "The *Indominable* has fired its first torpedo."

"He was supposed to wait until we were in position," Dalton complained, taking a couple of steps forward. He watched a thin blue streak make a wide arc around several of the larger faction ships.

"Tomplie and Alvadorian attack craft are moving to intercept," Captain Gulch announced, pausing. "Sir, the *Maxell* has fired a torpedo as well… Colonian fighters are moving to intercept."

As the *Tarzana* drew closer, several streaks of tracer fire flashed in a wide spread toward the approaching torpedoes on both sides of the alien ship. The Ratronian torpedo was almost instantly neutralized, exploding in a brilliant flash well clear of the alien ship. Turning their attention to the other torpedo, Dalton and Captain Gulch watched as the inexperienced Alvadorian and Tomplie pilots chased the Albion ordnance, spraying tracer fire indiscriminately. As the torpedo closed on its target, it appeared there would be no stopping it, until it suddenly erupted in a ball of flame, enveloping the entire bow of the alien ship.

Dalton studied the tactical screens in front of the weapons control pit. A couple of older Colonian ships were moving into position near the *Maxell*. He recognized the *Trax* and *Realistic* with their support ships. Looking back at the alien ship, he noted that the torpedo had been destroyed before it struck its target. The bow now sported a stark black blemish, but remained undamaged.

"Sir, the *Indominable* has fired five more torpedoes.

"Time to impact?" Captain Gulch inquired.

"Two minutes!"

"Helm, execute course change," Dalton called without hesitation.

"What course change?" Captain Gulch inquired quietly from behind.

"Plausible deniability, Captain," Dalton replied, looking over his shoulder. "I am responsible..."

They looked forward as *Tarzana's* bow veered to the alien ship's port side, rather than its starboard. At flank speed they would be fully broadside the alien ship before the *Indominable's* torpedoes reached it.

"Target those torpedoes..."

"Sir, you're aware the Thug can only be destroyed from the rear? Its forward race shielding is far too thick for conventional ordnance, and energy weapons are deflected."

"Do we have any Tigers or Hoppers in their flight path?"

"Negative, Sir."

Dalton let his gaze turn to the alien ship as they passed broadside; close enough it appeared as though he could have reached out and touched it.

"Emergency stop!" Dalton bellowed.

"Sir, the tracking on a Thug is a bit less than pitiful," the Captain said. "If we don't get out of the way, those torpedoes will hit us."

"Emergency evacuation of the starboard hull compartments," Dalton barked, turning to the opposite window as the great ship came to a stop directly beside the alien ship. "Officer of the deck, stand by emergency power to the thrusters.

Captain Gulch stood in stunned silence, watching the Thug's thruster flares coming directly at them.

"Brace for impact," he called out suddenly.

Moments later, the great battlecruiser shook violently, pitching hard as all five torpedoes struck the *Tarzana* on her broadside. A wave of brilliant flashes enveloped the bridge and for a moment, chaos ensued. Once he had regained his wits, Dalton got back to his feet and steadied himself against a railing as the ship slowly came back to level.

*   *   *   *

As the Wamar wrapped up his comments, the Pintar hall became a buzz of muffled voices, chattering about the revelations surrounding the return of the Oxlind heir. Jana remained seated in the lavish throne chair situated directly in front of the speaking platform. The last hour had been a whirlwind, but now as the proceedings were about to wrap up, she was feeling a little more relaxed and confident of the outcome. She glanced up at Diord and Blinda as they stepped back to her side.

"And now, Ladies and Gentlemen," the Wamar concluded. "We will proceed with the final portion of the Rite of Pintar; the Allegiance Ceremony. Each of the representatives chosen by their factions to pledge their allegiance, will approach in the order assigned, kneel before the Queen and recite the pledge as follows. 'As a duly appointed representative of the House of, state your faction, I, state your name, do hereby pledge my devotion and fidelity to the House of Pintar and her royal majesty, Jana Tilee Oxlind, Queen of Hadrian.' The representative will then kiss both Oxlind rings and return to their seats. My associates and I will oversee each vow taken and record it. The recorders will provide witness to the audience on the personal displays provided when you came in. You'll also be able to see it on the overhead display. If there are any factions that would like to withhold their allegiance to her Highness, they will not approach, but may indicate so to either myself or one of my representatives. We will then meet separately after these proceedings to discuss the matter and resolve any issues presented." The Grande Wamar looked down at Jana and smiled. "Are you ready, your Highness?"

Jana nodded confidently.

"Administrator Vandmire, Blinda Koss?" the Wamar motioned. "If you'll step away, we'll get started."

Diord bowed and took a few steps back, but Blinda remained unmoved. There was an awkward pause that only those close to the podium detected.

"Miss Koss?" the Wamar whispered. "You need to back up a little so the recorders and witnesses can observe."

"I will not move from her majesty's side." Blinda's demeanor was stoic as she stared straight ahead.

The Wamar looked to Diord who quickly stepped forward.

"Blinda? What are you doing?"

"Protecting."

"You can protect a few steps back," Diord instructed quietly. "Come on, we just have to get through this last segment."

"Speaking from experience, that's when things usually start to unravel. I will remain here until we can move the Queen to a more secure location."

"You don't think all the security surrounding us is enough to keep her safe?"

"They're not under my control," Blinda responded, glancing in the King Commander's direction.

Diord looked toward Theone Dismon as well, then nodded and stepped back.

"She is the Queen's personal guardian. She is required to remain where she is," Diord instructed the Wamar. "You'll just have to work around her."

The Wamar gave the Thane a wary look, then nodded tentatively. He stepped back to the podium and touched several controls on the panel built into the rostrum. A large overhead screen dropped down behind the stage with a magnified image of the immediate throne area. The Wamar then activated the faction list so everyone could see it.

"The Faction of Abatures," the Wamar started, motioning to a line forming to the side of the stage. The representative walked across the stage, stopped in front of Jana and knelt. They then recited the pledge, kissed her rings and moved quickly off.

"The Faction of Acura," the Wamar called out looking over to the line. The representative quickly moved across and repeated their pledge to the Queen, then quietly moved off. Several iterations repeated themselves when the Wamar finally announced the Albion representative. Theone Dismon came to his feet and stepped toward the front of the line, but before he made his approach to the Queen, he paused and motioned to the Duchess. Stephanie abruptly rose from her seat and glided to the King Commander, taking his extended elbow. Stunned, the Wamar hesitated, not sure how to proceed through an unscripted procedure. Theone eyed the Wamar, finally tipping his head slightly.

"The Faction of Albia... will... escort the Duchess Benetar... to present their pledges," the Wamar announced tentatively.

Jana detected Blinda shifting her stance. The Young Queen glanced back as the Thane flexed her gloved hands. Jana looked at the uncertain expression developing on Diord's face as he eyed the Duchess. Stephanie and Theone walked arm in arm across the platform, stopping in front of Jana. After a short bow from both, the King Commander helped the Duchess melt gingerly to her knees. The hall fell silent as the Duchess bowed nearly to the floor, then carefully looked up at the young Queen.

"Your Highness," she spoke softly. "I bow in humility at your greatness and hereby pledge my undying devotion and fidelity to the

House of Pintar and to the only true Oxlind heir, Jana Tilee Oxlind, Queen of Hadrian."

Jana watched the Duchess lower her head back down, then gave Diord another glance. With no indications from anyone, Jana came to her feet and gestured to the Duchess.

"Rise, Duchess Benetar." Jana's eyes shifted momentarily to Theone as Stephanie slowly came to her feet. Keeping her eyes lowered, the Duchess pulled her hands beneath the lavish material of her sleeves. Theone smiled, nodding to the Duchess, then faced the young Queen and sank to his knees, bowing low.

"Your Highness," he spoke boldly. "As the King Commander of the House of Albia, I Theone Dismon do bow to your greatness and do hereby pledge my devotion and fidelity to the House of Pintar and her royal majesty, Jana Tilee Oxlind, Queen of Hadrian."

Theone remained prone, finally looking up at the young Queen. Jana shifted uncomfortably as the King Commander slowly came back to his feet, putting his hand under Stephanie's arm.

"It is our sincerest hope that you know we bear you no ill will and wish your rein to be long and prosperous," Theone slithered quietly.

"Yes, your Highness," Stephanie said carefully. "If we can be of service to you in any way..."

Jana held out her hands so the two could finish their pledge. Both bowed and reaching for her fingertips, leaned in to kiss the rings. Before they could touch the Queen's hands, there came a sudden snapping noise followed instantly by a flash of light. Startled, Stephanie instinctively pulled back. Theone did the same. The circular radium blade of Blinda's Balkrum singed the Duchess' bangs, dropping a long lock of her hair to the floor at Jana's feet. A quick twist from Blinda's other arm and the young Queen was spun out of the way. In a blur, the Albion Thane was standing with her Balkrum extended between the Queen, the King Commander and Duchess.

"Blinda!" Diord shouted, bolting forward.

"What is the meaning of this?" the Wamar bellowed, motioning for the Pintar guards.

"Stand back!" Blinda blazed, bringing Diord to an instant halt. She separated her rings, winding them into an attack stance. Despite her drawn weapons, the guards surrounded the small group with blaster rifles ready.

"Blinda?" Jana demanded. "What are you doing?" She looked at Stephanie and Theone, then around at the guards. "You must stand down or someone is going to get hurt."

"Blinda, put away your weapons," Diord insisted.

"Surrender your weapons immediately," the Wamar demanded.

Blinda's eyes narrowed, looking around, then back at Jana. She carefully relaxed a little, lowering her weapons.

"The Duchess conceals a hypo sleeve under her index finger," the Thane declared quietly. "I suspect either Kodiac Blu or Tulane. Either is lethal."

Diord stepped tentatively forward, stopping in front of the Duchess.

"Your hands," he demanded.

Stephanie slowly raised her hands up, revealing a thin silver sleeve on the bottom of her index finger. A thread-like needle protruded to just beyond the end of her fingertip.

"Duchess?" Diord gestured holding his hand out. Stephanie hesitated, her eyes shifting momentarily in the King Commander's direction. She slowly removed the hypo and held it out to the Administrator. Diord carefully took it and held it up in full view for the witnesses and cameras to see.

"Your Highness," Stephanie interrupted. "If I may be permitted to..."

"To what?" Jana demanded, curtly. "Explain?" She stepped out from behind Blinda and stood next to Diord, her eyes fixed on the Duchess. She took the book from Diord. "Shall I play my father's journal here and now, for everyone to hear and witness? Will you try to *explain* what he clearly documented?"

Stephanie hesitated, hedging another look in the King Commander's direction.

"I would not presume to interpret your father's words," Stephanie offered nervously. She leaned in a little closer. "The King Commander," she whispered barely audible. "This whole affair is his doing. I have recorded proof of his..."

"Silence, Duchess," Theone growled, pulling something from under his cloak. A quiet pop accompanied a ruffling to the back of Stephanie's gown. She teetered a moment then dropped to her knees. Instinctively, Jana held her hands out to steady her. Theone pushed Diord aside and brought his weapon up to Jana's head. There came a momentary look of glee slither across his face as he squeezed on the trigger again, but before the pistol went off, the gun abruptly dropped to the floor in front of him, his severed hands still clutching the pistol grip. Amputated at the wrists, his flesh was fused by the searing heat of the razor sharp blades of Blinda's Balkrums. Theone reeled back, turning around holding his stumps up. His expression of complete astonishment mixed with escalating pain, riddled across his face. Blinda once again took a defensive stance in front of the Queen. Astonished, the guards stood frozen, their weapons aimed at nothing. Diord pulled the small blaster pistol from Theone's severed hand and pointed it at the King Commander. Now screaming in pain, Theone staggered into the arms of one of the guards. Stephanie looked up at Blinda, then around her at Jana, her breathing labored.

"Yes, I killed your family and poisoned your father." She looked back at a groaning Theone as he cradled his stumps in opposing armpits. "But I had nothing to do with sending you into exile in the Reako biosphere or Terminal Twenty-One. That was all orchestrated by the King Commander."

"Shut up you worthless coward!" Theone screeched.

Jana put a hand to Blinda's shoulder, bringing the Thane to relax somewhat as the Queen took a step past her. She stood above the severely wounded Duchess. Dropping carefully to one knee in front of the Duchess, Jana gazed into her tortured eyes.

"Are you prepared to take responsibility for what you have done?" Jana asked quietly.

"For what I have done, I take full responsibility and will expect no mercy, as I don't deserve any."

"Why wouldn't you expect mercy?"

"Because it's not something I would have given."

"That's what separates you and the King Commander from me," Jana said after a moment of thought. She came back to her feet and held out her hand. Stephanie looked at it a moment, bewildered. Wincing painfully, she took the Queen's hand and with great effort, came back to her feet. Jana motioned to the Wamar, who in turn directed a number of the guards to attend to the Duchess. As Stephanie was escorted out, Jana stepped over to the King Commander. Several guards had come to his aide, holding him up. The trauma to his nervous system was starting to make itself manifest. Jana's gaze burrowed into him. She finally motioned to Diord, holding out her hand as he stepped up beside her. Worried, he handed the small blaster pistol to the young Queen. Examining it for a moment, Jana finally pointed it at Theone's forehead. The King Commander tried to show his continued defiance by standing on his own, but his body was in extreme shock, requiring the guards to hold him up.

"Check his pockets," Jana demanded.

Diord moved carefully forward while Blinda stood next to the Queen with her weapons still activated. As the Administrator went through the King Commander's pockets, he pulled a tiny device out and held it up.

"It's a transmitter," Diord announced.

"My personal... communications..." Theone panted.

"He's been calling the shots from right here in the hall," Diord said, pointing out the window at the mass of ships outside.

Jana eyed Theone carefully, a scary expression developing as she reaffirmed her grip on the pistol.

"You and the Duchess are the cause of so much suffering going on in Hadrian right now," Jana said tightly.

Diord looked back at the Queen. She appeared to be wrestling with inner demons.

"Your Highness?" Diord inquired, moving away from the King Commander.

"The House of Pintar…" Theone puffed arrogantly. "Had become weak… It required, cleansing. The Oxlinds had to be… purged… in order to make way… for a new, stronger order."

Jana touched a control on the side of the pistol. The power pack immediately lit up, an indicator light blinking on the short barrel of the weapon.

"You're as weak as your father," Theone grunted defiantly. "You don't have the instinct to rule a galaxy as I do."

Jana firmed her grip on the gun and started to squeeze on the trigger until a gloved hand from behind took hold of the barrel.

"Your Highness," Blinda spoke softly. "I can speak from long, sad experience. He isn't worth it. You are as strong, if not stronger than your father. Now is the time to show that strength."

Jana looked at Blinda, then back at Diord, then Theone, who remained defiant. She took a deep breath, finally letting go of the pistol and stepping back.

"The right to execute you myself is mine and I would be justified, but that would make me no better than you and no one is above Hadrian law of trial. As Queen of Hadrian, I will rule as my parents did; by the laws that have governed this people for eons with justice and mercy."

Several tense moments slipped by. Diord smiled, watching the royal guards around the platform area shifting their stance, looking at Theone, then at the Queen. As if by some silent order, they all turned to her and came to attention.

"Take him away," she muttered. As Theone was carried away, Jana stepped back over to her seat and sat down, shaking.

"Well, this is off to a good start," Diord smiled at Blinda as she took her place beside the throne chair. "Your Highness," Diord said leaning down next to Jana. "General Niker had to leave because of the conflict outside that the King Commander was directing during these proceedings. You must act now; decisively to stop it. The King Commander has many allies and forces still loyal to him, most of them have followed blindly, but words from the Queen of Hadrian could sway them at this critical time. I suggest you make a fleetwide announcement, ordering a cease fire."

APEX

As if by some silent order, the four fighters that had been harassing the shuttle Rick and BachTL were flying, suddenly blew past them and disappeared into the myriad of tiny dots ahead of them. Rick abruptly stopped the shuttle to assess the activity surrounding the attack on the Albion Battlecruiser, *Tarzana.* Varying sizes of internal fires and explosion flares continued to manifest themselves on the starboard broadside where the five torpedoes had impacted.

"That was unexpected," Rick said, sitting back. "What happens next?"

BachTL leaned forward, trying to size up the situation.

"Not sure what to think. Normally, the *Tarzana* would respond with return fire and send at least half of her fighters and bombers out at whomever fired on them. But as they were hit by one of their own…" BachTL puzzled for a moment. "It doesn't make any sense. The *Tarzana* deliberately put herself in the path of those torpedoes. As if trying to…"

"Protect *Calypso*," Rick finished quietly. He leaned over for a look at the scopes in front of BachTL, then brought his communicator up.

"*Athena*, are you still with me?"

"Yes, General," Dakota responded. "Any idea what that was all about?"

"If I didn't know any better, I'd say we have an ally aboard that ship. Did the *Maxell* fire anything?"

"They got one torpedo off, but the fighters from the *Trax* took it out. Would you like us to teleport you aboard?"

"I'm not close enough," Rick replied.

"You would be if we weren't holed up inside *Calypso*."

"You being holed up is what's keeping this whole situation from blowing up in our faces. We're going to try to make it to the *Executioner*, then see if they can get close enough for you to bring us over." Rick turned to BachTL. "Can you get us clearance to the *Executioner*?"

"I can try," the Castellian replied, going to work with the communications equipment.

Rick looked around outside; still no fighters. A quick glance at the scanning equipment revealed a myriad of targets moving in clusters ahead of them, but nothing in their immediate vicinity. As he gave the shuttle throttle, he steered toward the large box shaped Tomplie carrier.

"They didn't sound very happy about it, but we've got clearance to land," BachTL said leaning over. "They're supposed to be sending a couple of escorts, but I doubt any will show up with all this going on."

Rick gave the scanner another look as they drew closer to the flotilla. Several flights of fighters and bombers streaked across their path moving in different directions. Moments later, a couple of odd looking fighters swooped in close.

"Transmitting landing instructions. Follow them into bay number two," a voice from the com system ordered. "It's pretty hot out here, so any deviation from that flight path could get you blasted to bits."

"We've got your landing instructions," BachTL responded. "We need to speak with General Aurora as soon as we land."

"That'll be your problem. We're just supposed to get you in there in one piece."

"Friendly bunch," Rick mumbled as the opening to the carrier's landing bay came into view. As he maneuvered into the approach path, he was able to get a closer view of the damage the *Tarzana* had taken, as well as a better look at *Calypso's* bow. There were large pit craters in the blackened port side bow of the great ship, but no breaches in the hull itself. *Tarzana* had not fared as well. Her starboard side sported several large punctures, exposing decks and internal machinery, some decks were swarming with emergency crews in space suits working with the fire control systems and making repairs. A few of the torpedo impact points had struck close enough together as to join with the damage of the next impact zone.

As Rick set the shuttle down in their assigned area of the landing bay, they were surrounded by a platoon of security guards with weapons pointed at them.

"Nice greeting," BachTL said, raising his hands before they even made it to the shuttle exit.

"Just stay calm and let me do the talking," Rick said, stepping out the door with his hands up. The two were quickly stripped of any weapons, taken from the shuttle bay and escorted to a holding cell just off the main hangar. "We need to talk to General Aurora," Rick yelled as the guards walked off.

"This wasn't the plan," BachTL complained, leaning against the cell wall.

"I suspect he's a little busy at the moment," Rick suggested.

"Now what?"

"Now we go see General Aurora," Rick said, closing his eyes and slowly exhaling.

BachTL gave the Kalamarion General a quick glance as Rick brought a hand up in front of him.

"Oh, that's right..." BachTL grinned. "You can do the whole Thane thing. Forgot about that."

The energy beams barring their exit from the cell suddenly disappeared, but Rick kept his hand up as he stepped out and waited for the Castellian to clear the emitters. The beams suddenly reappeared as the two began to make their way out of the holding area and to an elevator.

"Are you as good at that as Blinda?" BachTL asked as the door closed behind them.

Rick looked at the symbols on the control panel for a moment, then pressed one of the buttons. As the elevator began to move, he took a deep breath and faced BachTL.

"Blinda has been a Thane far longer than I have..."

"So not quite as good?"

"She has enormous potential to be one of the greats if she can learn to focus and channel her energy in a more controlled fashion. I suspect, up until now, she's had no need to concentrate to accomplish the things she's been doing."

"You're saying she could kick your butt if she really practiced hard?"

"Yeah, probably," Rick answered after a short pause.

"How did you know which floor to go to?"

Rick pointed to the panel and the button that had just stopped glowing.

"This panel was made by Diord's company. I can read it."

BachTL chuckled as the door opened up to an enormous control room. Making it look like they knew exactly what they were doing, the two walked directly in and recognizing General Aurora, headed quickly toward him before security could intercept them.

"General Niker," Lou exclaimed. "I thought you were on Calliope helping the Princess? How did you get out here?"

"We borrowed one of Diord's shuttles."

"Shuttles in a hot zone like this aren't a good combination. So how did...?" Lou turned to the security master. "How is it I wasn't informed about General Niker's arrival?"

"We requested entry and your approach control people gave us clearance, even an escort," Rick said, taking a deep breath.

"Then your security herded us off to a holding cell," BachTL added.

"Well?" Lou pressed the security master.

"You were busy."

Exasperated, Lou waved him away.

"I'll deal with you later. I'm not going to ask how you got out. My apologies that you were ever put in there. I assure you I had no knowledge of your arrival." Lou came to his feet and stepped to the main bridge windows. "This is a huge mess, General Niker. I think if you have any weapons on that colossal boat of yours, now would be a good time to deploy them."

"The fact that they haven't been deployed is probably the only reason this hasn't turned into a bloodbath already."

"Not a bloodbath?" Lou asked, surprised. He stepped over to a strategic viewer at the side of the bridge and pointed at the images on the screen.

"I've lost nearly a quarter of my fighters and bombers to Albion and Ratronian action. I think the only thing saving the rest of them is the Ratronians, Albions and Colonians have no idea who's fighting who." Lou swirled around and pointed out at the Albion battlecruiser. "And if that wasn't enough, the *Tarzana* just took five direct hits from the *Indominable*, and they're on the same side!"

"General," Rick started quietly. "I deploy anything from *Calypso* and all hell is going to break loose."

"I'm moving my flotilla to a safer zone, out of the way, and I'm advising Diord's ships to do the same," Lou growled. "I didn't sign up for this..."

"No, you signed up to help repatriate Princess Jana," Rick flared. "And turning tale and running at the first sign of a fight isn't doing her any good. Did you really think this was going to be easy? That you were just going to sail in here and knock on the doors of the factions and they would welcome you with open arms?"

"I assume you're standing here for two reasons," Lou said, standing up to Rick. "Number one; because you succeeded in getting Princess Jana into the Rite of Pintar. Awesome; mission accomplished. And number two; now you're trying to save yourself at our expense. Well, I have no devotion to you or anyone associated with you. Now, if you want to save that big pile of junk floating out there, then I suggest you deploy whatever defensive countermeasures you have right now or there'll be two burning hulks out here."

"General," one of his aides called from the main control console. "The *Tarzana* is attempting to turn."

"Turn?" Lou repeated. "To where?" He looked out at the damaged battlecruiser as it began to pivot in their direction. "Back us out of this mess. Give her room. If she gets too close, put a couple more torpedoes into her."

"She just saved *Calypso*!" Rick cried.

"...and now she's a floating time bomb," Lou cut back. "Get everyone clear of her!"

"The *Indominable* is launching more torpedoes."

"Emergency reverse," Lou ordered. "Get us out of here!"

"General," Rick stepped forward. "It's clear that the *Tarzana* is trying to do the same thing we are; protect *Calypso*. The commander of that vessel is moving to intercept the *Indominable's* torpedoes. Don't you think whatever their intentions are, they could use our help?"

"Nothing is going to stop this fight unless you deploy whatever weapons you have in that ship."

Rick considered, looking out at the *Tarzana* sweeping past the front of the *Executioner*. It was heading directly at the incoming torpedoes. As Lou's carrier continued to pull back, the *Indominable* came into view. Several clusters of Black Tiger fighter craft buzzed angrily ahead of the *Tarzana*, firing at the incoming ordnance. The Albion battlecruiser seemed undeterred as it swept through a cluster of larger ships, heading directly at the enormous dreadnaught. There came a sudden flare from midship as one of the torpedoes struck the *Tarzana's* lower superstructure.

"What are they doing?" Lou asked, looking over at the bridge command officer.

*　　*　　*　　*

"Helm, hard to starboard, three twenty-one mark six," Dalton shouted over the noise of sirens and the crew's panicked voices. "Flank speed," he bellowed.

"Where are you taking us?" Captain Gulch inquired, wiping blood from a cut on his temple and holding his arm.

"We can't allow any more torpedoes to be launched," Dalton answered.

"But your orders..."

"I am in command of this vessel, Captain," Dalton responded curtly.

"I understand that, Sir..."

"What's happening here has to stop. There has been no provocative action taken by that alien ship. If we don't end this right now, the other ships here will follow suit and nothing will prevent an all-out war from developing."

"Are we to fire on the *Indominable*?"

"Only if necessary," Dalton responded with absolute resolve.

The Captain turned to carry out his orders, but returned a moment later with a communication device and handed it to Dalton.

"The Admiral, Sir."

"Commander SoKnack," Admiral Hilton croaked. "You were ordered to keep the *Maxell* occupied. Explain yourself!"

"This is wrong, Admiral," Dalton replied calmly. "You need to stand down."

"You and I are under orders from the King Commander."

"I will not participate in something we both know is wrong," Dalton insisted.

Embarrassed, Admiral Hilton stuttered, caught off guard by the response of a lower ranking officer. His ego ruffled, he became more defiant.

"You are ordered to turn your ship around and fire at that ship."

"And I am refusing that order, Admiral," Dalton maintained. "I will not facilitate or escalate an unprovoked attack."

"If you do not turn your ship around and follow your orders, I will be forced to fire on your ship as well," Admiral Hilton warned.

"I will do the right thing, Admiral," Dalton countered, dropping the communication device and stomping on the display. He looked ahead of the ship. They were just clearing most of the other faction ships, heading straight at the *Indominable.*

"They can fire more torpedoes than we can," Captain Gulch remarked from behind.

"They're a bigger target," Dalton replied.

"True, but wouldn't it be helpful if you could keep them from firing torpedoes in the first place?"

"There are about three dozen torpedo launchers on that dreadnaught." Dalton turned to the Captain. "I am open to suggestions."

"The *Indominable* has a central launch control center located toward the bow in the crook of the forward sensor arrays. It's armor shielding is considerable for obvious reasons, but there are other ways to get through."

Dalton turned to the Captain, sensing an allegiance to his command.

"Damage report," Dalton inquired, stepping back down into the navigation pit with Captain Gulch directly behind him.

"Reports are still coming in," the Captain said, trying to keep up. "We took all five Thug torpedoes to the starboard broadside right below the Snuff gun banks. We took another torpedo to the lower conning tower. Decks five, seven and nine to the third tier of the interior have been sealed off. Casualty count is unknown."

"Are the Snuff guns still operational and are there crews to man them?"

"Yes, Sir. We only lost a few automated Spike turrets on that side."

Dalton turned to the navigation officer.

"Bring up the *Indominable's* base prints," he ordered quietly.

"What do you have in mind, Sir," Captain Gulch inquired.

"Gonna put your idea to the test," Dalton said, examining a dimensional readout of the enormous dreadnaught. "Right there," he said, pointing to the front of the ship. "Give those coordinates to helm control." Dalton nearly ran Captain Gulch over as he sprang back up to the command overview deck. "Helm, navigation has just given you a new set of course coordinates, adjust your course now."

Captain Gulch looked up at the overhead tactical display.

"Have the bow evacuated and prepare for damage control," Dalton ordered.

"You're going to ram it."

"There's no amount of shielding that can keep the nose of this ship from going right down its throat," Dalton responded resolutely.

Both officers focused on the *Indominable* as the *Tarzana* zeroed in on its target. An instant later, they began taking and returning fire as the gap between the two ships rapidly closed. A flurry of torpedoes streaked away from the dreadnaught and the battlecruiser as the ships held steadfast on a collision course with each other. Rattle and Snuff guns on both ships came alive as the projectiles came into range. The rapid fire guns swung with lightning precision at the inbound torpedoes. Much of the destructive ordnance was dispatched in a flashing storm of explosions before it could reach their targets, but some made it through the barrage of rapid fire guns, slamming into turrets and streaking toward the superstructures.

"All hands, brace for impact," Captain Gulch's voice echoed ship wide. Torpedoes grazed off the bow of the battlecruiser as they sliced through space. A few managed to explode against the superstructures, causing considerable damage.

Pitching violently, the bridge crew of the Tarzana struggled to remain at their posts. Trying to brace himself against the nearest handrail, Dalton's eyes widened as a sea of the *Indominable's* hull filled his vision. A split second later... it came! There was such terrific force that he was thrown against the forward bay window bulkhead. His head spinning, he managed to lift himself up high enough to see ahead of them. The bow of the *Tarzana* was tearing into the flesh of the *Indominable*. Likewise, the Albion dreadnaught was peeling back the metal skin and bulkheads of the battlecruiser. It took minutes for the *Tarzana* to come to a stop amid the carnage of the collision.

Pandemonium filled Dalton's ears as the bridge crew called out in panic and injury. Trying to get up amid the blaring sirens in his ears, he found he was unable. A pain in his leg stabbed at him whenever he tried to move it. Rolling onto his side, he felt something warm spilling across his face. He wiped whatever it was away only to find it quickly return. A number of utility lines and conduits had ruptured directly above his position. Wiping his neck and shoulder again, he checked his hand. In the flickering bridge lighting, he recognized... blood. He looked around, seeing Captain Gulch staggering toward him holding his shoulder and sporting a bloody eye socket, more blood streaming from an injury from somewhere behind his collar.

"Mission accomplished, Sir," the Captain grunted, helping Dalton sit against the wall directly below the main bridge window.

"Do we still have thruster control?" Dalton asked, trying to figure out where all the blood was coming from. "Directional control?"

"Yes, provided we can get the crew to refocus."

"Back us out and move her off to a safe distance."

"Are you going to be all right?" the Captain asked, helping his commander to one of the command chairs. He started to check him for injuries as there seemed to be a lot of blood coming from somewhere, but the situation wouldn't allow for a positive identification.

Dalton blinked several times, feeling strange. "Better get a medic up here as soon as you have the ship moving again." Dalton waved the Captain away and tried to settle into a comfortable position. He watched as the Captain turned and started barking orders. The bridge crew calmed down as the sirens finally went silent. He looked out ahead of them as the *Tarzana* began to pull itself free of the *Indominable*. Great pieces of ship parts dislodged themselves and floated free as the torn bow of the battlecruiser separated from the mangled front end of the *Indominable*.

"Try firing torpedoes now," Dalton mumbled defiantly, feeling sick. He again tried to figure out where all the blood was coming from, but his hands and arms were losing feeling. He scanned the bridge, watching as the crew worked to maneuver the great ship back and away from Calliope. As the *Tarzana* pivoted in space, the faction clusters came into view. Most had given *Calypso* a wide berth. Only the Alvadorian and Tomplie ships remained close by.

His thoughts turned to Colonel Barker and the conversation with her before she left him in command. He blinked sporadically. His sight was blurring and clouds of blackness were creeping into his vision. He turned his gaze to Calliope, then to the *Indominable*.

"Yeah," he drowsed to himself. "It was the right thing to do..." He looked down at the floor next to him. "That's a lot of blood..."

*   *   *   *

Everyone on the bridge of the *Executioner* watched in stunned silence as the *Tarzana* pulled back from the wrecked front end of the dreadnaught. Several minor explosions flashed from the mangled bow of the battlecruiser as it limped away. As the *Executioner* continued to maneuver from the fray, Rick observed several destroyer class vessels moving in closer to where the collision had occurred. He shifted his attention to the *Merc*, then the *Trax* and *Realistic*. They were still holding their positions around *Calypso*. Rick turned back to Lou.

"General..."

"General," Lou countered before he could get the words out. "Deploy your weapons."

Rick opened his mouth, but hesitated. He finally raised his communicator up but before he could speak, a communications officer approached.

"General, we're receiving a fleet wide transmission from Calliope."

"Who's it from?"

"It's encoded from..."

"Well, come on man; out with it!"

"Her royal majesty, Queen Jana Oxlind. Ruler of Hadrian."

Lou's expression shifted. He gave Rick and BachTL a glance. They were both smiling, relieved. The Tomplie General opened his mouth to say something, but hesitated. He tried again to speak, but his thoughts were in direct contrast to what he had wanted to say. He finally nodded.

"Bring the flotilla back in close to *Calypso*." He turned to a large viewing screen. "Let's see the message."

*　　*　　*　　*

"Fellow patriots of Hadrian," Jana began, sounding nervous. She paused, glancing back at Diord and the majority of faction leaders standing behind her. Several of the royal guards had taken up positions around her, but Blinda remained by her side.

"I stand before you as one who would rather be sitting next to you. Right now, I feel the weight of all your hopes and aspirations on my shoulders. My lineage reaches back into the eons when my Oxlind ancestors first colonized Hadrian. Certainly, it was a smaller place then."

"Some of you remember my parents. They were good and kind people that understood the enormous responsibility of governing a galaxy. My father wrote most of his thoughts to me here," she said, holding her book up. "Many of those thoughts were about the hard decisions that had to be made in order to preserve a peaceful way of life."

She set the book back down, running her fingers around its edges and thinking.

"Monarchs do not share their thoughts easily, but my parents realized that as Hadrian continued to grow into other regions, equity would become impossible. Greed and corruption in the government, military, corporate and private sectors would usurp royal power and authority into a twisted hungry monster that could only consume

everything in its path. It happened so fast that my father could do nothing to prevent his family's demise. My mother, my brothers and sisters, and finally my father, were murdered for their beliefs, their stations in life, and for the power they were responsible for."

Jana looked down the row of dignitaries at the Duchess's and King Commander's empty seats.

"As a child, I was to suffer a similar fate." She turned and looked at Blinda. "But through the heroic acts of those assigned to protect me and care for me as a child, I was spared."

Jana paused, thinking of her rescue. A melancholy smile drifted across her face.

"I've made many friends on my journey here. Many of them are not even from Hadrian. Yet, they have risked everything to see that I am able to stand before you now. They didn't ask to come to Hadrian, yet they've seen to my safety and I owe them an endless debt of gratitude." She paused, turning and looking out the walled window at the flotilla of ships surrounding *Calypso*. She then turned back and stood as tall as she could. "My friends need our help. Even now, their ship is under siege outside Calliope by those who would subvert Hadrian law and authority, trying to take what doesn't belong to them. As your Queen, I'm ordering the Factions of Colonia, Ratronia and Albia to stand down their forces everywhere. Instead of trying to destroy my friends and take by force what doesn't belong to us, we should be trying to help them figure out how to get back to their homeland. Would anyone of us want anything different were we in their situation?"

"My friends, I am one of you and like you, I believe no one person should be esteemed above another; for we are all equal unto each other. Our faction governments must be reorganized, starting with Colonia and Albia. Checks and balances must be put back in place that will keep greed and corruption from overrunning the people of this great civilization again. I am young and have much to learn so I implore you to be patient with me. I am ever your humble servant, Jana Tilee Oxlind, fifth daughter and seventh child to their Majesties, King Kiev and Queen Mila Oxlind, Queen of Hadrian."

*   *   *   *

While the *Executioner* pulled in close to *Calypso*, Rick folded his arms watching Lou slowly turn from the viewer.

"My apologies," the Tomplie General admitted sheepishly. "My faith in this actually succeeding was not as strong as it should have been."

"The cause of right always seems to be against overwhelming odds until everyone involved can see clearly," Rick said.

"We could have run and hidden in the Spartus and ignored all this, and probably no one would have been the wiser for it," Lou admitted. "At least not in our lifetime. I have to admit I got into this for the woman… Err, I mean, the android, Tonnie."

"It was a place to start," Rick chuckled lightly. "But you must have believed the Princess could be repatriated at some point."

"I'm an opportunist, General," Lou smiled. "I diverted for a rescue, hoping to get a payout of some kind and keep my ships out of harm's way. The idea all along was to let Diord take the brunt of whatever happened here." Lou looked genuinely disappointed in himself. "I guess some things never change."

"There's always what you do next," Rick suggested. "You and Blinda Koss now share something quite unique. You two get to alter your course."

"Is that Thane teaching?" Lou asked.

Rick thought a moment, a pleasant look forming.

"I think just good words to live by." Rick looked back at BachTL. "We need to take our leave. Thanks for getting us close to *Calypso*."

Lou shook his hand firmly, then Rick brought his communicator up.

"*Athena*, we're in range. Two to bring over."

"Right away, General," Laura responded.

*   *   *   *

Rick recognized the buzzing sound surrounding him and the sparkles starting to fly through his vision as the teleporter stream was activated. Moments later, they were standing in the teleporter room of the *Athena* looking at Captain Korack.

"Welcome back aboard, General," the Interceptor pilot said, coming out from behind the control console.

"Status?" Rick inquired, stepping into the hall and looking into the engine room.

"Fully operational. We're sitting right next to the *Intrepid* just inside the hangar doors ready to deploy. Though I'm not sure where we're going to deploy to."

"Then you saw the Princess…" Rick turned up the hall toward the bridge. "I mean, the Queen's speech?"

"We sure did. Impressive young lady. I gotta say, we're all a little relieved."

"Damage to *Calypso*?"

"Dãsha and Mister Dacey indicate only minor problems. That was a close call."

"Yes, you can thank the *Realistic* and *Trax* for that."

"Sir, what do you make of what happened with the *Tarzana*?"

"I can only assume they didn't agree with what the *Indominable* was trying to do and took the only action they could without drawing the rest of the factions into an all-out slug fest."

As Rick entered the bridge, Captain Abrams instantly came out of the command chair, snapping to attention.

"As you were, Captain, it's just me." Rick smiled, stepping up next to Captain Dayton at the helm. "Who's in command of the *Constellation*?"

"Engineer Moon is in command over there right now," Dakota responded. "He's just finishing the Asium refit."

"And Doctors Mantose and Yamoto?"

"Still in surgery," Dakota answered in a more somber tone.

"I'm heading over to check on their progress. You're still in command here, Captain."

They quickly made their way out of the ship and over to where the *Constellation* sat, away from the other five Starbirds. Rick took note of Engineer Moon standing at the engineering console in the engine room and Alex 7001 hovering in front of the Asium chamber. Not stopping, he turned up the main corridor to sickbay. He tried several times to enter, but each time the activators blinked a locked message. Impatiently pacing in front of the door, he paused a moment looking at BachTL.

"How am I supposed to know how things are going?" he asked, snapping his fingers repeatedly and continuing to pace.

"Didn't Caidin indicate this would take some time?"

"Yeah, but how long?"

"Simple math," BachTL suggested. "How long did it take to repair and revive Commander Atlanta?"

"Something like twelve... eighteen hours," Rick fumbled clumsily. "I don't know. I can't even think straight right now."

"Let's say fourteen..."

Rick stopped again, looking at BachTL.

"Just throwing a dart at it," BachTL tried to reassure him. "If the Colonel is as complex as your doctors say he is, then wouldn't it be logical to think that it would take at least twice as long, if not longer?"

Rick took a deep breath, thinking things through.

"Yeah," he finally agreed quietly. He considered for a moment, then stepped back to engineering. "Alex 7001?"

"Yes, General?" The little droid turned his head, but not his body. Billy remained glued to what he was doing.

"Alex, are you with me?"

The floating droid remained silent for a moment.

"Why would you ask that? I'm right here in front of you."

"Sorry, force of habit with CORA 500," Rick said, rolling his eyes. "Is it possible to get a status report on Colonel Conrad's condition?"

"No."

"What?"

"No, Sir?"

"What do you mean, no?"

"The medical bay is on a procedural override. No admittance due to surgery."

"Yes, I know why it's locked," Rick retorted, becoming irritated. "Can you get me any information about how things are going with the surgery?"

"Only to a certain point," the levitating AI responded.

"A certain point?" Rick repeated.

"I cannot access the Colonel's vitals as there is no current link to the Med assist or his stasis pod. I have a visual of the interior of sickbay, but most of what's happening is being done at the molecular level by Doctors Mantose and Yamoto."

"Have they said anything?"

"Very little. MedTech Gantrie has been tending to the doctors as they have worked, but has not provided any additional information. Based on my unfamiliarity with the procedural techniques being used, it is impossible to attempt any kind of estimate of how much longer they will be in surgery."

Rick rubbed a fist into the palm of his other hand.

"Let CORA know the moment you get something," Rick demanded.

"Why wouldn't I?"

"An AI's timing to cop an attitude is impeccable," Rick grumbled, heading back over to the *Athena*. He thought about a nap, but in this situation, he knew he would only lay on his bed staring at the ceiling, thinking about Gunnar.

After trying for hours to escape the realities of Gunnar's condition using Thane meditation, Rick finally found solace with intense study of the star charts from *Calypso's* data banks. Hour after hour he remained immersed. So much so, he didn't even hear the summons from his communicator or the calls from the door. He didn't notice it open or detect someone enter the room. It wasn't until Jayda sat down on the desk next to the display that his concentration shifted.

"Aren't you supposed to be on the *Montego*?" he asked, still riveted to the display.

"It's not going anywhere," Jayda replied quietly.

"Those ships need to remain battle ready," Rick countered, remaining engulfed.

"They're ready enough if something happens."

"You're not very good at following..." Rick looked up, noticing there were others in the room. "Orders..."

"No, we are not," Dãsha said with her fists resting on her hips.

"Sometimes you have to ignore an order for a greater good," Dakota said standing next to BachTL.

Rick shifted, trying to see if anyone else had come in. As the General turned his chair around, BachTL and Dakota motioned Jayda and Dãsha to the couch.

"Ok, what's this all about?" Rick asked as Dakota took a seat by the door.

"This time," Jayda began, "it's about you."

"Me? What about me?"

"You are a particularly strong individual," Dãsha started. "But even strong people require support from others."

"Support?"

"Yes," Jayda said. "The only person who knows you as well as I do, is Gunnar and he's not here right now. You've been a pillar for all of us. But who's your pillar?"

Rick rolled his eyes toward BachTL, then leaned forward.

"I appreciate your concern, but I'm doing just fine."

"Just fine? Really?" Jayda asked, leaning forward as well. "This is what you've always called *just fine*," she said, gesturing toward the display behind him. "We've been sealed for how long and you're going to sit there and try to tell me, tell us, you're *just fine*? Sweetheart, when you're upset and you can't resolve the issue, the first thing you run for is anything technical. The harder and the deeper you can dive into it, the better."

"Ok," Rick said, holding a hand up. He paused a moment, reading the looks on everyone's faces. They needed to be there for him just as much as he needed them to be there. "My best friend is lying dead on a surgeon's table. So yeah, I'm a little upset."

"It is clear that we all are," Dãsha said. Her voice was uncharacteristically soft. "I have not had the opportunity to interact with Gunnar, but I can see the profound effect he has had on your people."

"Not to make this sound like a eulogy but, he was one of us," Dakota spoke up. "I don't think he ever put himself above anyone else. Whether it was a new cadet trying to figure out how to properly navigate an orbit or any veteran fighter pilot in the Ministry. He was always anxious to help anyone needing a leg up. I recall a Mid-cadet that had just destroyed a Thumper with a bad landing. The kid barely got out of the wreck in one piece. He was in the process of washing out to an orbital station or something planet bound, but Gunnar wouldn't have it. He made the kid his wingman and taught him how to fly circles around the other Mid-cadets. Of course, no one could fly like the Colonel."

"Noone knows that better than me," Rick admitted smiling. "I started flying before I was eight years old. I've flown with some of the best pilots in the galaxy, but none of them ever came close to what Gunnar could do when he was at the controls. There was something about his talent in any cockpit that was unworldly. Flying came naturally to Gunnar; it was his first language."

As the room fell momentarily quiet, the door summons sounded, then opened. Audra looked up as she stepped in, stopping short when she saw everyone.

"Sorry, I thought this was a one on one. Am I interrupting?"

"Not at all, Commander," Rick beckoned. "Come in."

Dakota sprang from his seat so Audra could sit. After she got comfortable, she smiled at the looks everyone was giving her.

"What?" she asked, wondering.

"We were just talking about what an amazing pilot Gunnar was," Jayda said.

"Fascinating that he came by his talents naturally," Dãsha said, inflecting her voice. "Richard describes it as his 'first language'."

"First language?" Audra sighed. "Most of the time, it was his only language."

"You understood it," Jayda pointed out, remembering Audra's service background with traffic control.

"Sure, as a controller, I had to. Doesn't mean I liked it; at least not when he was supposed to be speaking my language."

"Yeah, we know what language that was," Jayda teased.

Everyone but Dãsha chuckled softly.

"I don't understand the reference. What language was it?"

"Reach back to Danis," Rick suggested.

Dãsha tipped her head, thinking for a moment. Her expression suddenly brightened and she smiled.

"The language between couples that precedes..."

"Ok, I think you've got it figured out," Jayda spoke up quickly.

Everyone snickered at Dãsha's expression. A long period of casual conversation ensued, disconnecting everyone from what they had experienced for the last several days. Rick and Jayda served drinks and snacks as they chatted. The conversations had nothing to do with mechanics, physics or military rank and protocol; only friends talking with friends about nothing in particular. After several hours of relaxation, BachTL and Dakota trickled out.

"We'd better get back to the *Intrepid*," Audra said, looking at Dãsha and starting for the door. "Thanks for this. I think we all really needed it."

"Yes," Jayda echoed. "It's been a while since we've had a chance to disconnect and unwind. This was wonderful."

"It will take me some time to fully process why this activity has been so therapeutic," Dãsha agreed.

"I have to admit, I've really enjoyed this too," Rick said following them to the boarding ramp. "Funny how we get tunnel vision and can't see the need to detach for a while."

They stopped at the bottom of the boarding ramp and looked around at the Starbirds positioned in *Calypso's* bay. Remaining silent for several moments, they hesitated to reconnect with their responsibilities.

"I'll walk you ladies back to your ships," Rick volunteered.

"Thank you," Jayda smiled, taking his arm.

"How do you think things are going?" Audra asked, looking over at the *Constellation*.

"I don't know," Rick said. "Alex and CORA said they would let me know as soon as there was something to report."

"How long have they been at it?" Jayda asked.

"Going on eighteen hours now. I don't know how they can go for that long with only five minute breaks every couple of hours."

"I witnessed Caidin reenergize himself when he was repairing Audra," Dãsha announced quietly. "I have to think he can do the same for Doctor Yamoto."

"That's a lot for one person to take on alone," Rick said, stopping at the *Intrepid's* boarding ramp.

"I have the utmost confidence in my husband," Dãsha asserted.

"Me too," Rick agreed.

"Let us know when you hear from Alex and CORA?" Audra requested as she followed Dãsha into the ship.

"As soon as I know anything," Rick reassured them as he and Jayda turned toward the *Montego*. Walking arm in arm, they passed several of the Kendalon officers, greeting them warmly.

"You gonna head right back to your star charts and math equations?" Jayda asked as they stopped at the *Montego's* ramp.

"I don't think so," Rick said after a brief pause. "Might just go take a nap."

"You can thank BachTL for the last couple of hours."

"Yeah," Rick smiled. "Sometimes it takes someone from the outside looking in to see what we can't."

"I love you," Jayda whispered, then kissed him.

"I love you too," Rick responded, holding her hands, then letting go and turning toward the *Athena*. After a quick check of engineering, he headed for his quarters and stepped inside. He looked at the charts still up on his desk display, then turned it off, shut off the lights in the living area and went into his sleeping chambers. Laying down, he quickly found sleep.

*   *   *   *

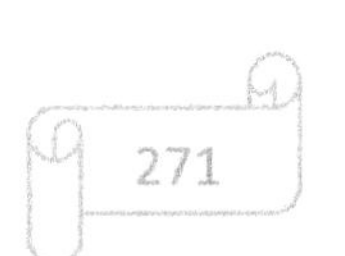

"Richard," a voice echoed through a dream. "Richard," the voice
repeated several times.

Rick finally opened his eyes and blinked, trying to clear his vision.
How long was I asleep?

"Richard," the voice persisted.

He recognized it.

"CORA?"

"Yes, Richard."

"How long was I asleep?"

"Three hours, fourteen minutes."

Rick sat up and stretched.

"Alex 7001 has indicated that Doctors Mantose and Yamoto have
finished with Colonel Conrad.

Rick was on his feet and moving in an instant.

"How long ago?"

"Just now."

"Have BachTL meet me..." Rick nearly ran into the Castellian as he
bolted from his quarters. Together, they rushed down the hall, down
the boarding ramp and across to the *Constellation*. The sickbay door
couldn't open fast enough. Upon entering, he found Caidin leaning
over Gunnar's stasis pod checking the readouts on the panel.

"Caidin," Rick asked anxiously. "How did it go?"

"We've done all we can," Caidin responded without looking up.

"What does that mean? Where's Fuji?"

Caidin made several adjustments to the controls on the panel, then
straightened up, turning to Rick. Rick and BachTL instantly came to
his aide as the older gentleman teetered, barely able to stand.
Darkened circles surrounded his tired eyes. After helping him into a
nearby chair, Rick turned to the pod and examined the panel.

"I had a couple of the crew help get her to her quarters."

"Her quarters?" Rick repeated. "What happened?"

"I told you going into this that there were inherent dangers with
what we were attempting to do." Caidin closed his eyes and leaned
back. "While she is a gifted doctor and surgeon, her frontal cortex
doesn't have the expanded abilities of a Thane. The process of taking
us both down to Gunnar's cellular level was extremely taxing on her. I
can act as a guide, but when all is said and done, she had to do much
of the work. Toward the end, she began to fade and I had to finish up
without her. I won't know her condition until she either wakes up or I
can go check on her. Right now, I need rest."

"Were you successful?" Rick asked, letting BachTL examine the
stasis pod. Rick turned to Caidin, waiting for a response.

"I don't know," he finally sighed.

"How can you not know?" Rick ramped.

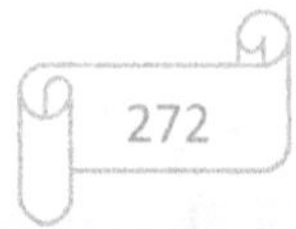

Caidin cracked an eye, looking at Rick's expression.

"We repaired everything we could to the best of our knowledge. His body is functioning..."

"Then why is he still in the pod?"

"Because it's the only thing sustaining him."

"Caidin..." Rick elevated even more, turning back to Gunnar's pod.

"Richard," Caidin burst. "You open that pod now and he's dead for sure."

"So how are we supposed to know when it's safe to open it?"

"When there's some kind of indication that he's actually alive. We've done everything we can. There were several things neither of us knew what to do about, so we had to... guess." Caidin settled back and closed his eyes again. "For now, his body is functioning... with the help of the pod."

Rick stared blankly at the tired gentleman. He tried his best to push out the obvious thoughts that Caidin was alluding to. Fighting the anger boiling inside, Rick finally realized there was nothing more that could have been done. Caidin and Fuji had stretched themselves beyond their capabilities.

"He looks like he's just sleeping," BachTL suggested as Rick turned back to the pod.

"He always slept with his left eyelid cracked," Rick said quietly. "I used to tease him about it. Never seemed to bother him though."

"It doesn't look cracked now," BachTL observed, looking a little closer.

"No, I don't imagine it would," Rick agreed barely audible. He peered through the glass at his friend, searching for anything that would signal that Caidin was wrong and maybe he would open the left eye a little and smile or yawn; something... anything.

"I think we need to get Caidin to bed," BachTL observed, stepping over to help the older gentleman.

Rick assisted as well, then turned back to the stasis pod. *Have we really gone through all this only to lose you? Audra is here! Shouldn't that count for something?* Rick remained frozen, staring at his motionless friend. He hoped against the inevitable that maybe Gunnar would open an eye and give him a wink. He'd pull him out of the pod and punch his lights out while hugging him.

"General," BachTL spoke in a low voice. "We need to let Caidin get some rest. I'm sure the others will be anxious to hear how things went."

Rick looked back at Caidin laying on the bed. He put his hand to the glass of the pod for several moments, then nodded and left sickbay.

"BachTL, what's your clearance status aboard ship?" Rick asked heading in the direction of the engine room.

"Captain Abrams had me back here with Tiana most of the time."

"Good enough," Rick said, examining the Mallory doors, then stopping next to Billy Moon. "Is she ready to go?"

Billy looked up, a little startled, but his demeanor shifted when he recognized the General.

"Ready to go?" The engineer shook his head. "Not by a long shot." Billy looked around at the engine room's interior. "This poor ship has had the living daylights beat out of her. She needs to spend about a year in spacedock with a dozen bricks of Nanomech."

"I need an answer to the question, Billy," Rick maintained, straight-faced.

Billy frowned and looked back down at the displays on the glass in front of him.

"The math says yes," he finally said without looking up. "But my gut tells me she isn't as strong as the others and will come apart under combat conditions or unnatural travel."

Rick looked at Alex 7001 as the small droid hovered over the console.

"Alex?"

"I share the engineer's concerns, General. Mathematically, everything is in place and functioning as it should. But this ship is in a fragile state. The corruption present in its computer operating systems has so many patches and work-arounds causing most of its systems to be sluggish if not inoperative, at least in a combat environment."

"Understood," Rick said, looking around again. "Thank you. I'm headed to the bridge. Have the crew meet me there." He turned and left, heading back up the main corridor with BachTL following behind. Entering the bridge, he found the crew already at their stations.

"General on the bridge," Lana called out, noticing them come in. Everyone came to attention as Rick sat down in the command chair.

"Knock it off people," Rick grumbled. "I'm not any different than Gun... the Colonel." Rick looked around as the crew relaxed a little. Alex 7001 and Billy quietly entered the bridge and took their positions at the engineering station. "I'm going to ask for a report from each of you individually. Please have a seat and be ready."

"Sir?" Mister Pippin spoke up. "About the Colonel?"

"Yes..." Rick looked around at everyone, considering. "Ask your questions."

"We heard you found him and that he's down in the medical bay. How bad is he?"

Rick read the anxiousness etched in their faces. He had hoped time would give him a clearer picture as to what he could tell Gunnar's crew. He understood how it all worked. You live in close quarters, breathe the same air, get to know people, laugh with them, cry with them and even watch some of them die. You develop a bond that is typical of a close knit family. All these people were family; even to him. They had a right to know exactly what happened to a member of their family. Rick took a deep breath and let it out slowly.

"You saw the transmission the Queen of Hadrian gave a short time ago?"

"Sir, the Queen looked a lot like..."

"Janox..." Rick answered quickly. "Her real name is Jana Tilee Oxlind. A Princess of Hadrian. She had been marooned inside the planet Reako by the Colonian Duchess Benetar and the Albion King Commander Thoene Dismon in hopes of hiding her from the factions of this galaxy. They were trying to usurp power and take over the throne of Hadrian. Gunnar and Alex 7001 helped to liberate her and find the proof she needed to claim her right as the royal heir to the throne. While trying to help collect her proof, Gunnar was taken captive by the Albions. You're aware that he escaped and helped free the *Constellation* from the Albions. He kept Blinda Koss occupied so you could make a run for it. They got into a firefight and both were mortally wounded. Doctors Mantose and Yamoto have tried to repair his injuries, but it appears that it's too late. There was too much damage. His body is in sickbay, in a stasis pod."

The bridge remained silent, only the soft humming of operating circuitry echoed around the consoles and displays. Somber looks dominated the crew, their gazes pointed at nothing in particular.

"What happens now, Sir?" Lana asked.

"Arrangements will have to be made for his burial as soon as I can manage it."

"How is Tiana doing?" Billy asked.

"She's recovering in the *Athena's* medical unit. MedTech Gantrie is looking after her."

"What about the ship, Sir?" Pippin asked.

"The good news is, Billy and Alex 7001 have you fully operational again, although there shouldn't be any plans to run out trying to blast your way through anymore Albion or Colonian blockades. She's still going to need a lot of work."

"Who's going to take over command?" Doran asked from the weapons console. "Is Captain Abrams going to take the Colonel's place permanently?"

"For now, that's the plan," Rick answered quickly.

"Where did those other Starbirds come from, General?" Pippin asked.

"That's still a bit of a mystery. For now, all I know is they were being transported in those Kendalon freighters in the other hangar. When we get back home, we should be in a better position to figure all that out. For now, they're combat ready."

"Whose manning them?" Pippin asked.

"Just the Kendalons for now. I'm hoping once things cool down a little more outside, I can get the new recruits from Aster command over here. These four ships were intended to be a bluff in case we had to launch to protect *Calypso*."

"General," Billy said, leaning against his console and folding his arms. "You need to tell them."

Rick gave the engineer and Alex 7001 a long look, then brought his wrist communicator up.

"Commander of the *Intrepid*. Can you teleport directly over to the bridge of the *Constellation*?"

"Right away, Sir."

There was something familiar in the voice, but no one made any connection. Rick came to his feet and started for the door.

"General, what about the individual status briefings?" Pippin asked.

Hearing the familiar crackle of a teleporter beam forming, Rick turned sideways and smiled.

"You can give your reports to the Commander of the *Intrepid*."

From the corner of his eye, he glimpsed a short form materialize next to the command chair as he exited the bridge. Before the doors could close, there came a collective gasp from the crew.

Aftermath

Calypso remained in position near Calliope after the Rite of Pintar had concluded. The *Trax* and the *Realistic* trolled slowly in and out of the vicinity, keeping watch on the mammoth alien ship. The other faction ships began to disperse a day or so after her Majesty, Jana Oxlind ordered an end to all hostilities. Normal faction traffic resumed shortly afterwards, most of it commercial. It had taken some time to remove the *Tarzana* and *Indominable* to separate facilities for lengthy repairs. The Starbirds remained inside *Calypso*, with the *Athena* being on active alert. With hostilities quelled, the crew personnel of all the Starbirds were able to mingle and relax in regular shifts.

Several weeks after the Rite of Pintar, Fuji Yamato sat at her desk on the *Constellation* studying several medical reports on Gunnar's condition. A grimace twisted across her expression as there seemed to be nothing to give her hope for his recovery. The stasis pod was keeping his body functioning, but there were no indications of a bio-signature or any other signs that he was alive. Audra had been coming into sickbay several times a day, usually just leaning against his pod. During those times, Fuji tried to quietly excuse herself from the room or sit back and occupy her mind with other things. She still didn't quite feel up to par yet. The miracle of Caidin taking her down to Gunnar's molecular level to repair the extensive damage to his structures had taken a heavy toll on her and she was only now feeling somewhat recuperated.

Discouraged by the data in front of her, Fuji turned to a different screen and studied it for a moment. Tiana Mantose had been placed in a Med-assisted coma directly after her surgery and had only been removed from it a short time ago. She looked at the display above Tiana's bed, feeling pretty good about the readouts. As she studied the information further, she noticed movement at the foot of the bed. Slowly coming to her feet, she watched the blankets shifting. Picking up a small hand scanner, she stepped over to the bedside and held it over Tiana's chest. After looking at the tiny readout on the end of the device, she pushed it into her pocket and glanced at the overhead display. Looking down, Fuji smiled as Tiana opened her eyes.

"Welcome back," Fuji whispered, pulling up a chair.

Tiana developed a confused look, slowly scanning the room, then looking back at the doctor.

"You've been unconscious for a couple of weeks," Fuji said quietly.

"Couple of weeks..." Tiana finally repeated hoarsely.

"Do you remember anything?"

"Remember anything? I don't even know where I am."

"*Constellation's* sickbay."

"Oh... yeah... Ok..." Tiana blinked several times, looking around. "Let's see... what do I remember... We were giving the Albions what for... then Engineer Brandon activated the emergency release control values on the power couplings. I tried to stop him when he went for the escape hatch, but he pulled a blaster on me. We fought over it for a moment, but it went off. After that, I don't remember too much. There was some arguing about getting me out of engineering, but the couplings valves had to be shut off... After that, everything is a blur."

"You basically saved the ship," Fuji said squeezing her hand.

"How did I get out of there?"

"Mister Pippin teleported you out. Captain Abrams brought you in here and we performed emergency surgery on you. It was touch and go for a while, but I think you're going to pull through.

"So what happens next?"

Fuji gave Gunnar's pod a glance, then forced a smile and looked back at Tiana.

"Since you're doing so well, I'm going to turn your care over to another physician as I have other matters to attend to."

"Just like that?"

"Just like that," Fuji repeated.

"So is Alder going to be the one taking care of me now?"

"No, I'm going to transfer you over to the *Athena*. The medical team there will take good care of you. In fact, the chief medical officer should be here any moment to make his assessment."

"Rather have you be my doctor," Tiana grumbled.

"That's about the sweetest thing a patient has ever said to me," Fuji snickered. "Oh, don't worry. I know this guy really well. He's pretty much amazing."

"If you say so." Tiana shifted her attention to the pod on the bed next to her. "What's going on with the pod?"

"Just checking its systems out," Fuji replied sounding elusive as the hall door snapped open. "Ah, here we are. Tiana, I'd like you to meet your new attending physician."

"He doesn't look very new to me," Tiana responded giving the older gentleman a glance as he shuffled into the room. Turning away as he sat down next to her, she watched the panel on Gunnar's pod flash information across its display.

"Tiana?" a gentle voice beckoned quietly. "I'm Doctor Mantose. I'll be taking over your care until you're well enough to return to active duty."

"Technically, I'm not part of the crew. I'm an Independent Contractor."

"Do you feel well enough to be moved over to the *Athena*?"

"Maybe in a while. I'm still really tired." She lay quiet for a moment, waiting for some form of encouragement from the new physician. Something wasn't right... She couldn't hear any movement behind her. Fuji should have stepped in with some kind of reassurance. Tiana thought a moment, trying to figure out what was wrong. She finally shifted, slowly turning back over. Fuji was standing a few steps behind the smiling bearded gentleman.

"What?" Tiana asked bewildered. She stared at the older gentleman for a moment. "Wait... Who did you say you were?"

"Doctor Mantose," came the calm reply.

"Doctor Caidin Mantose," Fuji reiterated from behind.

Tiana gazed at Caidin for a long moment.

"You're my... my father?"

"Yes, I am." Caidin smiled, waiting for recognition that couldn't come. "You and Taron were still quite young when I disappeared, so you wouldn't remember me, except by the Chyropaz necklaces I gave you and your brother; maybe what your step-mother might have told you."

Tiana shifted her gaze to Fuji who nodded and produced a small hand display. Tiana sat up and studied it for a moment, giving Caidin a glance, then looking back at Fuji.

"It's your DNA samples. He really is your father."

Tiana gazed at the information on the display again, finally setting it down. She stared off into space trying to fathom the gravity of the revelation. A myriad of questions suddenly swirled through her thoughts so fast she could scarcely process them. Tiana looked back at Caidin trying to decide what she wanted to ask. She had rehearsed what she wanted to say to him many years ago when things started to go badly with her step-mother. A wave of emotion abruptly engulfed her as the fulfillment of her life long hope was realized. Her eyes pooled as she tried to speak. What finally came out wasn't a question at all.

"Daddy," she burst crying.

Caidin came to his feet, wrapping his arms around his sobbing daughter, tears coming to his eyes as well. Fuji quietly backed up and stepped out of sickbay, giving the father and daughter some privacy. She stood just outside for a minute, smiling. Somehow, she didn't think Tiana would be objecting to being transferred now.

Once Caidin had gotten Tiana settled in the *Athena's* sickbay, there were many long discussions covering all the questions they had for one another. Certainly her twin brother's death was an emotional topic to cover, as well as their birth mother and the woman who now bore her name, Dãsha. Then there was the account of life with their step-mother, Nora and what was going to happen now.

Under the careful eye of her father, Tiana's recovery was ongoing. As soon as she was up to it, Tiana made good on her threat to belt Billy a good one. But the blow was followed by a relieved embrace and finally, an impassioned kiss.

Work on the *Constellation* continued well over six months, most of it was being handled by Alex 7001 and an army of Nanomech borrowed from the *Athena* to aid the contingent already deployed. Even with the wonders of Nanomech, it would be sometime before the *Constellation* would be considered back to normal operational status. For now, Rick insisted on a rigorous training schedule for Kalamarion and Kendalon crews. While Rick made sure he took part in many of the training exercises, he spent a lot of his time in his quarters looking over instruction and repair reports. Of most interest to him were the star charts Dãsha had provided. There were days when Jayda would come in to find him asleep at his desk with the charts up. On other days, she would find him hyper focused on them. Today was a blend of both.

"Inspections are due," Jayda said touching him on the shoulder.

Startled, Rick rubbed his eyes as he came back to reality. Turning his chair around, he found Jayda standing next to Dãsha.

"What, no miniskirt?" he asked looking at Dãsha's uniform.

"Such attire is neither regulation nor appropriate for visiting dignitaries." The tall brunette raised her hands to her hips.

"While I appreciate your attitude toward what's appropriate, I feel the need to remind you that you're not in the military and therefore not subject to those dress codes." Rick noticed Jayda take a step back, grimacing and waving him off.

"Jayda and Audra have informed me so, several times," Dãsha responded taking a step closer. "Nevertheless, as I continue to try to integrate into normal human interaction and behaviors, it is important that I follow expected conventions for any of the different situations I may be confronted with. I realize that my figure can be a distraction and do not wish to draw attention to myself when I am not to be the point of focus. As I understand we will be visited by several dignitaries shortly, I wish to make a..." Dãsha looked back at Jayda. "How did you put it? Good impression."

"You look great," Rick smiled. "Just don't go overboard."

Dãsha developed a strange look and cocked her head.

"I do not understand the reference, 'overboard'."

"It means, too much," Jayda offered from behind.

"Ah, yes. I see." She cocked her head in the other direction. "But can't that go both ways?"

"Yes it can," Rick chuckled, getting up and heading for the door. He glanced at the time. "Is the center landing bay ready?"

"Yes," Dãsha answered before Jayda could speak. "The Kendalons are masters of logistics. The center bay is smaller than this one, but they made cleanup look routine."

As they headed down the hall, Rick noticed a couple of the new recruits for the *Athena*.

"How are the new people dealing with the real thing?" he asked, pausing at the engineering doors and looking in at Frank Cooper. He had several engineers gathered around the console giving them instructions.

"Better than I thought they would," Jayda answered. "They have a pretty big learning curve to overcome."

As they exited the ship and started for one of the elevators that would take them up to the causeway levels, Rick eyed the *Constellation* carefully.

"Yes," Jayda answered the unasked question. "She's with him."

"Richard," Dãsha said plainly. "It is not good for Audra to continue to mourn for her husband. He is lost. You should proceed with his burial so she can begin the healing process."

"That's one of the reasons for the dignitary visit," Jayda said. "The Queen is coming to pay her last respects before the burial."

"I am sorry for my plain words, Richard," Dãsha offered. "I had no knowledge that you had already made the arrangements."

"It's ok," Rick said, staring across the metal grated causeway as they exited the elevator. "This has been difficult for Audra. Ultimately, it has to be her decision. It was hard enough two years ago when it was her we were shooting out into space."

"Would it not be easier to bury him on his home planet?" Dãsha asked. "Then you would know where his body is and its condition. You could go and visit his burial site as often as you'd like."

"It's a comforting notion, but in our current circumstance, an uncertain one. While we know the way home, we have no practical means of getting there."

"*Calypso* is still a very capable ship. I have no doubt that it could make the journey."

"Agreed," Rick responded. "But what's the fastest she can go above lightspeed? Five, maybe six times faster? Provided you could maintain those speeds indefinitely, it would still take three quarters of a millennia to bridge that kind of space. We might survive in stasis pods, but what would we be going back to?" Rick looked down into the center landing bay as they crossed between. The hangar doors were just opening up. "No, better to let him go now, here, where all his friends can say goodbye to him too."

"That is not something I had considered," Dãsha admitted. "Of course it makes sense." As they exited the elevator to the main floor, she leaned closer to Jayda. "Will the ceremony require me to wear this formal attire, or can I be more casual?"

"Dignitaries will be in attendance," Jayda whispered back.

"Formal," Dãsha frowned momentarily. Her attention was drawn to a number of ships entering the landing bay. Several escorts entered first followed moments later by a larger shuttle craft. Several more escorts entered and set down behind the rest as the bay doors began to close. A few larger vessels passed by the closing doors with the bow of the *Trax* as a backdrop. Without any fanfare, the occupants of the shuttle began to disembark as Rick, Jayda and Dãsha approached.

Administrator Vandmire was first to spot the three coming toward them and smiled broadly.

"It's been a while, General. How are things progressing?"

"We're getting things organized," Rick responded, shaking Diord's hand.

CJ Barker took Diord's arm and smiled at Jayda and Dãsha.

"It's good to see everyone again," she commented.

"Well, I'm sorry we haven't been able to keep in better contact," Rick said. "We got a little busy over here trying to get things put back together. All this time passed before I even realized it."

"We were a bit preoccupied as well," a voice spoke up from behind them.

CJ and Diord took a step to one side, allowing the Queen of Hadrian to greet everyone.

"Hello, General," Jana said, smiling. "It's so good to see you again. It does seem like an eternity since we were last together."

"As I recall," Rick said, stepping forward and bowing. "You had just convinced the factions of your royal lineage."

"Thanks to you and your people."

"Good friends," Rick said quietly.

"Yes, good friends," Jana echoed.

"Why didn't you come in one of your larger ships?" Jayda asked looking at the shuttle craft as more people exited. "It looks a little crowded in there."

"We have several other landing bays much larger than this one," Dãsha pointed out.

"With a ship this size, I have no doubt," Jana said looking around. "But my head of security insisted on these arrangements."

"It's easier to control," Blinda said, stepping up behind the Queen. "This little event is a security nightmare. If *Calypso* had some sort of armament... anything."

"You could have had everyone teleported over," Rick teased.

Blinda developed an embarrassed look.

"Hadn't thought of that."

"Blinda, Rick," Jana interrupted. "It's ok, I feel perfectly safe with the security Blinda has brought with us and with the people Rick has aboard *Calypso*. There are several warships sitting right outside with a couple of groups of fighters flying around out there. I think we're good. Now, while I've had a full day so far, I'm still young and happy to take a tour of whatever you'd like to show me."

Rick smiled and bowed again, gesturing toward one of the Kendalon freighters. After a brief walk around, they crossed over into the adjacent bay and made a lengthy tour of the *Athena*. As everyone exited the bridge, they were met at the door of the conference room by Commander Atlanta, Captains Abrams, Fowler and Quayle.

"I apologize for my tardiness," Audra apologized. "I completely lost track of the time..."

"It's all right," Rick said, putting a hand up. "You're right on time." He gestured into the room. "Why don't you make sure everyone gets seated correctly."

"Do they not know how to seat themselves?" Dãsha asked, perplexed.

Audra snickered, stepping quickly through the door and pulling chairs back to assist individuals to their seats. As Jana took her seat near the front of the room, Blinda Koss quickly melted back against one of the walls where she could see everything. Audra gave her a long look, then turned to her own chair at the far end of the conference table. As the room settled, Caidin and Fuji came in and stood next to the Albion Thane. Audra tried to get Caidin to have a seat next to Dãsha and Jayda, but he refused.

"Well, it seems like the last time I was in this room, I got the surprise of my life."

"What surprise would that be?" Diord asked.

Rick considered for a moment, a smile drifting across his lips remembering the moment Audra had returned. He looked at the other end of the table, then at the others, stopping at Blinda.

"Surprises come in different forms."

"Yes they do," Jana agreed. "I would like to make sure everyone in the room understands how grateful the people of Hadrian are that your two ships came to us. I am sorry for everything you've been through and as Administrator Vandmire as already pledged, we stand ready to help you in any way we can."

"At this point, I'm not sure what more you can do," Rick said. "*Calypso* has provided us with the location of our home world, but as was discussed earlier, there isn't a good way for us to reach it."

"How did you get here in the first place?" Blinda asked with folded arms.

"Without a drawn out explanation that half of you wouldn't understand anyway, we came by way of an accelerated wormhole. I've tried to duplicate the equations that brought us here, but Hadrian doesn't have the essential quantum dynamics required to recreate those unique conditions."

"What do we lack?" CJ asked.

"For one, a class four blackhole."

"Hmmm, yes, that would present a problem."

"Closest thing we have is the Lyon," Diord said. "But it's a small black hole and a very long way from here."

"Wormhole travel is a very risky business anyway," Captain Abrams spoke up.

"Very risky," Audra echoed.

"The only thing that's even remotely viable is high end light speed travel," Rick announced.

"If I understand that option, we would be traveling for a long time," Captain Fowler observed.

"Unaided, it would take over seven centuries to reach home," Rick stated. "We could all go into stasis pods, but what would we be going back to? Anyone or anything we ever knew would be long gone and we would be considered a relic from a distant past that anyone from a world we used to call home would never be able to recognize. Additionally, we would need to consider the ship itself. We know that *Calypso* is of unknown origin and its actual age a mystery. I've been assured that it could make such a journey and I have no reason to doubt that information," Rick said looking at Dãsha. "But what if *Calypso* encountered trouble, either in its flightpath or a mechanical problem of some kind?"

"Couldn't you adapt the pod systems to wake the crew in the event the ship encounters something?" Diord asked. "They could make the adjustments or repairs and then go back into stasis."

"For that matter, couldn't we just tie *Calypso's* systems into the Starbirds and have Alex 7001 and CORA 500 do all the monitoring?" Captain Abrams inquired. "They could tie into Dãsha's pod, wake her up and she can assess the need to bring the rest of us out of stasis."

"Why Dãsha?" Jayda asked.

"She's the only one among us who can function both systems seamlessly."

"I would not categorize my familiarity with either ship as seamless," the tall beauty admitted. "There is much about *Calypso* that I don't understand or am conscious of. Both ships have complex systems that are nothing alike."

"But you wouldn't be subject to the rigors of the stasis process like we would," Dakota pointed out.

"It's true that my physical properties enable me to withstand many environments a normal human cannot, but I worry that you are placing too much on my technical abilities."

"Dãsha's right," Rick interjected. "And I would be uncomfortable asking her to babysit all of us in this manner. What if something went wrong during a wake up operation? There are enough pods for everyone, but if one should fail for any reason, there aren't any replacements or backups." Rick shook his head.

"If this is our best option," Audra spoke up. "Then I don't see it as an option at all. If it's going to take that much time to get home, then there is no use in putting a ship and our lives through such a journey for something that will be long gone by the time we get there. We've made good friends here and with a little work, we can help Hadrian heal from something that I understand we had a hand in messing up."

"While what you say may be true," Jana spoke quietly. "I think you've left out a very important detail. You're also the ones who were instrumental in bringing us back together. Yes, there is still a lot of work to be done, but the more help we can get, I think the better off we'll be. Of course you are always welcome and we will do all we can to find you a place you can call your own and forge even better friendships with the people you already know and those you will undoubtedly meet."

"Thank you, your Highness." Rick smiled, bowing slightly. "My command staff will meet later and come to a decision on what we're going to do."

"Not to pry into your internal affairs," Captain Abrams spoke up again. "But what happened with the *Tarzana* and the *Indominable*?"

"You mean where are they?" Diord asked.

"I mean, what happened? The *Tarzana* turned on its own ship. Clearly it was to protect *Calypso.* But I thought the Albion objectives were to flush out our ships. Don't get me wrong, we're grateful, it just didn't make any sense. They could have opened up *Calypso* with those torpedoes."

"You are correct," CJ began. "KC Dismon ordered them to flush out your Starbirds at all costs. Admiral Hilton, of the *Indominable,* ordered Commander SoKnack, of the *Tarzana,* to provide cover while the *Indominable* fired a salvo of torpedoes into *Calypso.* Commander SoKnack brought the *Tarzana* in close, but at the last second, made a deliberate course change and put his ship directly in the path of the salvo taking all five hits. He then rammed the *Indominable* in her targeting section to keep her from firing any more torpedoes. Commander SoKnack was trying to do the right thing."

"Where is Commander SoKnack?" Rick asked. "We'd love to give him our thanks."

"He deserves a promotion," Dakota added.

CJ developed a somber look.

"Dalton was mortally injured during the maneuver."

Everyone fell silent, the warriors in the room understanding the highest honor one could give in the line of duty.

"What of the King Commander and the Duchess?" Jayda asked.

"The Duchess is cooperating with authorities," Blinda announced. "Both of them face a long list of formal charges that, at the very least, will put them away for much longer than they will live."

"Most of the Colonian government remains intact," Jana said. "But is being carefully vetted to eliminate any remnants of the Duchess's command structure. I've ordered the governmental structure to be modified to provide a more equitable system of checks and balances. A system that will rely less on a monarchy and more of a federalized republic. The design is to prevent such a catastrophic mess from developing again. The Albion government is a little more complicated as the King Commander had many allies in the command structure. For now, most have pledged cooperation with their internal investigations and reorganizational modifications."

"It's a sure bet the Ratronians won't be any easier," CJ said.

"Were the Castellians able to make repairs to Tonnie?" Rick asked, turning to Jana.

"Sadly, no. Her interior is so complex and so far beyond anything we have any knowledge of, there is no way to bring her back online. Even with Alex 7001's help, we couldn't get her main power system to respond. For now, I've requested that she be transferred to a secure facility on Tintee for storage." Jana paused, thinking. "Perhaps someday..."

"I guess that brings us to the main reason we're all here," Rick said quietly. He looked to the other end of the table. Audra fingered her ring despondently.

"Are you sure there is nothing more that can be done?" CJ asked abruptly. "I feel like Gunnar did more to facilitate Hadrian's return to the road of sanity than the rest of us combined. Surely you, Blinda and Caidin can combine your powers to do something more?"

Rick's gaze remained on Audra, whose blank stare continued to bore a hole in the table top.

"As we have explained earlier," Caidin added. "It doesn't work that way."

"If he hadn't been lured into Calliope's tunnels and ambushed," Dakota simmered, looking at Blinda. "He might still be here."

"Captain Abrams," Rick flashed.

"No, General," Blinda held a hand up looking at Dakota. "It's all right."

"No it's not all right," Rick maintained glaring at Dakota. "Captain, another outburst like that and I'll excuse you from this meeting."

"Permission to be excused now," Dakota flared, coming to his feet.

"Sit down, Captain," Blinda instructed calmly. "Please..."

Several tense moments passed, even after Dakota had returned to his chair.

"The Captain is correct. While I was acting under the orders of the Queen Captain, I could have let him go after we crashed. Instead, I lured him to my location. I was blinded by his stubbornness and what I had become."

"You didn't kill him," Rick interjected. "If anyone is to blame, it rests squarely on me." Rick paused, thinking. "I was his protector and failed in my charge..."

"Please stop..." a quiet voice interjected from the far end of the table. All eyes turned to Audra. She looked around at everyone, her big brown eyes pooling. "It doesn't matter, none of it does." She looked at Rick, then Blinda. "It's no one's fault. Gunnar did what he did because it's who he was and it wouldn't have mattered what you did or didn't do, the result would have been the same. Dakota, don't blame another warrior for doing what they're supposed to do. Most of us in this room know all too well that there are many things we have done and have to do in the line of duty. Gunnar was no different. He understood the costs of being a warrior, as we all do. Because of that, I've decided it would be best to move forward with the burial here, at Calliope with one condition. Can we make sure he doesn't get shot in the direction of Acuity?"

A soft chuckle rolled around the room, breaking the tension in the air.

"I think that can be arranged," Rick agreed. "Are you sure this is what you want?"

"You know Gunnar," Audra said sniffling. "He was never one to wait around for anything. This way, we can all be here to say goodbye."

"Ok," Rick nodded after a brief pause. "Is an hour enough time?"

"Yes," Audra's voice cracked.

"I'll have the crew get everything ready," Dakota volunteered, rising with Audra. The room remained silent as Dakota left with her. Rick gave Jayda a quick head nod and she and Dãsha followed.

It took a little longer for Audra to find something appropriate to drape over the stasis pod. At first she thought about just using his shoulder sash that held his awards, but when she reached for it, her attention was drawn to the garment next to it. She carefully pulled his service jacket from the closet. She loved this on him. She recalled the first time he wore it after being promoted to the rank of Colonel. He looked so dashing and even though he didn't care so much for rankings and more about being in a position to get the job done, he loved the way he felt when he was wearing it. She almost wished they had something like this for Lieutenant Commanders. She slowly sank to the bed and for several minutes held it close. Fighting her emotions, she finally took a deep breath and left her quarters. By the time she reached the *Constellation's* sickbay, the command staff had assembled in the hall.

"Can you give me a couple of minutes, Captain?" Audra requested holding the jacket close.

"We'll do you one better, Commander," Dakota responded quietly. He turned to the rest. "Everyone will form an honor guard just outside the boarding door." Dakota turned back to Audra. "You take as much time as you need."

"Thank you, Captain," Audra smiled politely. She waited patiently for everyone to leave, then activated the door and stepped inside. The room was dark, except for a light directly over the stasis pod. Setting Gunnar's jacket on the table next to it, she checked the control panel, then leaned against the wall. Closing her eyes, she listened to the sound of the ship and the silence of the medical bay. A sudden flow of rage griped her as she glanced back at the pod. *Why couldn't you have been more careful? Why did you have to stick your neck out and get yourself killed?* She clinched her fists and throttled up a scream, rushing the pod and pounding on it. Audra's fits of anger quickly gave way to a mournful lamentation.

"Please don't leave me like this," she whispered, fighting to regain her composure. She wiped her eyes and took a deep breath, looking across the surface of the pod. Being a wonder of technological engineering, the stasis pod could keep a person alive for as long as it had power. It could also preserve the body indefinitely, making a perfect coffin. Gently touching the glass, she peered in at the form of her husband, wishing she could touch him one last time.

Fighting a constant tsunami of emotion sweeping through her, Audra stared at him. For several minutes she wrestled internally with what to do. *If I open this pod, he's dead for sure. But this will be my only opportunity to touch him one last time. At this point, it's not going to matter.* She finally dropped her hand to the control panel, slowly shifting her gaze to the display. She watched part of it race with information while the rest remained stationary. Audra looked back in at Gunnar, fingering one of the dials. After slowly turning it, she pressed a short sequence of buttons. The panel began to flash and a split second later the lid to the pod hissed open.

Apprehensive, she held her hand over his chest. Tears once again began to flow. She hoped that he would suddenly open his eyes and grab her. Leaning close, she took hold of his hand. It felt pliable, but lifeless. She kissed it, then brought it to her cheek. In all the time she had known of his demise, this was the one thing she longed for; to touch him one last time. Holding his hand, she laid her head next to his, nuzzling as close as she could.

"I'm here," she whispered, her eyes again engulfed in tears. "Please come back..." She wept softly for several minutes, holding onto him, not wanting to let go. Her communicator went off, but she ignored it, swallowed in grief. As she lay next to him, she became aware of an odd sensation. A rhythm; slow and steady. Puzzled, she raised her head. Trying to clear her eyes, she blinked and came upright. She could hear air quietly moving. *His left eyelid...* Stunned, she took his hand again, feeling a grip growing. Audra carefully popped the limb restraints designed to keep the body in place should the pod experience external forces. She suddenly jumped back, stumbling against the adjacent bed as Gunnar turned his head and looked at her.

"You wouldn't believe the dream I was having," he groaned, astonished. "The crappiest thing ever." Stiffly sitting up, he looked around the room.

"I think... I'm having the same dream," Audra stuttered. "Gunnar," she sniffled in shock. "You're... alive!"

"Pretty sure that's what this is..." He tried to stretch, rubbing the back of his head and neck. "Every muscle in my body is in knots..." Gunnar abruptly held a hand up in front of Audra, looking at her. "Hold it!" His mind became a whirl of flooding thoughts. *Rick in the cave, him and Blinda shooting each other above the hole. The chase in the assault craft, leaving the observation level, Doctor San Sann. Blinda with the mind probe, dinner with CJ and Drax, speeding across the dry lakebed, tossing out the Hallavertor chip...*

"Where am I?"

"On... the *Constellation*," Audra answered bewildered. "Sickbay." Gunnar stared at her as she wiped her eyes.

"You're not real." His expression chiseled sharply. "You're a program. But... the chip... How are you functioning?"

"No... Gunnar," Audra stammered. "I'm not a program. I'm real."

"Audra is dead." Gunnar pushed the cradle out of the way and swung his legs painfully over the side of the bed. "Computer, end program."

"Gunnar, there's no Hallavertor in sickbay!"

"Clearly there is!"

"Gunnar," Audra pleaded.

"Computer! End program!" Frustrated, he staggered from the bed. "Where's Alex? He'll get this idiot thing shut off!"

Bursting through the door, he nearly ran into Dakota and Willis with Audra fast after him. Shocked, they pressed themselves against the wall, watching the Colonel stumble to the open boarding door and down into the hangar bay. Still somewhat stunned, Dakota and Willis looked at each other, then took off after Audra.

Not realizing what was happening, everyone assembled at the bottom of the boarding ramp made way for the chase. Once in the open, Gunnar abruptly stopped and looked up. *What's going on?* Two Starbirds sat in front of him. He looked to either side. The *Constellation* was on his left, the *Athena* on his right. He turned around. Looking past Audra and his astonished crew, he counted two more Starbirds.

"Alex!" he yelled looking around, counting the ships again. "Alex!"

"Gunnar," Audra called. "It's ok. Just calm down. I can explain."

"You can't explain anything. You're nothing more than a computer program that shouldn't even be active. Alex!"

"Colonel Conrad," Alex called, coming to a stop directly in front of him. "You're alive."

"Of course I'm alive, you tin-plated eggplant," Gunnar growled. "Why does that seem to be in doubt? If I weren't alive, I'd be dead."

"Precisely," the droid responded, pivoting his head to Audra, then spinning it completely around as more of the crew began to assemble.

"I need a little clarification here."

"What would you like to know?"

"I just came from sickbay. Is there something wrong with me, because I'm seeing multiple ships right now?" Gunnar asked, looking around again.

"You are correct. There are six ships in the hangar bay."

"Why are there six Starbirds in the hangar bay?"

"Because there are not seven?" Alex responded.

"Funny droid," Gunnar fumed. "Why is that program active?" He pointed behind him at Audra.

"Colonel, clearly you're a little disoriented..."

"A little?"

"If you will take a couple of deep breaths, I can explain everything. You are currently standing in one of *Calypso's* hangar bays stationed over Calliope."

"Ok," Gunnar said, trying to calm himself. "Calliope I get. Don't know what *Calypso* is... Clearly this is a ship of some kind."

"The Starbirds you see came from a couple of Kendalon freighters General Niker rescued while retrieving Commander Jayda from the Spartus quadrant."

Gunnar looked to his right. The *Athena* was sitting right there. He looked around, then stepped over to the ship and touched its hull. *Feels real enough.*

"How do I know this isn't all just an elaborate Hallavertor program? You've got that stupid thing running," he said pointing back at Audra.

Alex remained silent as more crew assembled around the focal point.

"Because the Hallavertor can't function outside the ship," Audra said quietly from behind. "And even if it could, there are some things even it can't duplicate."

Gunnar whirled around and took a tentative step back. He looked into Audra's brown eyes, instantly feeling something. Audra stepped closer, keeping her eyes on his. She carefully reached out and took one of his hands.

"A program can't do this," she whispered, drawing close and taking his other hand.

Gunnar felt it. The program was able to duplicate touch, even warmth of another being, but it couldn't duplicate this. A warm surge of something that only one person could produce. He slipped his hands around to the small of her back and pulled her close. Audra kept her eyes locked on his, sliding her arms up around his shoulders.

"How?" he whispered, but could say nothing more as Audra pressed her lips to his. The surge of energy blossomed. All his doubts suddenly vanished. *Audra!* For several moments, they remained locked together in an energizing embrace, until they became aware of the thunderous applause surrounding them.

*   *   *   *

Rick and Jayda stood with the Hadrian delegation in front of the *Montego* inspecting the Pin Missile launchers and torpedo tubes.

"That sounds like applause," Dãsha announced, moving between the *Montego* and the *Intrepid*. As the Starbirds were parked in pairs perpendicular to the others, it was impossible to see what was going on.

"Captain Abrams," Rick spoke into his wrist com.

"General, you need to come back over to the *Constellation*."

"What's going on?"

"You better come see this for yourself."

Rick trotted off, leaving the others to follow at a slower pace. Jana looked at Jayda and Fuji.

"I'm sure it's just a crew game," Jayda said. "We've been cooped up for a long time. They have inventive ways of blowing off steam."

"Yes, I've seen how they blow off steam," Jana smiled as they made their way back to the *Constellation* and *Athena*.

"What's the excitement all about?" Rick puffed as he approached the crowd.

Captain Dayton turned, her eyes wet.

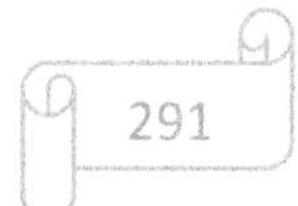

"I'm not much of a romantic, but…"

"But what?" he asked, pushing past Toby Mavis and Laura Habba. "This is supposed to be a…" He stopped short, recognizing the man standing in front of him, holding onto Audra. Rick shook his head slightly, not believing what he was seeing. He looked at Audra, her smile beaming under wet cheeks. She sputtered briefly, trying to hold back her emotions.

"He came back to me," she laughed and cried at the same time.

"Gunnar?" Rick still didn't believe it. He took a couple of steps toward him, but hesitated. The last time he was this close, Gunnar nearly killed him.

"Once again, you've saved my sorry butt," Gunnar said with a grin as he stepped forward and threw his arms around his friend.

Rick instantly let go of his doubts. He held his friend as though he would never let him go. After a few moments, he realized he was crying, then heard a familiar gasp from behind. He pulled away and turned, seeing Jayda standing a few feet away with her hands to her mouth, her eyes quickly pooling. Gunnar chuckled, for once loving all the attention. He let go of Rick and threw his arms around Jayda, spinning her around.

By now, the rest of the Hadrian delegation had reached the crowd. Not relying on royal decorum, Jana burst toward Gunnar as soon as she saw him.

"You're alive!" she beamed jumping into his arms. "I knew it! Somehow, I just knew it." She took a step back. "And look at you. Your Sunggo, it's bright and the color is right."

"What about Tonnie?" Gunnar asked, looking around for the nanny. "She should be able to see it too."

Jana's expression suddenly saddened.

"Gunnar, Tonnie no longer functions."

Gunnar wrapped consoling arms around the young Queen.

"I'm so sorry."

"Like you, she died protecting me."

"I hope you've found someone else to keep track of you," Gunnar smiled. "I have other things I need to get done."

"You're still with us!" CJ rushed forward, grabbing him, hugging and kissing him, but pulled quickly away when Diord cleared his throat. Diord smiled and took Gunnar's hand, then pulled him in close for a hug. As Gunnar returned the embrace, he looked over the Administrator's shoulder at a familiar form standing back away from the crowd. Gunnar instantly pulled back, his guard up.

"What is she doing here?" he sharpened, releasing Diord.

"Now, let's all stay calm," Rick cautioned, surging forward to hold onto his friend. "She's a little different than you remember."

"Looks the same to me," he replied snidely.

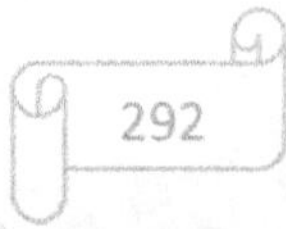

"Seems we share a membership in a fairly exclusive club, Colonel," Blinda mused calmly.

"Yeah, it would seem so," Gunnar relaxed, keeping his eye on the Thane.

"Gunnar," Rick said, standing ready. "She's the Queen's Thane, Jana's protector."

Gunnar looked at Rick for a moment, reading the sincerity in his eyes. He then looked back at Blinda. She took a couple of steps toward him, but remained just out of reach. Gunnar looked back at Rick, then at Jana.

"It's true," Jana said with anxious eyes.

Gunnar narrowed his gaze, looking back at Blinda.

"I don't expect you to give me a hug, but we are on the same side," Blinda smiled.

"Well, given our history, you'll have to excuse my general lack of enthusiasm at seeing you." Gunnar looked back at Jana again. "There's a lot of new stuff here to get used to." He forced a smile and turned back to Audra and the crew assembled.

Blinda lowered her eyes and stepped back, letting out a sigh.

"It'll take some time," Rick said, reading Blinda's body language.

"It's not that," Blinda said quietly, looking after Gunnar as he greeted the others. "I've never felt this way toward someone I've tried to... and actually killed."

"Like I said, give it some time. He'll warm up to you." Rick turned but then paused. "And to be clear, he killed you..."

Blinda grinned broadly, watching Rick turn back to catch up with his friends. She felt strangely; completely out of her element. *What's changed? Me?*

And Discuss...

"I think I might have been better off back in the pod," Gunnar complained. A small army of medical people hovered around him as he lay in the *Constellation's* medical bay. "Where did all these people come from?"

"Try to relax, Colonel," Fuji said looking at the readout on her portable scanner. "The more patient you are, the faster I can finish up."

"You mean you're about done?"

"Just relax."

"Where's Audra?"

"I'm here."

He could hear her, but there were so many people around his examination table, he couldn't see her.

"Seriously, Doc. How many people does it take? It used to be just you and Sindee..."

Fuji gave the monitoring system on the wall above Gunnar's head a glance, then looked down at her patient.

"All right, Colonel. As long as you promise not to complain when it's time to finish the rest of the exam."

"I promise."

"Everyone, I need the room. Caidin... Audra, please remain."

The quiet sounded good as the door finally closed and they were left alone.

"Better?" Fuji asked, still looking at her scanner.

"Much." Gunnar smiled and took Audra's hand as she came to his side.

"There, right there!" Fuji pointed excitedly at the readouts.

Caidin studied the monitor on the wall, then looked at Fuji's scanner.

"It's no wonder we couldn't get any kind of a reading on him," Caidin said, focused. "This pod isn't calibrated for anything even close to wavelengths like this. Is this even his pod?"

Fuji gave the stasis pod a quick look, then tapped several points on a glass display.

"No, it's General Niker's," she said.

"That explains a lot. This would be my fault," Caidin admitted. "I was in such a rush, I didn't check it."

"What's this all about, Doc?" Gunnar asked. "Not that I'm complaining, but was I dead or not?"

"Yes, you were definitely dead," Fuji answered, thinking. "But the pod Caidin put you in was the General's. It wasn't calibrated to your unique physiology and biosignature.

"But that doesn't explain why I couldn't sense him," Caidin perplexed. "I checked on him several times after the surgery but detected nothing. There was body energy, but no biosignature."

"I don't think it would have mattered," Fuji said. "Audra, let go of his hand."

As Audra let go of Gunnar's hand, Fuji and Caidin watched the readouts on the wall monitor drop off.

"I suspect without Audra's influence, he wasn't producing enough energy to detect through the incorrect stasis telemetry. Grab his hand again," Fuji instructed, studying her scanner with an excited grin.

"Whatever the case," Caidin admitted, "she's sure having a profound effect on his biosignature right now. It's a flaming torch."

"Hear that?" Audra grinned. "Profound effect."

"Yeah," Gunnar agreed. "I'm a *flaming torch.*"

"What have I done?" Caidin grinned looking back at Fuji.

"Good luck living with her now," Fuji chuckled, looking at Gunnar. "...and I might add, serves you right."

"His preliminary scans are just fine," Caidin said smiling. "I think you can release him."

"Ok, Colonel," Fuji agreed. "I'm not seeing any signs of the Kodiac Blue or any side effects from it, so I'm going to clear you for active duty. But, I want to see you every morning. Audra, I'm looking to you and your magic touch to keep him in line."

"Not to worry, Fuji, he's not leaving my side for a very long time," Audra said, helping Gunnar off the table.

"Thanks, Doc," Gunnar said whole heartedly as he and Audra walked out arm in arm.

Fuji and Caidin watched after them as they turned up the hall toward their living quarters.

"This is the part of medicine I really love," Caidin said.

"Yes," Fuji agreed. "And now for even more fun. Let's have you lay back on this bed here. I want to do a deep scan of your frontal cortex?"

"Should I be worried?" Caidin asked, apprehensively.

*   *   *   *

As the door to their quarters closed behind them, Gunnar and Audra were instantly locked in another embrace. After several moments, they came up for air, breathing heavily.

"This all seems like a dream," Gunnar whispered, holding her close.

"A dream I no longer want to be a part of," Audra said softly, pressing herself to his chest.

"The last two years have been a black hole," Gunnar said.

"Poor word choice."

Gunnar chuckled.

"You were asleep for most of it."

"I think Caidin would take issue with that characterization," Audra responded.

"Caidin Mantose," Gunnar repeated. "Can you even imagine the odds of connecting with him here?"

"I don't care about odds," Audra said, holding him tighter. "We've just beaten the odds because he and Fuji were here."

"Let's give credit where credit is due." Gunnar held her back, looking into her brown eyes. "You're the one that brought me back. How did you know to open the pod? If I wasn't dead before, it should have killed me."

"I had to touch you before we fired you out into space."

Gunnar smiled broadly.

"Only you have the magic touch."

Audra carefully pulled back, looking away from him.

"What happened to you?" she asked.

"What do you mean, what happened?"

"I've heard things about what you went through when I died. I know it had to have been unbearable, but I've also heard and seen things."

"You're talking about the Hallavertor program?"

"Not just that, but yeah, let's start with that one."

"What's this all about?"

Audra fidgeted a moment, looking away.

"Being in hyper stasis for so long did something to me. I'm not sure I'm back to normal yet. My emotions are all over the place."

Gunnar considered for a moment, then nodded.

"I think I understand how you're feeling. After my little bout with death, I'm not sure I totally feel like my old self either. I can only imagine what you're going through."

Audra smiled softly, trying to open her mind.

"Hopefully, Rick and Jayda have given you a good view of how things happened from their perspectives."

"They did, but neither of them had a complete picture."

"Fuji tried to help, but I'm afraid I wasn't very cooperative. I did try."

Audra gave Gunnar a skeptical look.

"Really, I did," Gunnar defended. "But, it seemed like the slightest things would set me off. I remember having to practically bite my tongue or lip through, or put my fist into a wall to keep from doing something stupid. Because of that, I spent a lot of time alone. It's hard to explain what was going on. Rick and Fuji finally came up with the Hallavertor program. I assume you've seen it."

"Only the beta version aboard the *Athena*," Audra said, nodding patiently.

"It's an amazing piece of tech."

"Yes, it is..."

"I kept my version on a fob. It learned everything about me and remembered all the conversations we had together. At first it really helped with the mourning process. But after time, it became less and less effective. I reached deeper, relying more and more on it to find relief from what was happening inside me. It wasn't until Fuji caught me embracing the program, that I realized things were progressing in the wrong direction. I forced myself out more, trying to use the program as little as possible. At one point, I was drawn to Tiana, but there was something missing."

"What about Lieutenant Starman?" Audra asked.

Gunnar rolled his eyes.

"I did follow regulations after all." He turned his desk chair to Audra and sat down. "Then I got mixed up with Casey."

"Colonel Barker?"

"Yeah. Initially, we were attracted to each other and there seemed to be something there, but over time, it turned to a friendship rather than something romantic. We were in a lot of situations together. Not sure how we could have gone through all we did and not come out of it anything but friends. But nothing else ever happened with her. When I finally got myself aboard the *Tarzana*, I was paired up with a Malina Cass. She was all right, but married."

"Married?"

"Yeah, I sort of killed her husband during the incident on Dither. Some nice folks on Albia helped me assume his identity in order to get aboard the *Tarzana*. Apparently Malina and her husband weren't on the best of terms, but she was willing to help me in exchange for passage to wherever I was going, as long as it was away from the *Tarzana*. We played a married couple for the time we were aboard ship."

"Played a married couple?" Audra repeated.

"Nothing happened..." Gunnar stopped short, thinking. "Well, nothing important."

"What?"

"Well, we sort of made out on her couch." Gunnar cringed as he said it, watching Audra's face.

"You made out with her?"

"We had to," Gunnar defended. "Security was checking our idents. It was the only way to get rid of them. Sort of the, *I haven't seen my spouse for a long time and now we're making up for it,* kind of thing."

Audra sank to the couch and cradled her head in her hands.

"What's the current definition of 'making out'?"

"We're going to use the one you and I know," Gunnar assured her. "Hey, nothing else happened. If it makes you feel any better, when the guards left, she said it was like she was kissing her father. Sick of hearing things like that," Gunnar grumbled, shaking his head. "Am I really that old?"

"We're not as young as you'd like to think."

"Then there was Blinda Koss..."

Audra looked up at him.

"Gee whiz, honey! Did you ever stop?"

"Hey, that was all her. She was coming on to me. I didn't return any of it."

Slowly shaking her head, she looked back at the floor. Her mind was messing with her emotions. Gunnar knelt down, taking her hands.

"I appreciate what you're going through here, but you have to understand that I thought you were dead. I'm not trying to throw a pity party, but it took me over two years to finally accept the fact that you were gone. I'm sure you have an idea of what something like that does to a person; what it did to me."

"No," Audra whispered, shaking her head. "I... I get it. It's just..." She fumbled, trying to figure out what she wanted to say. "Most of what I've pieced together about what's happened since the wormhole points to the same thing that we agreed to have you in the Starbird program for in the first place. Fighter duty was too risky and your predisposition for recklessness was going to get you killed. But the last two years or so have proved that being out of the cockpit and in a command chair hasn't changed much, only the situations you put yourself in. You're just as careless now as you were before, if not more so. Having to deal with your death has only amplified my feelings on the matter. I can't go on like this."

"So, what are you suggesting here? You want to quit?" Gunnar's tone sharpened. "You think that's going to change anything?"

"It will for me. Once we've settled wherever we're going to be, I'm resigning my commission.

Gunnar just stared at her. Her determined expression defined her resolve.

"You want me to quit too, don't you?"

Audra remained silent.

"Audra, I'm a warrior; it's what I am. I can't remember a time when I wasn't and now you want me to stop being what I am?"

"I want you to be with me."

"I have been with you. From the moment we met, I've been with you."

"No, I've had to share you. Share you with a fighter; with this ship. You've treated this like it's some kind of game. Like cheating death is fun. Do you think I'm having fun when you're having fun? You think when you're chomping at the bit for a probe assignment in Valkyrie territory that I'm thinking, *Oh boy, this'll be fun…* ?   I love you, Gunnar. More deeply than anyone has a right to… But right now, I'm not enjoying it. Every time you take off, every time you leave a launch bay, I wait for that call to come. I go to bed sick and wake up scared. I hate being sick inside all the time. I don't like feeling like the next assignment is your last one. I can't stand being silent about it."

"Does Jayda feel the same way about Rick?"

"No, Rick has always been a by the book kind of pilot. He doesn't look for the scariest way to get something done. He thinks things through before he takes action."

"He makes up things on the fly," Gunnar defended.

"Not like you do. You don't plan anything out ahead of time. You react to what's happening around you."

"But that's what makes a great fighter pilot. The ability to react to any given situation. Audra, we've had this discussion before. Why are you bringing it all back up now?"

"Because things have changed… I've changed. I just can't do this anymore."

"So, what?" Gunnar stuttered, dumbfounded. "You want me to resign my commission too?"

"I was sure once you were out of the cockpit you'd settle down and I'd feel better about how things were going, but that hasn't happened. And after all we've gone through in the last two years, it's clear it's not going to get any better. I can't do it… I'm sorry… I just can't."

"I don't get this at all," Gunnar complained. He stepped to the port window next to the desk and looked out at the other Starbirds in *Calypso's* hangar bay. "You've been in that pod for over a year and a half. You've barely given me a chance to show that I've changed."

"Just look at the last two years. I don't have enough fingers to count all the scrapes you've been in, and that's just the ones I've been able to find out about. Do you even remember the promise you made to me? About not getting back into a fighter?"

"That's not fair! I was forced to fly, and it was never in offensive action. I was either running from hordes of people hell bent on catching me or trying to rescue someone."

"Rick told me what happened between you and Blinda. That sounded like offensive action to me."

"Did he also tell you who I was trying to rescue and why I was trying to keep Blinda occupied? Not to mention that I was shot not by gunfire from a fighter, but by Blinda's pistol?"

Audra remained silent with her head in her hands while Gunnar watched the goings on in the hangar bay.

"There's always going to be a reason and likely a scary outcome to go along with it," Audra finally spoke up.

"Then what's it going to matter if I'm out there trying to do the right thing, or sitting behind a desk staring at a wall?"

"Because I'll be there, and happy inside."

Gunnar looked back at Audra. Something suddenly felt different in him. They had always loved each other and felt like they stood out from other couples in that they had a connection that so few shared. Maybe because he was unique and they complemented one another in every way. Eldon and Rayna came to mind. He remembered how he felt watching them interact and how they felt towards one another. Rayna was truly happy and Eldon was happy in her happiness. Suddenly, he could see nothing but Audra and the life they had together. There came a swelling desire in him to see to her every need and that meant her happiness.

"Ok," Gunnar agreed quietly.

Audra looked up.

"You talked me out of being a fighter pilot and sitting in a command chair, and I was good with that. Now I'll give up the command chair if that's what you want me to do. But I have a couple of conditions."

"What conditions?" Audra asked carefully.

"I can't tell you. I have to show you."

*   *   *   *

Rick sat focused on a sea of information displayed in front of him.

"Gonna burn your retinas if you stare at that too long," Jayda suggested from the couch. She was working on a personal display device, inputting some logging information.

"Gotta get this figured out," Rick mumbled.

"What's to figure out? If you send this ship home at light speed, it will take over seven hundred and thirty-six years to reach Kalamar. We can all survive in our stasis pods for that length of time and distance. If a problem arises that Alex and CORA are unable to solve

on their own, they can wake us up. Once the problem is taken care of, we simply get back in our pods and continue on."

"Yes, I'm not worried about that, although there are a myriad of scenarios that could totally screw all that up. The problem is the time and distance."

"Don't worry, you'll still be older than me no matter how long it takes us to get there."

Rick looked back at Jayda's playful smile.

"I'm serious," Rick responded, scratching his forehead.

Jayda paused, reading the look on his face.

"Look, take a poll of the crews. Do they want to go back to something that no longer exists or would they rather remain here and live their lives among friends?"

"Are you suggesting we stay?"

"Not at all," Jayda responded without looking up. "I'm suggesting that in the absence of a clear cut ideal, you leave it to them to decide. You and I have nothing to go back to whether we could do it instantly or not. Even Dãsha and Caidin have nothing to go back to. I suspect Audra and Gunnar feel the same way. But we're the exceptions. Everyone else may have something they left behind. A loved one or family member. Maybe they would like to know what became of them. Why is this such a hard subject for you? It can't be me. You know I'll go anywhere with you, as long as we're together."

"Not sure what's driving me. I guess there are a lot of unresolved issues and unanswered questions at home."

"You're talking about those Starbirds out there," Jayda said, still focused on what she was doing.

"What were they doing in those freighters and why were they bound for Genis Two near the Androlus Quasar complex?"

"Questions that no longer have any meaning..."

"Maybe so," Rick replied. "But it's nagging on me more than it should. Like it's a history changing event."

Jayda fell silent while Rick turned back to his displays. After several silent minutes of study, he pushed back and came to his feet.

"When did Gunnar say they'd be back?"

"CJ said it was a day to get to Albia and a day back. I can't imagine there aren't some things they wanted to see while they were there, so at least three days."

"Given everything that's happened here, I can't help being a little nervous about that little trip."

"Gunnar seemed pretty determined to go, and of course Audra wasn't going to let him go alone. Don't worry, CJ will take care of them. They're in good hands."

"Remember who we're talking about here." Rick stretched and looked out the port window. "Want to go for a walk?"

Jayda looked up and smiled.

"And where will this little walk take us?"

"Outside..." Rick answered. "Anywhere but here. Dãsha told me about one of *Calypso's* observation decks. Might be a good place to think."

Smiling, Jayda took her husband's hand and followed him out of their quarters.

Maturity

"This is impressive," Audra said, scanning the skyline of Zeppelin as their shuttle swooped closer. "A lot like Kalamar."

"I've always thought it was one of the most beautiful cities in all of Hadrian," CJ commented looking outside as they flew over the vast expanse of the Albion city.

"There's something you don't see every day," Dakota said pointing out from the copilot's seat.

Gunnar and Audra pressed their faces to one of the shuttle windows. In the distance, the massive shape of the Albion battlecruiser, *Tarzana* was clearly distinguishable from a sprawling industrial area. Surrounded by girded structures, the enormous ship sat in a specially designed cradle, holding the ship up high enough to clear it's lower conning tower. A flock of utility craft buzzed all around the ship, moving parts and replacement panels to the damaged areas of the hull.

"Why didn't you just put it in an orbiting ship yard?" Gunnar asked. "Seems kind of big to have sitting on the ground."

"Her antigrav, inertial guidance and environmental systems were damaged in the collision," CJ responded. "The ship cradle was available, so it made sense to bring it down here. Logistically, it's a whole lot easier to make repairs on the surface than in an orbiting ship yard."

"Sure is impressive to see it like this," Gunnar commented quietly. "Gives you a little better perspective."

"Where's the other ship, the *Indominable*?" Audra inquired.

"Talus Mane. Albia maintains an orbiting repair facility for capital ships there. Both ships will be out of commission for some time." CJ tapped the shuttle pilot on the shoulder. "Willis, were you able to get clearance over the shipyard?"

Willis Ruston smiled and waved her arm across the windshield.

"We own the skies wherever we fly."

"How did you pull that off?" CJ asked.

"Apparently it's who you know," Dakota responded.

Moments later, a familiar voice blared through the communications system.

"Executive one, you are cleared over the Faylane shipyards directly to the Milner water works station. From there you are free to navigate without restrictions over lake Pantera."

"Talia, how did you end up at Albia Command?" CJ asked with a grin.

"Excellent references," came the smiling reply from the comm officer. "Part of the cultural exchange program Administrator Vandmire brokered with the Albion military. One guess who got picked first?"

"Is it actually a lake now?" Gunnar asked, looking back at CJ.

"It only took fifteen minutes to remove the hydro dams and channeling walls," CJ explained. "There was enough hydraulic energy created by the breach to wash everything else away. It only took a couple of days for most of the lake to fill back in. It'll take another week for the inlets to fill and about a month for the debris to clear from the shores. We'll have to run crews along the perimeter afterwards to remove anything that doesn't belong." CJ looked over at Gunnar as he and Audra watched the vast Faylane ship yard passing beneath them. "I still don't understand how you even knew about the Milner station, let alone that it was obsolete. Or why you asked for it to be removed. I mean it was scheduled for decommissioning anyway."

"I'm grateful for your indulgence," Gunnar replied without taking his eyes off the *Tarzana*. "I promise I'll show you when we get there."

CJ looked at Audra.

"Don't look at me," she shrugged. "He wouldn't tell me either."

After the *Tarzana* had disappeared beneath the shuttle, everyone settled back into their seats.

"So where's Diord off to these days?" Gunnar asked.

"Attending to matters of state," CJ replied. "Since the Queen made him a regional governor of the Spartus quadrant, he's been so busy we've hardly had time to see each other."

"Probably have to start making an appointment," Gunnar smirked.

"He's been spending a lot of time trying to bring the Drake and the Skillman together. The hope is to make them feel more a part of the galactic government and not so isolated. It's a big challenge "

"Coming up on Pantera," Willis announced from the front.

Everyone came to the windows again. Beneath them, a sprawling body of water had spread out where there once was none. The breaching of the Milner water works, not only released the vast unutilized body of water stored there, but also allowed all the tributaries feeding it to flow freely out across the vast dry plain of the Pantera flats, recreating Lake Pantera. Gunnar stepped up between Willis and Dakota.

"Take us to the far end," he said, pointing out ahead of them and to the left.

"What are we looking for?" Dakota asked.

"Small settlement near the edge."

"There are settlements all around the perimeter of Pantera," CJ informed him.

"This one is the furthest out."

"Isn't that the shoreline?" Audra asked, pointing at a distinct line below them.

"No," Dakota said, pointing further out. "I think that is."

"This is just as far as the water has made it so far," CJ observed.

"Perfect," Gunnar muttered, smiling.

"Is that it?" Willis asked, suddenly pointing to a spot near a defined tree line.

"Could be," Gunnar said looking closer. "Set it down in that clearing."

"What could you possibly want with anyone clear out here?" CJ asked, shaking her head as they circled over a tiny cluster of homes nestled in the trees.

Dakota pointed to the clearing as Willis brought the shuttle in for a near perfect landing. As the shuttle touched down, Gunnar was out the side hatchway before the steps could fully extend. Turning to the tree line, he could just make out the shapes of several small dwellings. One dwelling had an open deck facing the dry lakebed. Two figures were standing against the railing, squinting in the light of the twin suns. Gunnar waited for Audra and CJ to disembark, then headed toward the tree line. With Willis and Dakota bringing up the rear, the group approached the rustic dwelling situated in the trees. Looking up, the group ascended the outer stairs to the deck, but were stopped at the top by a firm male voice.

"State your business," came the gruff order from an older gentleman.

"Eldon, it's me, Gunnar." Gunnar held his hands up slightly, recognizing the man with a drawn pistol. Behind him, Rayna held tightly to her husband, peering around one of his shoulders.

"Gunnar?" Eldon repeated. The pistol came down with the recognition.

"It's Gunnar?" Rayna asked, coming out from behind Eldon. "Put that thing away. I told you, you didn't need it. Besides, I doubt you even know how to use it."

"I know how to use a blaster pistol," Eldon sniped back at his wife. "And you're the one that told me to get it." Eldon set the pistol down as Rayna opened the deck gate. No sooner had Gunnar stepped up onto the deck than the older woman threw her arms around him.

"It is so good to see you again."

"You too," Gunnar grunted from the hearty squeezing.

"For heaven's sake woman," Eldon said tugging on his wife's arm. "Let the poor man breathe. You're gonna crush the life out of him."

"Oh hush," Rayna said, releasing him and looking around his shoulder. "Introduce us to all your friends."

"This is CJ Barker," Gunnar said, gesturing toward the Albion redhead. "She's made this little reunion possible."

"Does the CJ stand for something?" Rayna asked.

"Casey Janae," CJ smiled, taking her hand.

"This is Dakota Abrams and Willis Ruston," Gunnar said. "Dakota is a close friend and Willis is... well, we have high hopes for her."

"Pretty sure there's no need for hoping," Rayna smirked looking down at the couple's tightly clasped hands.

"And this," he said pulling Audra close, "is my better half; Audra Atlanta."

Rayna and Eldon developed an odd look, glancing at Gunnar then back at the short brunette.

"I don't know about better half," Audra grinned brightly.

Rayna carefully reached for Audra's hand.

"I'm pleased to..." She looked back at Gunnar. "Gunnar, I don't understand. You told us Audra was dead."

"Bald face lie if you ask me," Eldon smiled, stepping in to take Audra's hand. "This girl is very much alive."

"Gunnar?" Rayna repeated, looking at Audra.

"Why don't we all sit down?" Willis suggested looking around at the deck chairs and benches.

"Where are my manners?" Rayna said heading inside. "We weren't expecting company, but I'm sure I can get us all something to drink."

"We'll help you," Willis offered, pulling Dakota back off the bench and disappearing after their hostess.

"So what brings you back to our little corner of Albia?" Eldon asked.

"Unfinished business," Gunnar replied, getting comfortable next to Audra.

"What could be more unfinished? Clearly you were successful in getting the Princess to the throne."

"I can't take much credit for that," Gunnar responded. "It was a group effort and I had only a small part in it. In fact, I missed most of it."

"Literally," CJ added.

"Last we saw him, he was boarding a military transport for the *Tarzana*."

"You'll forgive us if we seem a little inquisitive," CJ smiled. "But Gunnar hasn't really explained who you are or how he knows you."

Eldon grinned broadly, looking back at CJ.

"He didn't really articulate much about himself or his friends either."

"Hey, I was protecting everyone," Gunnar defended.

"Eldon and Rayna Grant. We found your boy here in the wreckage of a transport and patched him up. He spent a couple of days with us before he convinced us to help him get to the *Tarzana*."

"That must have been a trick," CJ chuckled. "Let me guess, retired military? Probably Special Forces or Intelligence?"

"Intelligence," Eldon answered.

"That explains a few things," CJ continued smiling. "May have to ask you to consult on a couple of things."

"CJ is a Colonel in the Albion military," Gunnar acknowledged quietly.

"Thought you were military," Eldon surmised.

"Hard to tell without a uniform," CJ said.

"She has been a big help in repatriating the Queen." A smile swept across Gunnar's face. "And she's a good friend."

"Everyone has to try my secret elixir of nirvana," Rayna said with glee as she reappeared holding a pitcher of liquid. Willis and Dakota followed quickly holding empty glasses. After everyone had a full glass, Rayna stood up.

"A toast." She looked at Gunnar and Audra. "To good friends, those present and those no longer with us." Everyone raised their glass and clinked them together, then sat back and took a drink. No sooner had Rayna swallowed her first sip, than she turned back to Gunnar and Audra.

"Gunnar, you were about to explain to us about Audra."

Gunnar smiled and gave Audra a glance. She grinned broadly, trying to hide it in her glass. Gunnar looked at the pink liquid in his glass.

"It's a very long story that obviously has a happy ending."

"You don't have time to tell it?" Rayna asked anxiously.

"Ray," Eldon scolded. "Did it ever occur to you that it might be a very private matter?"

"We're family," Rayna looked back at Eldon, then at Gunnar. "Aren't we?"

"Yes, of course," Gunnar chuckled. He thought a moment about what he had just said, then nodded. "Yes we are. I already told you we're not from this galaxy and that she died of her injuries during our wormhole operations getting here."

"Yes," Eldon recalled. "I assume in your military, you buried her in space?"

"Hardest thing I've ever had to do," Gunnar said quietly as Audra took his hand and squeezed it gently.

"You should know he was still a total wreck when we found him," Rayna whispered to Audra.

"Ray," Eldon scolded again.

"Well, he was. Honey, this boy was still pining for you, even after all this time."

There was a brief pause as everyone smiled.

"A good friend found her burial pod in the Spartus quadrant and…" Gunnar looked around, holding his breath, then looking back at Eldon and Rayna… "He repaired her injuries and revived her."

Eldon and Rayna froze, staring at Gunnar, waiting for more. When nothing came, Rayna twitched her head and leaned forward.

"And just brought her back to life…? Just like that…?"

"Well, there's a little more to it than just *that*," Gunnar admitted.

"Well, come on then," Eldon demanded. "Give us the rest of the, *that*!"

"My friend is what's called a Thane…"

"Never heard of it," Rayna blurted.

"I'm not surprised," Gunnar returned quickly. "They have the ability to manipulate matter at a molecular level. Depending on your talents and skills, they can use their abilities to accomplish tasks that would otherwise be impossible. Caidin is an accomplished medical doctor. Using his powers to manipulate matter, he repaired Audra's injuries and revived her."

"But how do you revive someone that's dead?" Rayna asked.

"This never gets old," Willis whispered to Dakota.

"When I was put in the burial pod, my life energy had not fully dissipated," Audra explained. "The burial pod is a stasis chamber. Everything is held in suspension, including that energy. Once Caidin had repaired my injuries, it was just a matter of reversing the stasis procedures."

"So was it a happy reunion?" Rayna asked, smiling at Audra.

"I wasn't exactly there for it," Gunnar announced. "I was off trying not to get killed."

"Tell them how all that worked out for you," CJ teased.

Willis and Dakota tried to hide their snickers as CJ put a finger to her lips, holding her grin back. Gunnar appeared a little embarrassed looking at Eldon and Rayna who remained riveted to the storyline.

"After I got the Princess's artifacts from the *Tarzana*, I was caught trying to escape. After the Albion voodoo doctors did some freaky experiments on me, I finally broke out. One of their assassins caught up to me and got off a lucky shot."

"Hold it," Eldon said, sitting forward. "I've been able to buy everything you've explained, till now. After all you went through before we found you, and now you're going to sit there and expect us to believe that an assassin took you down with one shot?"

"Yeah, I don't think so," Rayna agreed.

"Well," CJ cut in. "To be fair, he was pretty heavily drugged, then injured several times before crashing his ship. Then he was shot..."

"Again?"

"...and fell down a hole."

"That's when you... died," Rayna finished.

"Not exactly," Gunnar admitted.

"There's more?" Eldon asked.

"Just a bit. I fell a couple more times, then I... died."

"Takes quite a bit to knock you down, doesn't it?" Eldon chuckled.

"I'm sensing a pattern," Rayna admitted bewildered.

"A predictable one," Audra agreed.

"So did you have this friend of yours bring you back as well?"

"Yes," Gunnar said hesitating briefly, then looking back at Audra. "He did repair most of my injuries, but it was Audra that brought me back. I just needed her touch."

"Awh," Rayna said with heartfelt admiration. "So romantic."

"I think that story deserves another toast," Eldon said, raising his glass, bringing everyone else to raise theirs. "To Gunnar and Audra. It seems that even death can't separate them."

Everyone drank again, then fell silent. The quiet was interrupted by a chirping sound coming from Dakota's hip. Pulling the unit, Dakota looked at it for a moment then nodded to Gunnar.

"Why don't we all go for a walk," Gunnar suggested coming to his feet.

"What are you in the mood for?" Rayna asked, swallowing the last of her drink. "A bit of a hike, or maybe a stroll through the woods?" She gestured towards the hills, then at the surrounding forest.

"Maybe down along the shoreline?" Gunnar suggested.

"This is really good," Audra said, swallowing the last of her drink.

"Yeah, and she can cook like no other," Gunnar whispered, moving back to the deck gate.

"There isn't much along the shoreline but bleached rock and rotten logs," Eldon informed them.

"Eldon's right," Rayna confirmed. "We spend most of our time out here looking the other way." She looked back at the bleak lakebed and did a double take. "Eldon? What's that?"

Eldon shaded his eyes, squinting in the sunlight as he joined Rayna at the railing. Looking out beyond the lakebed, he struggled to identify a darkened swath steadily enveloping the bleached ground. Gunnar swung the gate open for the older couple as they scrambled down the stairs. Hurrying past the Talladega in the driveway, they quickly made their way down the lane toward the shoreline.

"How long have they lived here?" CJ asked following Gunnar and Audra.

"They didn't say, but I'm pretty sure as soon as his stint in the service was up, they moved out here and built this house," Gunnar said.

"It's a wonderful thing," Audra smiled. "I don't understand why it took your waterworks people so long to remove the dam."

"Well, to be fair," CJ admitted. "I haven't been in Zeppelin for quite some time, but from what I've been able to piece together, it boiled down to some underhanded profiteering on the part of one company that could have switched to a better source a long time ago. Once I pointed out the *error*, the company quickly corrected the problem and was happy to turn the water loose to avoid any catastrophic repercussions."

"I'm not even going to ask," Gunnar said as they rounded a stand of trees and looked toward the shoreline. Several other couples had observed what was happening and were gathering to watch near the shore as the water approached. Gunnar noticed Rayna and Eldon standing arm in arm a little ways off from the others and quietly stepped up behind them. Rayna's head was resting on her husband's shoulder as they watched the water rolling gently over the bleached ground that had once been the main roadway across the lakebed to the other side.

"Doubt you're gonna want to be wading or swimming in it for a while yet," Gunnar smiled lightly.

"Has it really come back to stay?" Rayna asked, looking back at Gunnar.

"Permanently," Gunnar responded.

"How did you manage it?" Eldon choked slightly, then cleared his throat. "We've been trying for so long to get them to let it go."

Gunnar looked back at Audra and CJ standing a little ways behind him.

"I know a good friend who knows people."

"I don't know what to say," Eldon's voice cracked again.

"We don't know how we can ever thank you," Rayna said, squeezing her emotional husband.

"There are no thanks needed," Gunnar said. "I'm just repaying an act of kindness."

"We just did what anyone else would have done." Rayna looked at Eldon. "Didn't we?"

"One can only hope," Gunnar agreed. "But you also showed me what the rest of my life can be like."

"Oh, brother," Eldon said, wiping his eyes. "Surely you're hoping not to be cursed by..."

"Don't say it," Rayna warned, sliding her hand up to his earlobe.

Gunnar laughed out loud and took Audra's hand as she stepped up next to him.

"I only hope we can be that way when we get that old," Gunnar whispered to her.

Rayna and Eldon turned to Gunnar and Audra.

"Did you just call us old?" they asked collectively.

Gunnar and Audra's eyes widened and they quickly pointed at a couple standing nearby.

Math Class

Rick leaned over the *Athena's* conference table, looking at everyone gathered. Her Majesty, Queen Jana Oxlind sat at the far end with Blinda Koss leaning against the wall directly behind her. Diord and CJ sat together next to Captain Abrams and Willis Ruston. Gunnar and Audra sat together across from Dãsha and Caidin. Frank Cooper, Billy Moon and Tiana sat next to Jayda with Fuji, Mister Pippin and Mavis standing behind them.

"I've listened to many of the crew explain their feelings about making the journey back home or staying here in Hadrian." He paused, smiling at Jayda. "There were some surprising as well as compelling explanations given for both options. I appreciate everyone who has expressed their feelings on the matter. But as the Officer in Command, I have to make the final decision. After a lot of consultations with upper staff and personal soul searching, I've decided that it would be unwise and serve little purpose to return to our own galaxy."

"So we're going to be staying here in Hadrian?" Gunnar asked, looking at the smiling faces around the table.

"If the Queen is able to provide us with a place to be and the means for some kind of security," Rick looked at Jana, "then it seems like the only logical choice."

"Governor Vandmire has informed me that he and Admiral Aurora have identified a planet that should provide the isolation you require," Jana responded.

"Please tell me it isn't Aster," Gunnar mumbled, leaning closer to Audra.

"I've never been there," Audra whispered back. "How bad can it be?"

"How much do you enjoy the dark?"

Diord pointed at the display behind Rick.

"It's an exoplanet located here, near the Jurass quadrant. The system is small, only four other planets. One has an Indigenous population of humans in the early stages of development."

"Not technologically advanced?"

"Hardly out of the stone age," Diord replied.

"Hadrian law has protected this system so that the population can develop on its own," Jana explained. "If we move you to this exoplanet, you'll be required to follow those same restrictions."

"What about these ships?" Dakota asked. "What's to prevent someone from coming along and trying to access our technology? And what's to protect Hadrian from us?"

"As this system is protected by Hadrian law, you would fall under the same protection as the people developing in this system," Diord said. "As for protecting Hadrian from you?" Diord looked back to Jana who turned to Rick and Gunnar. "I think I know you well enough to trust that won't happen."

"Then it's all settled," Rick agreed, smiling. "We'd like to send a site survey expedition to find a suitable area for a base of operations."

"We'll have personnel and transportation made ready for you," Diord said.

"I can see keeping your Starbirds hidden on the surface," Dãsha said. "We may even be able to dismantle the freighters and use their parts as living spaces, but what of *Calypso*? It is far too large to hide. What would become of her?"

Rick stood up straight, thinking.

"We could blow it up," Dakota suggested. "Leave nothing for anyone to find or use."

"Certainly that is not acceptable," Dãsha objected sharply.

"I know how you feel about *Calypso,* my dear," Caidin said, putting a hand on her arm to calm her. "But perhaps its destruction would be in the best interests of everyone."

"Destroying such a wonder would be unacceptable," a voice stated calmly from a corner of the room.

Blinda instantly had her pistol out of its holster pointed at a young woman standing next to her. *Where did she come from?* Blinda hesitated.

Half of her long copper colored hair flowed down the front of her robes, the other half disappeared behind her.

"Ona," Blinda whispered, lowering her blaster.

"I am grateful you still remember me," the Master Thane smiled. She stepped over to Blinda taking her hand. "It's important that you know, I never lost hope in you."

Blinda dropped to her knees, but Ona was quick to bring her back up.

"No one need kneel to me."

Ona turned to the other occupants of the room. Seeming to ignore Jana, she moved around the conference table as everyone came to their feet. She smiled through an emotional expression as she stopped in front of Audra and Gunnar.

"You two have endured so much. You've become an amazing testament of what love can transcend." She looked down at their clasped hands. "Cherish each other." She looked at Rick. "It was never my intention that anyone should get hurt," she said looking back at Audra and Gunnar.

Gunnar glanced at Rick.

"Is this the Ona lady you were telling us about?" he asked quietly.

Audra pushed her elbow into his ribs.

"Pretty sure she can hear you," she said, hypnotized by the Master Thane's expression.

Ona's smile broadened as she stepped past the couple toward CJ and Diord.

"Thank you for believing in a cause that had so much set against it. There are still some difficult times ahead, but the right people are in the right place to see them through."

CJ curtsied awkwardly, not knowing how to respond. Diord snickered and bowed as Ona moved to the front of the room, facing Rick and Jayda.

"Everyone aboard this great ship was brought here for a grand purpose, each of you with a part to play in averting a chaos that would have devastated this galaxy. You have my thanks for helping to bring balance back to Hadrian." Ona looked to the other end of the table at Jana. "Your Majesty, you have my eternal gratitude for teaching one who needed to learn what it is to be human."

"I don't understand..." Jana puzzled.

Ona gestured to the open conference room door as a figure appeared. Letting out a shrill squeal, Jana was out of her seat before anyone recognized the Royal Au Pair with outstretched arms. The young Queen nearly knocked Tonnie over as they embraced. Sniffling through her tears, Tonnie stroked Jana's hair as the young Queen cried into her shoulder.

"How?" Jana asked. "Did Caidin and Doctor Yamoto heal you?"

Tonnie shook her head.

"This would be far beyond my abilities," Caidin admitted with a grin.

"This is an android," Rick said looking at the Royal Nanny. "I witnessed the sizable blaster holes in her chest. How did...?"

"Not just an android," Ona corrected. "Tonnie is a life form. She operates on a similar plane as Dãsha, with one notable difference. Dãsha is human with artificial substrates. Tonnie is artificial with human substrates."

"A human android?" Rick asked.

"So to speak, but like all things around us, living."

"Are there more of her here?

"No. Like yourselves, she is from another place. They are called the Odecka."

"Why was she here?"

"All things evolve, but as pure androids, the Odecka were having difficulty progressing any further in their evolution without an environment that would foster teaching of a unique characteristic. You call it humanity. A trait only a human is born with. Because of the inherent sterility of their artificial makeup, the Odecka were incapable of developing a moral compass on their own, so I brought Tonnie here to learn and grow." Ona looked at Jana. "You provided her not just the opportunity to learn how to love, but the chance to demonstrate her capacity to love another enough to sacrifice herself on your behalf."

"What happens to her now? Will she stay with us; with me?" Jana asked, hopeful.

Ona stepped over to the young Queen and took her hands, smiling.

"No. Her place is back with the Odecka. Now she can help them learn by her experience here, so they can continue their evolutionary track."

"Hello, Tonnie," Gunnar said, stepping next to them with Audra on his arm.

"You survived!" Tonnie exclaimed.

"Takes more than a little fall to knock me down."

"Look at your Sunggo!" Tonnie exclaimed, looking him up and down. "Nice and bright; and the color. This can only mean one thing," she said smiling at Audra. "This must be Audra."

"Yes," Gunnar grinned broadly. "This is Audra."

Tonnie looked over at Ona, who raised her eyebrows.

"It wasn't me," Ona said. The Thane Master turned to Caidin and Dãsha.

"Well done, Caidin," Ona said. "You are a true Master Thane of medicine."

"I still have much to learn," Caidin admitted humbly.

Ona shifted her gaze to the tall beauty on his arm.

"Like so many things that surround you, you are a true marvel. Thank you for taking such good care of *Calypso*."

"I still know so little about it," Dãsha said, unsure how to process the significance of this greeting.

"There was no other better to figure out what was necessary for the purpose of *Calypso*."

"What is its purpose?" Dãsha asked.

Ona looked at Rick and Jayda as she made her way back to the front of the room.

"To take you home."

Crickets would have been deafening.

"We've made the calculations," Rick finally replied. "There is no logic to such a journey. The spacetime is too great to be meaningful."

"You would be correct," Ona agreed, stepping around them to the wall display. She waved her hand across the star chart, the images instantly changing to an enormous mathematical equation and a representation of two galaxies. She stepped back and studied the information for a moment then turned to Rick.

"Richard, your math is spot on. At *Calypso's* fastest sustainable light speed operations, it would take entirely too much spacetime for such a journey to have any meaning to any of you. Based on this, your current decision to remain here is the only correct one." Ona turned back to the display, thinking.

"But...?" Rick hedged.

Ona turned to everyone in the room, looking at each one for a moment.

"You don't belong here. It's not your home."

"There's no home to go back to...," Rick stated almost inaudibly.

"In the time it would take you to get home using the means you're proposing, you would be correct. It's a very long way back to Kalamar. But in your absence, this is what has become of your worlds." She swung her arm in a wide arc across the display, the image instantly changing to scenes of utter destruction of many cities and worlds in their own galaxy. Notably, Kalamar and the beautiful city that was once a shining beacon to the rest of the galaxy lay in burnt out ruins, most of its occupants dead. Gutted out hulks of great ships and debris floated aimlessly in space along the main travel routes, some stuck in eternal orbits around many of the planets.

The ships roaming the trade routes, prowling and destroying everything they came upon, were frighteningly familiar. Starbirds, ghastly red strips and symbols painted over their grey hulls, traveled in packs, others escorted larger capital ships. Their markings were that of the Valkyrie.

Horrified, several in the room had come to their feet, while the others remained seated with their head in their hands, some fighting back tears. Gunnar stood next to Rick, watching the scenes change before them, both stunned.

"What happened?" Rick asked, watching the terrible scenes on the display.

"Much of the information about your Starbirds has become a tool for Croft Heckla and the Valkyrie."

"So there really is nothing for us to go back to," Rick stated, his voice heavy. "Even if we could somehow recreate the conditions that brought us here."

"Did we just help save this galaxy from itself only to lose our own?" Gunnar asked, looking at the Master Thane.

"Yes and no," Ona responded. "These events have happened in your absence. The question is, do you have the courage to do something about it? Based on your time here in Hadrian, I have no doubt of the answer. Everything you need to correct this wrong is here."

"I don't understand," Rick said.

"If you change the perspective of your understanding of the universe, you will find your answers."

"This is the part where she doesn't tell you anything useful, then disappears, isn't it?" Gunnar grumbled.

"I'm not going to spoon feed it to you," Ona smirked. "However... What would happen if you were to modify your equation, here?" she suggested flicking two of her fingers across the screen. The lines of numbers and symbols instantly vanished, giving way to other symbols and numbers that took up only half the space as the first equation. Rick stared at the display for several moments, then stepped closer. Ona folded her arms, her eyes following his as they scanned the equation. Rick's lips were moving as he worked through the math. Halfway through the equation, he glanced at Ona.

"This formula..."

"It's time to go, Tonnie," Ona said, turning to the door.

"Will I ever see you again?" Jana asked, holding on to the nanny.

"I hope so, some day," Tonnie said, stroking Jana's hair.

They held each other tightly for several moments, then Tonnie pulled free and moved next to Ona.

"I trust you'll use what you have learned here in Hadrian to make your home a better place," Ona said, raising her arms out and flicking her wrists. In an instant, they were gone, leaving the room silent. Gunnar turned to Rick, who was transfixed on the new equation in front of him.

"Wait!" Gunnar objected. "Just like that, she's really gonna leave without a little more explanation?"

"That's Ona," Caidin said. "Not much for conversation."

"Talk about no help at all," Gunnar said looking back at where Ona and Tonnie had been standing. He turned back to the display. "So what's all this gibberish?"

"The way home," Rick whispered, his eyes still fixed on the numbers and symbols.

"Exactly how is a math problem the way home?"

"It solves the problem of folding space."

"Folding space? There's no such thing."

"Actually," Rick said, continuing to work through the equation in hypnotic fashion. "There is such a thing." For several more moments, he cruised through the rest of the math, then turned to Jayda. "But no mechanism or interface."

"You can't bend space," Gunnar pointed out. "Naturally occurring space bodies can; quasars, black holes, suns, planets; stuff like that, but it can't be artificially manipulated. All our attempts ended really bad."

Rick held a finger up.

Ona wouldn't have provided the equation unless there was a way to make it work. Rick searched his thoughts. *Something recent...*

"*Calypso...*"

"What about *Calypso*?" Dãsha asked.

"The engine room," he said, recalling his meditation session. It had taken him outside the ship during lightspeed operations. When he returned, he came in through *Calypso's* engine room. There was another room containing a strange apparatus with no visible means of operation or way to interface with it. "In a small room just off the starboard engine compartment."

"Slow down, Skippy," Gunnar said. "Are you talking about Hyperspace or a Warp bubble? Even if you had something that actually worked, it's still going to take too long to make a journey like that. Modifying a math equation doesn't solve the problem of getting from one side of this galaxy to the other side of a different one."

"Hyperspace and Warp bubbles have their advantages to be sure," Rick agreed. "But this equation relies on something else entirely." He turned back to the display and swiped across the screen. The symbols and numbers moved to the side and the images of the two galaxies flattened out. "Lightspeed relies on moving along a straight line from point to point. A warp bubble does as well, just by a different means that allows for greater speed. Hyperspace passes through the space existing above or below the dimensional space we currently occupy. Again, greater speed and distance along this flat, point to point line. Now imagine what would happen if you could move these two points closer together?" Rick slid the image around with two fingers. This time, the two galaxies curled until the two points came together.

Gunnar stared at the image as Jayda stepped closer.

"This seems rather simplistic," Dãsha commented, cocking her head at the image, then looking at Rick. "Does this change your decision to stay?"

"I think it has to," Rick said.

"How?" Gunnar asked. "We'd still be going back to something that no longer exists. Not to mention we'd be flying right into the middle of an apocalyptic reality. Two ships against a fleet of our own and Croft's."

"Six ships," Jayda corrected.

"Whatever," Gunnar blurted. "Even if it's possible, I see nothing good coming from this. We've made the decision to stay here. Everyone is good with that. Home seems a lot less inviting now."

"What if we weren't going back to what exists now?" Rick asked.

"I would love to hear this explanation... and actually understand it," Gunnar frowned skeptically.

Rick turned to Gunnar.

"Colonel, let's continue this discussion privately." He glanced around at everyone else in the room. "Please excuse us."

The conference room remained silent as both officers walked across the hall and the door closed. Audra and Jayda looked at each other, then turned to everyone else.

"Anyone know any fun games?" Audra asked, smiling.

*   *   *   *

Rick sat down at his desk and turned on the display. As the equation appeared, he made several adjustments to the galaxy images then turned back around to Gunnar.

"I know you." Rick started.

"...and I know you," Gunnar shot back curtly.

"In all the time we've been together, we've had only three arguments."

"Only three? We argue all the time."

"No, you argue and I listen."

"Pretty high opinion of yourself."

"Gunnar, you and I have always seen the bigger picture in any operation we've undertaken together. I'm asking you to see that now."

"And I'm asking you to consider what we've all been through for the past two and half years," Gunnar countered. "We're all tired. Tired of getting chased, shot at, captured, tortured, injured or killed. Here, we have a place where we can slow down and maybe settle down. But you want to change all that and take us across two galaxies in the blink of an eye to start a two and half year fight all over again. Not to mention putting the girls and all those people out there into harm's way. I've gotta say, you're going to have to do a little more to sell this idea."

"I know we're all tired," Rick said after a brief pause. "I'm just as weary of the chase. I have to admit I was looking forward to arriving home to a hero's welcome. We'd all walk into Admiral Mandell's office to high praise and a hardy thank you for a job well done for whatever kick-ass story we make up. Everyone will get their choice of extended shore leave or reassignment to wherever they want. Every member of these crews deserve that much and more. But if there's been one thing our time here in Hadrian has taught me is that life isn't about stopping or sitting it out when it comes to moving forward, and all the things we love or hold dear are never just handed to us. We have to fight for what's right. Why would our galaxy be any different than this one? We thought we came here by accident. Turns out there was a purpose, so we could help correct an evolutionary track that had veered off course. You and I have done this kind of thing in one form or another our entire careers. Now, in order to go home, we're going to have to correct a course deviation it took in our absence. Ona was right, we don't belong here; this isn't our home. We belong in our own galaxy."

"Hate it when your lectures make sense." Gunnar looked over Rick's shoulder at the display. "I just can't see how a math problem is going to solve the problem of folding space."

"It's not just a math problem, they're coordinates for the location of Genis Two and the correct time for arriving there. The actual mechanism is on *Calypso*. But there's a bigger question here."

"Something back home went wrong when we left." Gunnar thought a moment, looking at the image.

"It probably happened before we ever got to the Oneida." Rick postulated.

"Clearly." Gunnar's eyes widened.

"I was referring to the Starbird initiative."

"You're talking about the four you found on those Kendalon freighters."

"Yes, what were they doing there? Constance and Manx said they were to rendezvous with the *Zebra Phar* over Genis Two. That's in a system just out of the influence of the Oneida, on the fringe of both Kalamarion and Valkyrie space. The *Zebra Phar* is a Benton class light cruiser operated by the Filon. Commodities brokers, mostly third party dealings in small to mid-sized ships of all kinds. They contract with independent freight companies to move their merchandise around so they don't attract attention."

"Are you suggesting the Kendalons are part of some grand conspiracy?"

"Not at all. They didn't even know what they were carrying until I opened the holds. Why?" Rick shook his head slowly. "You and I were part of the Starbird program from the outset."

"You mean you were," Gunnar corrected.

"I guess that's true. I brought you in shortly after the project got under way. The point is, we've known everything, or at least I thought we did. Somewhere in the command chain, something was going on. Kalamar Command would never purposely send uncommissioned Starbirds anywhere near that area."

"So who made the order?" Gunnar asked. "How is it, when the freighters didn't arrive, Croft was able to take down Kalamar Command and lay waste to the entire region?"

"That's what we need to figure out. Otherwise, we're going to be second guessing every move we try to make."

"Ok," Gunnar paused. "Let me make sure I understand where you're going with this. We're going to use this math equation to fold space and take us back to our galaxy to a specific location... and time?"

"This is amazing science," Rick said. "We're going to recreate the same conditions that brought us here, only this time we can control both time and space." Rick silently considered the ramifications. Glancing back at the equation, he did a double take, focusing in on several lines of symbols and numbers.

"What's wrong?" Gunnar asked, looking at the display.

"This isn't good," Rick mumbled.

"Information please," Gunnar pressed.

"There's a time limit on this equation."

"A time limit?" Gunnar complained.

"This is some complicated stuff," Rick said looking up at his friend.

"Yes, yes, amazing science..."

"In a nutshell, Folding operations have to be right on the money because of the parameters you're dealing with. Arriving at a given spot at an exact moment in time, requires you to align your folding points perfectly. As the universe is in constant motion, meaning galaxies move, just the slightest shift will throw everything off and the Folding operation won't work. So, we only have a short window of opportunity in which to use these coordinates."

"Define, short."

Rick looked back at the screen.

"If I'm reading this right, about sixteen hours."

"You're joking," Gunnar said. "Maybe you added this number to that number wrong?" he suggested pointing at the characters on the display. "Any other fun-sicles in this equation?"

"There's some other stuff tacked onto the end here," Rick acknowledged. "But it doesn't appear to pertain to our destination."

Gunnar chuckled and widened his eyes.

"...And I thought I was the reckless one."

"You've used up all your reckless while we've been here."

Gunnar let a heavy sigh go.

"Ok, I've still got some crazy left in my pocket. If you say we can do this, then I'm with you. But there are still a few questions that stand in the way. We'll start with who's behind all this and how are they executing these plans without command knowing about it? Even if you have the answers to those questions, you still haven't addressed how we're going to keep the girls out of harm's way. If we end up going up against Croft and the Valkyrie, those Starbirds are likely to get chewed up as a snack?"

"I have a couple of ideas as to the who and with what," Rick said, turning to his friend. "It's the why that has me baffled. Those freighters never arrived to make the exchange, so it seems logical that something critical happened there."

Gunnar's expression chiseled as he thought through the possibilities.

"If we reload the four new Starbirds back in the holds they were originally stored in, then launch the freighters when we arrive, seeing how the *Zebra Phar* responds could really shed some light on what's going on, or at least point us in a definite direction."

"That still puts the girls in harm's way," Rick reminded him, remaining silent for several passing moments, thinking. He slowly looked back up at Gunnar, his expression blossoming.

"I know that look," Gunnar smirked. "Usually a great idea behind it, but always a little scary."

"No, this one has the potential to be a *lot* scary."

"Ok, out with it..."

Unhappy

Rick and Gunnar remained in closed conference for some time. When they finally emerged, Rick stepped back into the conference room, scanning the occupants as all eyes turned to him. He hesitated a moment, looking at Caidin, then Jayda.

"In light of the information that Ona has provided, it's been decided that we need to return to Kalamar immediately. Starbird commanders need to have their ships loaded back into the freighters they were originally found in, but remain launch ready. *Calypso* will fold space to Genis Two where we'll launch the *Stark*, *Huntly*, *Merc*, and *Tommy-Lee*. We'll then assess the situation with the *Zebra Phar* and act on those conditions. Are there any questions?"

"Any questions?" Dakota burst.

"I'd say there are at least a couple," Jayda chimed in.

"Captain Abrams," Rick said purposely ignoring the protests. "You will join Colonel Conrad with the logistical tasks of getting the Starbirds loaded back in their holds and completing the assigned modifications to their control systems and weapons inventory."

"Modifications?" Dakota inquired, visibly irritated, "Weapons inventory? What the...?"

Rick turned to Jayda.

"Commander Niker, you'll be in charge of seeing that the *Constellation* and *Athena* are staged at *Calypso's* bay doors ready to launch. Caidin, you need to remain here on the *Athena* as ship's doctor. Dãsha, you'll need to be on the bridge of *Calypso* with Mister Dacey. I'll join you shortly to discuss the controls for folding space." The room remained still until Rick clapped his hands together. "That's an order people, let's go!"

Dakota slowly got to his feet and followed Dãsha and Jayda from the room. Caidin lingered a moment, looking at the equation on the display. Shuffling to the door, he turned in the threshold.

"There are better ways of getting things done, Richard."

"Not in this situation," Rick responded quietly.

Caidin opened his mouth to say something, but ultimately left the room without another word.

Rick remained silent, turning to the display. He detected soft footsteps behind him, but kept his eyes fixed on the display.

"What can we do to help?" Jana asked.

"I'm sorry," Rick responded after a moment. "Just about every one of our crews were looking forward to staying here in Hadrian. I think even I was looking forward to the prospect of not having to fight a fight or solve a problem at every turn." He looked back at the other individuals left in the room.

"You and Gunnar have to do what you do best," CJ spoke up as she stepped around the table. "You make things right."

"When is it going to be time to make things right for ourselves?" he asked.

"That's what you're doing now. If you don't go home, you'll never have rest here or any place you go. You're always going to wonder what you could have done and why you didn't do it. Being at peace isn't going to mean anything if you don't make peace with yourself and those who mean the most to you."

Diord put his hands on CJ's shoulders.

"Anything you need, your friends are here to help."

"Anything," Jana echoed as she left the room.

As Willis followed CJ and Diord out of the conference room, Rick watched them disappear around the corner. Again, he detected someone behind him and turned to Blinda Koss.

"If I didn't have responsibilities to the Queen, I think it might be fun to come with you. I feel like we've sucked all the excitement out of being in Hadrian with everything set right."

Rick smiled.

"There will always be a need for an unruly Thane to keep the peace around the throne."

"Unruly?" Blinda repeated. "Never thought of myself as unruly. Maybe I ought to change my hairstyle to match the description?" Blinda flicked her fingers. Her jet black hair suddenly turned platinum, then shortened to a straight cut just covering her ears. She stroked her dark velvet body suit up her waist. "What do you think?"

"I think too many women in this galaxy ask me for fashion advice," he chuckled slightly. "Keep looking for the other perspective, Blinda."

"There are always better ways of solving a problem, right?"

"Exactly," Rick agreed with a smile.

Blinda left quickly to catch up to Jana and the others, leaving Rick alone in the conference room. As he contemplated what might lay ahead of them, he turned back to the wall display and studied the information Ona had left them and the images of the two galaxies folded together.

"Why are we in such a hurry?"

Rick turned around. Audra Atlanta stared at her fingers as she twiddled them silently on the conference table. Now Rick wished Gunnar were still here.

"The folding equation has a narrow time window. If we don't fold during that window, the entire equation is thrown out of balance and we could end up someplace we don't want to be."

"Like here?"

"We didn't want to be here," Rick responded. "We were brought here."

"And now you want to take us back the way we came."

"Sort of..."

"What bridges the exit and entry points?"

"An energy conduit."

"Is that its technical term?"

"No."

"It's a wormhole, isn't it?"

"Technically, yes."

"You understand my extreme apprehension about such an operation?"

"This wormhole will be far shorter with a much smoother ride."

"It's a wormhole! So you'll have to forgive me if I seem a little less than excited about it. Doesn't matter how safe you think it's going to be."

"I understand."

"No, I don't think you do," Audra maintained. "I don't want to do this anymore."

"I think everyone feels the same way."

"I don't care how everyone else feels. I care about one thing, Gunnar. We lost each other because of a wormhole. No matter how safe you say it's going to be, I can't get over what happened the last time and what these last two and a half years has done to us.

"What do you want me to do?" Rick asked.

"Let it go."

"You know I can't do that."

"Yes, you can." Audra came to her feet. "You're going to dive headlong into this and Gunnar is going to dive right in with you. Then a whole new nightmare will start." Audra stepped over in front of Rick. "Let it go. Stay here and make a new life; a quiet one, where there's no more war or galactic conflict."

Rick looked into her eyes. Maybe it was the side effects of hibernation sickness that shaded them a little darker than normal. Her face was etched with exhaustion. For a moment, he flashed back to a time at a concert gathering where he and Jayda had plotted to get Gunnar and Audra together. They were so young and dumb. It was a wonder things had worked out. His and Gunnar's careers were based on doing the right thing, even when it wasn't popular with the men and women they served with.

"Report to your duty station, Commander. Be ready to launch when the *Tommy-Lee* opens its doors."

Audra stared up at him for a long moment, her eyes burrowing into him. She finally nodded.

"Aye, General." She brushed past him without another word.

Rick knew what she was feeling as he turned and watched her go. It was one of the hardest things about being the Officer in Command. Ordering friends to do things they didn't want to or worse; ordering them into a situation that could get them killed. But this situation, while dangerous for everyone, was going to put him and Gunnar at the forefront of whatever was to come.

"CORA, have you sent the equation to Dãsha?"

"It's being uploaded to *Calypso's* navigational computers right now."

"Once you've finished the upload, I want you to delete it from the ship's memory banks and your own."

"Rick, are you sure?"

"I don't want it getting into the wrong hands if something happens. Have the crews go to yellow alert. I'm going to *Calypso's* bridge for the trip home."

"Genis Two is hardly home."

"No it isn't, but it's the only route to get there."

"Are you sure about this?"

"I have to be," Rick muttered as he turned to the teleporter room.

The Last Word

Gunnar had his hand up in an access panel of the *Constellation* when Alex 7001 swooped in close.

"Colonel, may I inquire as to what you are doing?"

"She's seen a lot of action," Gunnar replied, checking a readout inside the panel. He closed the door and took a couple of steps back, looking up at the nose of the Starbird. "If we're going to take her into battle again, I want to give her a good looking over."

"While I admire your homage to a pilot's affections to his craft, I can assure you that the engineers and I have already checked and rechecked every system on this ship. It is as functional as it can be, considering what it has been through."

"I would expect nothing less," Gunnar said, walking around and examining the Pin missile snouts and the torpedo launch doors. "So you won't begrudge a well-seasoned fighter pilot his ritualistic constitutional."

The little droid watched Gunnar for a moment longer, then turned as a pair of individuals approached.

"Gunnar, we've come to say our goodbyes," Jana said smiling broadly.

Gunnar stopped what he was doing and turned to the young Queen. He hesitated a moment, catching sight of Blinda standing a few steps behind her.

"Thank goodness I haven't had time to really think about leaving you guys behind."

"But you are going to miss me, aren't you?"

"How could I not?" Gunnar smirked. "You've been like that little sister that keeps getting under foot. Just as soon kick you out of the way as look at you."

Jana giggled and threw her arms around him.

"This is almost as hard as it was leaving you on Dither."

Gunnar detected her shaking.

"Hey, none of this," he said, pulling back to look into her tear filled eyes. "Where's that tough girl CJ and I rescued from Reako? You are the Queen of Hadrian. Royalty has to keep a straight face so everyone can see strength. You've got a lot of work ahead and you can't do it if you're crying your eyes out." He wiped her cheeks with his thumbs and kissed her forehead.

"I never knew my family, but I like to think of you as my big brother."

"Hooray," Gunnar rolled his eyes. "Finally someone thinks I'm young enough to be a big brother instead of their father."

"Big, dumb brother," Jana snickered, flicking his nose with her finger.

"I'll never forget you, baby sister." Gunnar hugged her again. As he let go, he looked up at a straight faced Blinda Koss. Jana stepped back, watching as the two former foes faced each other. Blinda shifted, letting her hands drop to her hips. Gunnar stepped forward.

"Did you change your hair again?"

"I'm a woman of many styles." She furled her hair, running a hand under it.

"You see that you take care of this little girl here, or I'll have to come back and shoot you again."

"Yeah, never going to happen." Blinda smiled. "At least not if I can help it." She took his hand and squeezed gently. "Stay safe, Colonel."

Gunnar nodded and turned to go back to his inspections.

"Gunnar," Blinda spoke up. "Keep Audra close."

"I don't think she'll have it any other way."

"Goodbye, Gunnar," Jana called loudly as she and Blinda turned for their waiting transport.

Gunnar took a step back and jumped up onto the nose of the *Constellation* for a closer look at the various gun pods. As he was examining the laser emitters, Alex 7001 joined him again.

"Now what are you doing, Colonel?"

"I'm still doing what I was doing the last time you bothered me; inspecting."

"This really is unnecessary."

"Humor me."

"Governor Vandmire and Colonel Barker are here to speak with you."

"Couldn't you have just said that in the first place?" Gunnar sniped, dropping to the bay floor. He turned around to find CJ's gleaming smile directly in front of him.

"Alien or not, I'll never get used to seeing you do that."

"I don't think I'll ever get used to seeing so much red hair on an officer."

"Take care of yourself," Diord said, stepping up and shaking his hand.

Gunnar pulled him close and hugged him. They embraced for several moments, then pulled back.

"Take care of this silly redhead. She's pretty high maintenance."

"And see that you take better care of Audra."

Gunnar looked over his shoulder at the short brunette next to one of the Kendalon freighters directing the work of getting the *Intrepid* moved back into the freighter's hold. He smiled and turned back.

"Yeah," he agreed.

CJ gave Diord a glance.

"Can you give us a minute? I'll be right along."

"Make it quick." Diord nodded and stepped away. "The Queen is ready to leave."

CJ and Gunnar watched him walk toward Jana's shuttle as its engines started to wind up. Looking back at each other, Gunnar smiled and flipped her second lock of white hair.

"From all the excitement," CJ said, curling the lock with her finger. "It matches the one on the other side. I was really looking forward to checking up on you in your new digs."

"Yeah, I was too. Funny how things got turned around so fast."

"Probably just as well," CJ snickered gleefully. "Eventually, you and I would have gotten into some kind of trouble. Then you'd bail me out and I would have to save you from something else entirely."

"Come on," Gunnar defended. "We made a great team. Seems like whenever Rick and I were doing a joint operation, I was always having to pull his sorry butt out of a sling."

"Funny, he tells a slightly different version of the same story."

"Yeah, but you can never quite believe what that guy says."

"Sounds like you're going to be jumping into a hot zone when you get home. I have no doubt that you're up to it, but is your ship?"

"She's taken a pretty good beating, but I think she's still got something left in the tank."

CJ stepped back to get a wider view of the Starbird, then looked back at Gunnar.

"It's going to seem boring around here without you guys stirring things up."

"We didn't stir anything up," Gunnar smiled.

CJ turned, spreading her arms out at the other Starbirds being staged around them. "And this would be...?"

"Ok," Gunnar admitted goodheartedly. "Maybe a little."

"Yeah, just a little," she said, squinting through her thumb and index finger.

They both laughed, then fell silent, each shuffling awkwardly.

"I wish you weren't leaving," CJ finally said quietly. "Not sure what I'm going to do not having you around to get me into trouble."

"I could say the same about you," Gunnar responded.

CJ looked up at the hull of the *Constellation*, then around at the other Starbirds.

"I'm sorry about your sister," Gunnar said. "Why didn't you tell me?"

"It's not something I'm proud of..." CJ diverted her gaze to a torpedo being loaded into one of the other Starbirds.

"I can't even imagine what that must have been like."

"I wish..." CJ bit her lower lip, shifting her eyes to nothing.

"You had to make a choice. BachTL is alive and Hadrian has her Queen because of that choice. What would have happened if you hadn't acted?"

CJ looked back at Gunnar.

"I know, but she was my sister..."

The two fell silent for a moment, listening to the bustle all around them. CJ's crystal blue eyes finally brightened a bit.

"Oh, before I forget. Doctor San Sann destroyed all his research materials on you. Super Soldiers aren't something Hadrian has the wisdom to deal with."

"What? You don't want copies of me running around saving the day?" Gunnar grinned arrogantly.

"Tough enough keeping track of one, why would I want an army of them?"

"You never know," Gunnar laughed. "What's BachTL up to now?"

"As one of Diord's lead envoys, he's been assigned to foster better relations with the Ratronians as well as seeking out lesser known factions and inviting them to join in with the other factions."

"He's going to be one busy boy. What about you?"

"I've got my hands full seeing that the void my sister left is properly filled."

"QC Barker?"

"No, I've never had any aspirations to be a Queen Captain. I have way too much past with the last one, and I don't think anyone should have that much power."

"Think you can change the way your military works?"

"I'm gonna give it a good try."

"Thanks for letting Willis and the others come with us."

"Either she was going with you or Dakota was staying here."

"Young and in love, what ya gonna do?"

They laughed again as Jana's shuttle engines suddenly came alive, giving CJ cause to look over her shoulder.

"I don't like goodbyes..."

"Me either," Gunnar agreed, fidgeting a little.

"I want to say something grand, but all that's coming to mind is a sappy, gooey bunch of rubbish," CJ said, fighting her emotions.

"It's ok," Gunnar said, taking a step closer. There came a tingling to his ears. He couldn't help but smile, thinking back to when they first met on Carolon. The tingling was there for only a moment, then it was gone. "I've got the same ailment, so I'll go first. Thank you for seeing me through the hell I've put everyone through. I don't know if I would have survived without you."

"It's me that should be thanking you," CJ replied, her voice cracking. "You forced me... out of my shell... and showed me it's ok to love again."

Gunnar reached up and dabbed a tear rolling down her freckled cheek. As he did, CJ leaned her head into his palm, holding it. Gunnar could see more tears flowing from the other side and couldn't help but get misty-eyed as well. He carefully leaned closer and kissed her cheek.

"Thanks for being such a good friend. Now, you better get out of here before you get left behind," Gunnar said looking over her shoulder as Diord appeared at the shuttle door.

CJ let go and wiped her eyes, trying to smile.

"I hope we'll see each other again... someday."

She suddenly leaned forward and kissed him on the lips, then hurried off to the shuttle.

"Dang if she didn't get in the last word," Gunnar said watching as CJ and Diord climbed inside the shuttle and closed the door. He grinned broadly, letting a chuckle go. "Well..."

"Colonel," Alex 7001 said from behind. "We are ready to stage the ship as soon as the Queen's shuttle has cleared the main doors.

Gunnar started for the boarding ramp, giving the shuttle a quick glance as it rose from the hangar floor and turned for the open bay doors.

"Have Lieutenant Starman get her started up."

*   *   *   *

Dãsha and Dacey sat staring at a small display just off the main navigational consoles on the bridge of *Calypso*. There was an unremarkable control panel in front of it with an odd looking terminal to the side. A myriad of foreign symbols scrambled across the display in an erratic fashion, some stopping, scrolling down, then backing up before proceeding again.

"Is this what you were looking for?" Dãsha asked.

"I'm not sure," Rick replied. "It has the same characters and symbols as the equation Ona provided. Do you know how to work it?" Rick asked examining the information a little closer.

"As this equipment appeared to have no vital function to our existence while marooned in Acuity, I never tried to discover its purpose," Dãsha admitted.

"Well, now it's time to figure it out," Rick said.

Dãsha looked at a larger display as Dacey started working with the small control board in front of him.

"Careful, Mister Dacey," Dãsha said without looking down. "Unless you fully understand the symbols associated with those controls, you could end up sending *Calypso* someplace we would rather not be."

"I've got a handle on what most of these controls do. I even know the symbols. It's the math that's killing me," Dacey complained.

Moments later, Alex 7001 sparkled into existence directly behind them.

"Why is Colonel Conrad's droid here?" Dãsha asked without looking back.

"How did she even know that?" Dacey whispered.

"The reflection in her display," Rick whispered without looking up. "He's here because he understands these symbols and equations better and faster than we do."

"I don't understand the human reference to a droid," Dãsha commented.

"It's an affection humans develop when dealing with anything they have a vested interest in," Rick responded, looking up at Alex as he floated closer.

Dãsha turned her head, looking at Rick as if he had lost his marbles or something.

"Oh come on. Surely Danis knew something of this kind of thing?"

"Danis never had opportunity for any such affections to anything, human or mechanical."

"No, I don't suppose she did."

"Who's Danis," Dacey asked looking at Rick, then Dãsha.

"Have Caidin explain it to you sometime," Rick said, ignoring the Drake pilot.

Dãsha cocked her head, processing.

"Thank you, I will."

"You requested my presence, General?" Alex said.

"Yes, Alex. You have the equation file in your memory?"

"Yes, Sir."

"Can you give us a rundown on what we're looking at and in what order we need to do this?"

"It is surprisingly simple," the droid started.

"I indicated as much when Ona Tusk provided the equation," Dãsha announced, stepping back from the display to allow Alex closer access to the console.

"As I said," Alex restarted. "It's quite simple. The symbols represent the time component. The numbers in this column indicate the locations of not just the target point, but the origins and where the bending must take place. That last part of the equation is the most critical. Get it wrong and the other formulas won't match up and you'll be sent to who knows where, and you won't be able to get back to the starting point to make another try. Keep in mind that you're dealing with an interface to make the system function. I don't know who the author of your equation is, but I haven't been able to theorize the details of how it works, let alone what the mechanism is that facilitates the operation."

"Ok, so using this equation, we input our current location here?" Dacey asked, pointing to a column of numbers moving vertically through the display. "And we input our destination in right here?"

"Affirmative," Alex responded.

"Then these symbols are for the time component." Rick said pointing.

"And the Bend information must be entered right here," Dãsha directed, tapping at the console with her fingers. She stepped back, looking at the display above. Cocking her head slightly, she turned to the droid. "Does this information look correct to you, my friend, droid?"

Dacey and Rick looked at Dãsha. Alex remained silent, processing.

"What?" Dãsha asked.

"Yes, according to the information in my database, this is correct."

"How is it activated and what should we expect?" Rick asked.

"As this will be my initial folding operation," Alex admitted. "It will be difficult for me to provide you with any kind of description."

"I can see no reason to expect anything other than perhaps a slight jolt," Dãsha said. "If your calculations are anywhere close to accurate, we should arrive almost instantly."

"Input the information," Rick directed. He watched carefully as Dãsha and Dacey worked the controls while Alex hovered close by. After several checks and rechecks, Dacey turned to Rick.

"We're ready to go, General. Activation is this little button right here," he said pointing at the unremarkable control.

"You'd think they would have used a bigger button," Rick joked. "Something with big letters and maybe a cover guard with warning labels all around it. 'Warning! You're about to blast your reality into oblivion!'"

"Would that have made this operation any easier?" Dãsha asked stoic.

"It might," Rick retorted good-naturedly. He looked around for any indications of a problem. As this had never been done before, he was a bit apprehensive. Thinking for several moments, he finally gave a nod.

Dacey turned back to the controls and watching the screen in front of him, pressed the button. The displays suddenly became a flurry of information, impossible for the human mind to follow. There came a dull thumping sound from somewhere beyond the ship's bulkheads, then a sudden beam of light from outside glared through the bridge windows.

"Dãsha, where are we?" Rick asked, stepping to the pilot's station for a look. Outside, the bright orb of a sun burned bright, giving everyone cause to shield their eyes until Dãsha activated the window filters.

"Androlus Quasar complex," Dãsha answered after a quick check of her instruments. "There are two stars behind us, another one to our starboard and three more to the port."

"We're in the right place," Rick said, folding his arms. "Now are we at the right time?"

"If I'm reading this correctly," Dacey said looking at a screen near Dãsha. "There is a large ship in orbit around one of the nearby planets."

"Dãsha?" Rick turned back to the tall beauty.

"Patience, Richard," she responded without looking up from the vast instrument console. "There is a lot of information coming in. As *Calypso* has never been here before, it will take her a moment to collect and organize what's here."

"*Her*?" Rick repeated, stepping beside Dãsha.

Dãsha instantly turned her head, looking at Rick with her black eyes.

"Have I used the inference incorrectly?"

"No, not necessarily. Just wondering about your gender choice."

"It was the only logical choice to make. My observance of you and your crew indicates your preference to referring to inanimate objects in the form of the female gender. And while the reason still escapes logical thinking, considering *Calypso's* vast capabilities, the designation fits better than a male reference."

"I think we've just been insulted, Mister Dacey," Rick chuckled looking at the information on the displays.

"How so?" Dãsha asked.

"Not important."

Dãsha gazed at Rick for a moment longer, then looked back at the readouts and pointed at a picture.

"Benton class light Capital cruiser. *Zebra Phar*. In orbit over Genis Two," Alex 7001 announced from behind.

"Any Valkyrie T-Class ships in the area?" Rick asked.

"No other ships in the area," Alex replied.

"Have they scanned us?"

"Not yet, and if you want to keep it that way, I recommend you move *Calypso* further away, toward the fourth sun to help hide her profile," Alex suggested.

"Mister Dacey, will you get us to the other side of Shano?" Rick gestured to the pilot's station.

"Where is Shano?" Dacey asked moving over to the pilot controls.

"The fourth sun," Rick said. He noticed Dacey turn back as he stepped into position. "The one over there," he pointed at the large overhead display at one of the blazing orbs. "Ok, Alex, time for you and me to get back to our ships." Rick took a position next to the droid. "You two gonna be all right here by yourselves?"

"Does it matter?" Dãsha asked.

"Can't really go back now, can we?" Dacey asked as the surrounding Quasar clouds outside began to move across the massive windows.

"Technically, in this ship, you can go anywhere you want," Rick suggested as he felt the teleporter beam starting to sparkle around him. "Provided you know the equations to use."

"Rick?" Dãsha said, turning before he disappeared. "Let Caidin know that I am just fine."

"I can do that."

"Be careful."

Rick winked and was gone. Dãsha looked back at the Folding displays and several columns of numbers and symbols still racing across the readouts. Her mind shifted back to Ona Tusk. There was still so much that she didn't know about *Calypso*. The Folding controls had been a mystery to her, but looking at them now, she suddenly had a grim understanding of what the racing symbols and numbers meant. She stared at the displays, realizing the unit was still counting...

A *Viper* among us

Exiting the *Athena's* teleporter room, Rick was met by Caidin. Focused, Rick breezed past the older gentleman.

"Dãsha? Is she all right?"

"Yeah, why do you ask?" Rick suddenly stopped, wondering if he had overlooked something.

"No reason, I just felt the need to ask."

"She said to tell you she was fine; don't worry. This is all part of the plan. There is no one better equipped to care for *Calypso*. And no ship is better to take care of a woman like Dãsha." Rick waited for a response, but Caidin only nodded and shuffled into sickbay. As Rick stepped quickly through the bridge doors, Laura Habba turned to him.

"General, Colonel Conrad is waiting for you on com line one."

"I'll take it up front," he said, heading to the weapons station and sitting down. "Have you got Dãsha and Mister Dacey?" he asked, pushing his headset down into place.

"Yes, they are on com line two."

"How are we doing?" Rick asked Gunnar.

"That depends. When are we going to get this show on the road? You wait too much longer and you're going to have a mutiny on your hands, and I'll be the one leading it. Then you'll have to put those new recruits to work. Won't that be fun?"

"Folding operations are complete. We're here, next to Shano in the Androlus Quasar."

There was silence for a moment.

"Seriously?" Gunnar finally asked. "We didn't feel a thing. How long did it actually take?"

"We almost didn't see it," Rick said, checking his displays. "I'm sure Audra will be disappointed."

"I'll let her know she can pry her fingers out of her armrest. We need to figure out how that thing works before you return it."

"Agreed, first things first," Rick agreed. He turned his chair around and faced Toby Mavis at the science station. "*Calypso* reports the *Zebra Phar* in orbit over Genis Two. Are you able to confirm?"

"Not until those doors are open," Toby responded. "Genis Two is too far away."

"Time to get them open then. Contact the *Stark*. Have Lieutenant Gillespie tell the other freighters to launch and follow the plan. The *Athena* and *Constellation* will hold here next to *Calypso*."

The communications officer went to work and moments later, the enormous bay doors opened. One by one, the four Kendalon freighters exited the great ship. After they were clear and moving away, Captain Lisa Dayton took hold of her controls and guided the *Athena* out with the *Constellation* directly beside her.

"General, Commander Niker is calling on the open battle channel," Laura announced.

"Yes, Commander," Rick nodded.

"Richard... I mean, General. Just to confirm, once we're in orbit and make contact, what happens when we off load?"

"Lieutenant Gillespie will move the *Stark* and the other freighters back to the far side of Shano as soon as your ships clear the doors."

"Then what?"

"Then we'll see what comes out of the cracks."

"She sounds really nervous," Captain Dayton commented.

"She has every right to be," Rick agreed. "Aren't you nervous?"

"I feel like I should be wearing a diaper... Sir."

Rick cracked a smile and turned to Toby.

"Now, how does it look out there?"

"Very quiet. Only the *Zebra Phar* on the scopes."

"All we can do now is wait."

"Aren't you a little concerned about our four proteges?" Gunnar asked from his channel.

"Little doesn't even begin to cover it," Rick responded quietly. "What about you?"

"You could say I'm a little nervous," Gunnar said. "I'd rather be on a beach somewhere getting a sunburn... oh wait, I don't burn. The only reason I agreed to this part of your plan is because those Starbirds have the new AI units in them that can operate those ships without anyone onboard."

"Let's hope it's enough if this little operation goes the way I think it will."

"Sir," Laura called from communications. "I'm picking up chatter on the fleet channels as well as the emergency channels."

"Out here?" Rick asked, a little surprised. "The Quasar complex's proximity to the Caldron should be blitzing everything."

"Coming from the other direction," Laura responded, listening carefully.

"Are you hearing this?" Rick asked after listening closely.

"It would appear Miss Tusk wasn't making it up," Gunnar replied. "Sounds like half the fleet is under attack."

"Lieutenant, can you pinpoint their locations?" Rick asked.

"Fleet chatter is originating from the Marcos system," Laura replied.

"That's next door to Kalamar," Gunnar said.

"Yeah, a little too close for comfort," Rick agreed.

"The emergency beacons are coming from the Bellon trade routes," Laura said listening. "I count at least five. I'm also picking up the *Zebra Phar*. They're signaling the freighters."

"Are they using the command codes?"

"Yes, Sir."

"General," Jayda called from her channel. "We're in orbit just behind the *Zebra Phar*. They're ready to make the transfer."

"Stick to the plan," Rick said quietly, still listening to the fleet chatter.

"Something isn't right," Gunnar remarked, listening as well.

"What's not right?" Rick asked.

"Something about Assault Corsairs attacking the fleet?"

"That's got to be garbled transmissions," Laura said listening.

"Can you clean it up?" Rick asked.

They continued to listen until Rick suddenly sat up straight.

"Did you hear that?"

"The *Lexington*, *Tut* and *Sung* are attacking the fleet."

"What?" Lisa asked.

"That's got to be a mistake," Toby said, turning to his scopes as a warning came alive.

"Put me on the open fleet frequency," Rick ordered.

"You're on, General," Laura called.

"Mayday, Mayday. This is the *Athena* of the Royal Kalamarion Starbird fleet; General Richard Niker commanding. We are in general distress in orbit over Genis Two in the Androlus Complex. Can anyone assist?"

"That was kind of generic," Gunnar announced flatly.

"By design," Rick responded. "Let's see if it gets anyone's attention. Lieutenant Gillespie, get those freighters out of there now."

"Sir," Laura called out. "Message from Dãsha. She says the Folding readouts are still active and the unit is counting."

"We just left the unit on," Rick said preoccupied. "I'm sure it's nothing to worry about."

"We've got company," Gunnar called out. "Mister Pippin has four T class destroyers coming over the upper pole of Genis Two."

"Confirmed," Toby announced.

"Are the Starbirds in position?"

"Yes, they're holding just off the *Zebra Phar's* port bow."

"Destroyers are Valkyrie," Toby announced as the freighters turned and sped away. "They're targeting the *Zebra Phar*."

"Nuts!" Gunner yelped. "Let's get in there..."

"Hold it," Rick ordered, remaining calm.

"You hold it. Audra and Jayda are in there."

"Well aware of that, but if you go plowing in right now, it will tip them off that something's wrong and this could be all for naught."

"It looks to me like something is going wrong right now."

"Calm yourself," Rick reassured his friend. "They came here for those Starbirds. They're in no immediate danger. I can't say as much for the *Zebra Phar*."

From their vantage point partially hidden behind Shano, they continued watching the events unfold near Genis Two. The *Zebra Phar* seemed up to the task of defending itself as it maneuvered away from the four attacking Valkyrie destroyers. Ordnance began crisscrossing between the ships as they traded pulse cannon, conventional missile and torpedo fire.

"I'd love to put a couple of our Mark Vs into those destroyers," Gunnar spoke quietly into his pickup.

"From what Dakota tells me, that might not be such a good idea," Rick replied, scanning the small weapon's tactical displays in front of him.

"Ok, maybe just one?"

"You probably wouldn't even need to hit any of them," Rick said. "Just set it to proximity detonation. It would be enough to take all four of them out."

"Yeah, we'll just keep that idea in our pocket for now," Gunnar smirked. "How long are we going to let this go?"

"Until someone gets the upper hand."

"Such as?"

"Either the *Zebra Phar* is going to take out the four Valkyrie ships or vise-versa. That's when things will really start to happen."

"Yeah, it's called four Starbirds just sitting out there waiting to be scooped up."

Everyone watched and listened to the space chatter blasting from the bridge speakers and a flurry of tactical information flowing across the displays in front of them. For better than half an hour, the battle raged between the destroyers and the Capitol ship. When the skirmish was finally over, all four Valkyrie destroyers disengaged as they had been nearly incapacitated. Despite being severely damaged, the *Zebra Phar* moved back toward Genis Two and the four waiting Starbirds.

"They're signaling again," Lieutenant Habba announced.

"Let the AIs acknowledge," Rick said.

"Are we still sitting on our hands here?" Gunnar asked.

"Wait for it," Rick replied, his eyes frozen to the screen.

"Wait for what?" Gunnar sniped.

"AI transmission completed," Audra called from the com channel.

"Wait for it," Rick repeated, watching.

"*Zebra Phar* is confirming," Jayda announced.

"Their bay doors are starting to open," Toby commented from his chair. "They'll be sending tugs..."

The science officer tensed.

"Long range sensors picking up another ship coming in," Pippin announced.

"There it is," Rick breathed.

"Daliger class dreadnaught dropping out of lightspeed near their position," Toby called.

"Idents?" Rick asked.

"You're really going to ask?" Gunnar ramped. "How many times have we been up against that thing?"

"The *Viper*; Croft Heckla himself," Rick said, eyeing the tactical displays back at the science station.

"He's arming weapons," Toby announced.

"Of course he is," Gunnar growled. "Enough waiting for it! We've got to get in there."

"Wait!" Rick studied the tactical display.

"No more waiting," Gunnar insisted.

There came a sudden flash on the displays as one of the Valkyrie destroyers suddenly disappeared.

"He's firing on his own ships," Toby exclaimed.

"Ok, so that's a little unexpected," Gunnar admitted, surprised.

"Commanders," Rick spoke quietly as if trying to remain undetected. "You have to hold your positions. Don't move..."

"You're going to let them take us?" Audra cut in excitedly.

"Stay calm, Commander," Rick cautioned. "With the *Viper* here, you need to let the commander of the *Zebra Phar* deal with him."

"I'm no expert," Constance cut in, "but we're looking at a very large, very well armed ship with a very well-seasoned crew that, by reputation, has no problem cutting down anything or anyone that gets in the way."

"If I'm right, he's here for those four Starbirds," Rick responded, watching the dreadnaught on the tactical display.

"I sure hope you're right," Gunnar said bluntly. "You have four Corsairs out there loaded with Mark V torpedoes and four inexperienced commanders with less than skeleton crews, up against one of the most ruthless and cunning warriors the Valkyrie has ever produced."

"It wouldn't be my first time," Audra spoke, trying to sound unafraid.

"Oh... you're going to count your encounter over Commenor as experience enough to take on the *Viper*?"

Audra didn't answer.

"I believe Croft is here to take possession of those Starbirds before the *Zebra Phar* does. But someone has to be the broker," Rick postulated.

"The commander of the *Zebra Phar* is sending instructions to the AIs to form up on him," Jayda announced.

"Let the AIs follow," Rick ordered. "Be ready to override if Croft comes at you."

"Leave the *Zebra Phar*?" Jayda asked.

"As soon as Croft engages, I guarantee the *Zebra Phar* will turn tail and run."

"Then we're really dead meat."

"When he runs, that's when Gunnar and I will come in."

"Are we ending this out here?" Gunnar asked.

"One way or the other."

Everyone on the bridges of the *Athena* and *Constellation* watched anxiously as the four Starbirds maneuvered into position on either side of the Benton Cruiser and waited while the *Viper* made quick work of the remaining damaged destroyers and turned back to face off with the *Zebra Phar* and its newly acquired escort.

"All right," Gunnar spoke up. "So far, you've called every move. I sure hope you know exactly what the next move is going to be, cause I've got to tell ya, right now, I'm scared out of my pants."

"And cue Mister Heckla," Rick said pointing to Laura at communications.

"Croft is signaling the commander of the *Zebra Phar*." She put her finger to her headset. "He's demanding the Starbirds be transferred over to him or he'll destroy all of them." Laura paused, listening. "The commander is refusing. Something about not having received full payment."

Several tense moments ticked away as the standoff seemed to continue for an eternity. Rick gave Captain Dayton a glance. She held her hands poised directly over her glass controls, waiting.

"That's a big ship," Audra commented from the comms. "Can a Starbird really stand a chance against something like this?"

"We're going to find out pretty quick," Rick said, watching.

"Croft just fired!" Toby announced.

"Now we go," Rick called, charging the *Athena's* weapons. "Captain Dayton, get us in there now!"

Lisa pushed her throttles all the way to the stops.

"Launch the Interceptors," Gunnar ordered. "Captains Abrams and Dalley, you need to get in there before us and knock out the shield arrays, then draw off their fighters."

"Captains Zek and Zak," Rick called. "Cover their six. This party will turn nasty real fast if we get too many of their fighters on the outside. Once the shield arrays are down, provide cover for the other four Starbirds."

"On your wing," Gunnar called. "Wished I had a better idea of what was going on here."

"I suspect once we give this big dawg a taste of what a couple of Starbirds can do, the plot will reveal itself in quick order," Rick said, setting up the *Athena's* weapons as the two Starbirds sped away from Shano. "I'll take out his launch bays."

"The weapons control stacks and launchers are mine," Gunnar announced. "We'll meet up at the command and control decks. You have no idea how long I've waited for a chance like this," Gunnar paused. "Yeah, well, maybe you do." He looked up in time to see the four Interceptors flash overhead toward the large grey bulk directly ahead. He felt a twinge of jealousy, thinking of being in the cockpit of the big fighter heading at such a large target. He recalled leading his squadron against this ship many times. The Valkyrie fighters were formidable, but the T-6 Tempest had always held its own against them. In the hands of a skilled pilot, what the T-6 lacked in firepower, more than made up for in speed and maneuverability. Now, the Interceptor was rivaled by nothing, Kalamarion or Valkyrie.

"As advertised, the *Zebra Phar* is running," Jayda called.

"Stay with him," Rick ordered. "You'll have better cover, but if things get too hot, go into override mode and get clear."

"We using torpedoes here?" Gunnar asked as the *Viper* loomed before them. The Interceptors had already engaged their targets, working to destroy the shield arrays of the mammoth ship.

"Stick to Pin missiles for now. We may need the bigger stuff later."

Rick zeroed in on his targets as the *Athena* cut through the *Viper's* failing shields. Touching several points on the weapons control glass, the *Athena* bellowed deadly pulses of energy at the fighter launch tubes. As Captain Dayton swept the Assault Corsair past, a volley of Pin missiles rocketed away from the forward bridge pod and streaked directly into the opening. A plume of expanding metal and hot gases erupted from the launch bay sending large chunks of debris hurtling away from the ship. The *Athena* repeated this process for the seven launch bays all around the dreadnaught. Unfortunately, as the attack was analyzed and their purpose discovered, the Valkyrie fighters quickened their launch procedures and a good number made it out before the last bay was either disabled or destroyed. Focusing their attack, the Valkyrie fighters quickly set upon the two Assault Corsairs as they made vicious attacks on key locations around the enormous dreadnaught.

Both Starbirds began to take a beating, not only from the fighters swarming after them, but also from every cannon turret and Rattle gun mounted on the dreadnaught. The pursuing fighters didn't last very long as the top turret cannon of the Starbirds seemed to follow any incoming fighter with pinpoint accuracy. It wasn't until two large Kalamarion fighters streaked directly over the *Athena,* that the pounding from the pursuing Valkyrie fighters stopped. Captain Zek and Zak Korack made quick work of the enemy craft, then moved onto their next targets.

"Take her close along the starboard side," Rick ordered, without looking up from the weapons console. "We'll blast as many maneuvering thrusters as we can."

"Do you have a preferred speed?" Captain Dayton asked quietly.

Rick gave the Starbird pilot a glance, noticing how relaxed she appeared to be.

"I'll try to keep up with you," Rick said. "Gunnar, how you doing over there?"

"These Rattle guns are murder and this thing has a few more launchers than the last time we did an inspection this close."

"You gonna be able to handle it?"

"I choose to be insulted by that question," Gunnar replied. "We're already heading for the weapons control stacks. Don't make me wait for you at the command and control decks."

"Fighters killing you?"

"Still Insulted! Dakota and Logan made quick work of them, so they're headed back out to the *Zebra Phar.*"

"The *Zebra Phar* is getting hammered out here," Jayda called. "They're still trying to reengage the AIs for assistance."

"Captains Abrams and Dalley are headed your direction to help," Gunnar announced.

"Be there in two shakes, Commander," Dakota spoke up over the comm chatter. "So nice to be behind the controls again."

"Nice to have you back on my wing," Logan said.

"Yeah, let's see if I can remember how to fly wing."

"Get in my way and you'll find out just how good I am."

"To your left," Dakota called, making an adjustment to his targeting scope.

"Three coming up from behind," Logan responded, touching the firing control on the yoke.

"'Scuse me while I take care of some bidness back here," Dakota said, tapping the reverse thrusters.

In an instant, three Valkyrie fighters streaked past the cockpit windows, chasing after Logan's Interceptor. A moment later, Dakota had one in his targeting sights and fired his weapons. His powerful guns quickly tore into the thruster ports of the ship; a split second later, the fighter disintegrated. He shifted to pick off the other one, but it maneuvered away too fast. The Valkyrie fighter disappeared from his heads-up display, but he was able to follow quick enough to

keep him on his external scopes. Dakota throttled harder in pursuit.
If he didn't do something now, he'd only have to contend with him
later on. The Valkyrie fighter went through a series of complex turns,
strafing the *Zebra Phar*, the *Phalanx* and the *Montego* before he could
get close enough to pop off several shots. Evading effortlessly, the
enemy ship made a sudden turn and came right at Dakota, firing. His
Interceptor shuttered violently as the ordnance found its mark.
Holding down the firing button, he gave the sighting display a glance.
Brilliant flashes swept back from the fighter as it streaked directly past
him. He turned hard to reengage, but noticed the enemy fighter had
disappeared from his scopes.

"Oh, this guy is good," he grunted looking around and making
several adjustments to his tracking computer. His eyes widened when
something flashed across his displays. Before he could react, the
Interceptor was slammed by a wall of gun fire. Warning alarms blared
in his ears as he shoved the throttles all the way forward and turned
away. Barely able to follow, he kept his foe on his tracking scanner as
the fighter raced back toward the *Viper*. Pushing the throttle ahead to
catch up, he glanced at his targeting computer. Realizing he was
being pursued, the other pilot weaved erratically.

"Thought you were on my wing?" Logan called out.

Dakota glanced at his rear tracking screen, seeing his wingman.

"I was, until this smart guy decided to break things up a bit."

"Let him go, we need to get back down to those Starbirds."

Dakota observed the *Constellation* on his scanners, skimming along
the drab hull of the dreadnaught toward the command and control
decks. A moment later, he took note of the *Athena* converging on the
Constellation's position. Frowning, he started his turn when the
Valkyrie fighter he had been chasing suddenly turned and came right
at him. Angry that he had let himself be drawn away, he lined up on
the fighter as they closed on each other. As the fighter swung into his
sights, he pulled the dual triggers on the underside of the control
yoke, activating the powerful blaster cannons mounted on the leading
edges of the V-winged craft. The enemy fighter opened up as well.
They held their courses straight at each other. Dakota let out a drawn
out, enraged yell while his Interceptor started taking hits, but held his
firing buttons down. There came a brilliant flash at the same moment
the Interceptor was thrown violently into a wild tumble.

This is it, he thought. *Gonna smear all over the hull of the Viper.*

"You're welcome," a voice called out over his headset. "Now you
and Logan get back here with the rest of us."

It took a moment for Dakota to regain control of the Interceptor,
correcting the tumble in time to see Captain's Zek and Zak Korack
throttling toward the four Starbirds near the *Zebra Phar*.

"Oh, that was nice," Logan said calmly. "You were going to go out in a blaze of glory. And for what? A single Valkyrie ace? Yeah, that would have been worth it. Not to mention making Willis a widow before you even marry her."

"Yeah, sorry," Dakota said, feeling quite sheepish. "I guess I got a little carried away."

"Stay with me, Captain," Logan said as they steered toward several more fighters attacking the *Intrepid*. "There are plenty of hot targets right here."

Dakota looked out at the *Zebra Phar* as it took several hits. It was still throwing a blaze of fury at the *Viper*.

"Is this a good time for one of those torpedoes?" Gunnar asked, looking out at the enormous dreadnaught still sending a constant volley of fire at the cruiser behind them.

There came a sudden flash that lit up the entire grey surface of the *Viper*. A moment later, the dreadnaught stopped firing.

"The *Zebra Phar* is gone!" Toby called, still looking at the tactical display.

"It just exploded and crumbled to pieces," Pippin reported over com.

"All ships take up defensive positions next to the *Athena* and *Constellation*," General Niker called out.

"Commanders Atlanta and Quayle," Colonel Conrad called out. "Form up next to the *Constellation*. Commanders Niker and Fowler, next to the *Athena*."

As the *Constellation* and *Athena* maneuvered into position, the other four Starbirds came along side, facing off with the *Viper*.

"Another ship just dropped out of light speed," Toby called from his station.

Rick turned for a look as the tactical display flashed a familiar image.

"It's Admiral Mandell's flagship," Toby announced excitedly.

"Knew we'd get someone's attention," Rick said. "But what's the Admiral doing out here by himself?" Rick puzzled.

"Right now? Probably wondering the same about us," Gunnar commented. "We're supposed to be rescuing some freighters at the Caldron."

"I'm sure a distress signal from Genis Two got his attention," Rick said, thinking carefully. "Seems convenient the Admiral is in the area. Mister Mavis, what's his weapons status?"

"Pulse cannon batteries at full power, but his missile and torpedo magazines are empty."

Rick looked at Gunnar with raised eyebrows.

"Curious he'd be out here all alone without a load," Gunnar commented.

"Unless he's just used them," Rick postulated.

"We're being hailed on the battle frequency," Laura Habba announced.

"Patch it through on to the fleetwide channels," Rick ordered.

"Sir?" Laura questioned.

"Make sure the Admiral can see the Colonel as well."

"General," Toby objected. "The rest of the fleet will be able to monitor our transmissions, including the *Viper*."

"You are correct," Rick responded, stepping to the command chair and sitting down. Without hesitation, he turned to the communications display over Laura's head as Gunnar appeared in a separate picture next to the stoic image of Admiral Mandell.

"General Niker, Colonel Conrad," the Admiral began. "You're supposed to be at the Caldron. What are you doing out here?"

"Responding to a distress call from the *Zebra Phar*."

Admiral Mandell looked at his scanning equipment.

"I don't see it anywhere."

"That nut-job behind you sorted of... blew it up," Gunnar cut in.

Admiral Mandell sharpened a glance at Gunnar.

"Admiral," Rick said. "We've disabled the *Viper's* defensive capabilities and we're just about to put an end to this."

"You've taken out the *Viper's* shield generators and torpedo launchers?"

"...And its ability to launch fighters," Gunnar sniped impatiently. "So if you'll move aside, we'll finish the job."

"Have you offered negotiations for Croft's surrender?"

"Are you kidding?" Gunnar continued flippantly. "When a known Valkyrie master criminal is caught red handed destroying another ship in an outright act of war, surrender generally isn't an offer on the table."

"Colonel, I don't much care for your tone."

"And I'm not going to wait around for Croft to get away again."

"Well, I can't allow you to do anything to the *Viper*."

Gunnar fell silent, trying to figure out what the Admiral was suggesting.

"What do you mean, you can't allow it? We put an end to him now and the war with the Valkyrie is over."

Admiral Mandell stared at Rick and Gunnar for a moment and finally let go of a heavy sigh.

"That's what I can't allow."

"Why not?" Gunnar asked, angrily.

"For precisely the reason you just stated. It would end the war."

Dumbfounded, Gunnar looked at Rick. His friend was caught in a deep stupor.

"This is what countless have fought and died for," Gunnar retorted, trying to figure out the paradigm. "We've been at war with the Valkyrie since before I enlisted in the Flight Ministry. I've lost friends and great pilots to this war. Now, you better believe if I have a chance to finally end this madness, I'm certainly going to take advantage of it. Why wouldn't you want it to end?"

"Because if the war ends, we become obsolete," Rick stated coldly. "Isn't that right, Admiral?"

"The galaxy needs war in order to maintain balance," the Admiral nodded. "General, you shifted that balance when you presented the Starbird Initiative and subsequent program. To add insult to injury, you also came up with the basic design for the Mark V torpedo. Turns out it changed the paradigm of the war almost as much as the Starbird did."

"You and the other Starbirds attacked the fleets in the Marcos system and the Bellon trade routes," Rick announced as the puzzle pieces fell into place.

"With the Starbirds operational, there seemed little point to having so many large capital ships. You've proven that easily enough by rendering the *Viper* nearly helpless with just two. So, the size of Kalamar's fleet needed to be... downsized, just to make things a little more even."

"But Kalamar still has the advantage," Gunnar pointed out. "The Starbirds, the Mark Vs and the fleet, or what's left of it."

"That's not entirely true," the Admiral responded calmly.

"We've scanned Croft's ship," Gunnar countered. "He's got nothing but fighters and assault craft stuck in his launch bays. He can't fire any torpedoes or High Velocity mines either, unless he tries to toss them out a window. Even if he could, we would take them down before they had a chance to reach anything."

An awkward pause held time frozen as each combatant considered what should come next. Rick finally leaned forward.

"You couldn't deliver directly to Croft without raising suspicion," Rick revealed. "So you orchestrated the sale and delivery of these four Starbirds, making it look like they were going to one of Kalamar's allies by way of the Filon. They contracted with the Kendalons to transport them, marking them as spare parts. You leaked their route through the Oneida Caldron to the Valkyrie where a couple of Croft's ships ambushed them. But the Kendalons gave them the slip and got to Genis Two anyway. When Croft was informed of the failure at the Caldron, he ordered them here, sending two more destroyers. They had to either intercept the transfer to the *Zebra Phar* or at the very least, delay it until the *Viper* got here. But why would he destroy his own ships?"

"You'll have to take that up with him," Admiral Mandell answered. "Croft is a bit of a fanatic. Perhaps a low tolerance for failure or he didn't want anyone knowing the Valkyrie had made a deal with Kalamar?"

Rick and Gunnar eyed the Admiral with profound disappointment.

"Admiral, you've betrayed us all," Gunnar surmised.

"It was supposed to look like technological espionage which isn't easily traced," the Admiral admitted. "Croft would have the secrets of the Starbird by having four fully operational prototypes with the new AI installed, and the war would continue with a purpose. You gentlemen were never supposed to have any part in this. Croft was supposed to send a couple of his best destroyers to the Caldron to see that your rescue operation experienced complications."

"You didn't think a Starbird could outgun a Valkyrie T-class destroyer?" Gunnar asked.

"On the contrary, I knew they couldn't outgun you. To tell the truth, I was a little worried about them going up against two of the Flight Ministry's finest. But they only had to nudge you into a bad position. Clearly that didn't happen and the freighters arrived just fine. So things worked out anyway."

"Admiral," Rick spoke up. "Surely you know we'll make sure the other Starbird commanders know about this."

"By the time you can get through to them, the damage will have already been done. Croft will have these Starbirds and if you're smart, yours as well, which will more than even out the balance."

"You're suggesting we join you?"

"With or without you, he's getting these four. I wasn't going to give him mine, but since you two are here messing things up, I may have to alter the plan a bit."

"Well, you have a bit of a problem then," Gunnar snarled. "We have control of these four Starbirds and we're not letting you or Croft have them. How does that fit into your grand plan?"

"Well, to be honest, it doesn't fit in with the plan at all. But I had to make the offer to have you join me and Mister Heckla back there, knowing you wouldn't take a liking to the idea." The Admiral looked over his shoulder at the science station displays. "Now, due to a tactical blunder on your part, you've made it fairly simple for me to finish our business here. I noticed that you have personnel manning those birds. Somewhat minimal, but the new AI is designed to be autonomous. A partial load of Mark Vs. Funny, they weren't included with the order. So much the better. So here's how this is going to work. You're going to hand over control of all four Starbirds so I can get them loaded into one of the *Viper's* cargo bays. Now, listen carefully because I'm only going to say this once." The Admiral leaned forward. "Both of you are going to make sure nothing goes wrong, because if something does, my fault, your fault, nobody's fault, using my command override codes, I'm going to blow them up, with your crews in them. Let's see, whose transponders am I reading? Captains Manx Quayle and Constance Fowler…" Admiral Mandell looked back from the display. "Commanders Jayda Niker and Audra Atlanta. Very nice, a little extra insurance."

"You wouldn't dare," Gunnar fumed.

"Give me a reason," the Admiral hissed back.

"Can he override our control?" Toby whispered to Rick.

Rick slowly shook his head.

"No, but he can use his command codes to order the ships to self-destruct."

"You can't let him take these ships," Jayda called from her comm channel.

"General," Constance called. "We'll blow them up ourselves before we let him have control."

"An offer to blow themselves up," the Admiral mussed. "Quite noble and certainly worthy of hero status. But I suspect you won't let that happen for exactly the reason I let you make your wives your first officers. While a marriage in the same command has never worked, in this case it works exactly the way I need it to."

"He wouldn't dare," Audra piped up from her comm channel.

"Everyone stay off the channel," Rick ordered.

"Don't kid yourself, Commander Atlanta," the Admiral warned. "You think I didn't study General Niker and Colonel Conrad's profiles enough to know they would never put you in harm's way?"

"He's bluffing," Audra maintained.

"I said off the channel, Commander," Rick barked.

"I see Commander Atlanta still has a little trouble with knowing her place," Admiral Mandell grinned. "Why don't we start with her?"

"Don't do it, General," Audra objected.

Rick thought carefully, looking at Gunnar. His friend remained quiet, a steely expression held firm on his face. He finally looked out at the four Starbirds on either side of the *Athena*.

"Captains Quayle and Fowler. I'm ordering you to turn over control of your ships to Admiral Mandell. Commanders Niker and Atlanta, you are ordered to do the same." Rick looked back at his viewer. "What guarantees will you give my crews, Admiral?"

"I'm not without heart," the Admiral replied, smiling. "Once I have those ships safely secured in the *Viper's* cargo bay, your crews can teleport over here. Once the *Viper* leaves, I'll have them teleported back over to you."

Rick nodded, looking at Gunnar who remained unmoved. Admiral Mandell turned away momentarily to direct the control of the four Starbirds.

"It's going to take a couple of minutes here," the Admiral announced stoic. "Just enough time to try and talk some sense into you boys."

"What? Joining you in betraying everything we stand for?" Gunnar asserted, a little calmer.

"We are all warriors here. Is it really going to matter what side we fight on? So long as we still have a purpose. War is that purpose."

"War isn't a purpose," Gunnar ramped. "War is a byproduct of someone with a narrow mind. And it *does* matter what side a warrior fights on, so long as he or she stays true to their convictions."

"Are you going to offer me a towel with all that dribble?" the Admiral responded, smiling. "So am I to understand that once you get your crews back, you'll want to continue with what you've started here?"

"Something like that," Gunnar maintained.

"Too bad. You are some of the best Kalamar has ever produced. You would have been a valuable asset in bringing up the next generation of warriors." The Admiral let out a heavy sigh. "Disappointing." He glanced back again, seeing the last of the four Starbirds land in the *Viper's* cargo bay and the door close. "I guess this is where we start shooting at each other."

"This ain't over until it's over," Gunnar fumed, feeling helpless. He took note of the remaining fighters entering the landing bays of the *Viper* as it prepared to leave.

"It will be soon enough."

"What happens to the other Starbirds?" Rick asked, seemingly caught in an endless loop of thought. He gave the sensor displays a glance.

"I'm confident once they've been given additional orders and explain the options available to them, they'll make the right decision. If not, they'll be faced with the same consequences the two of you are facing."

"And what additional information would you be referring to?" Gunnar asked.

"The other Starbird commanders are mindless simpletons; easily manipulated. They've acted on orders believing they were squashing an internal revolt by several rogue fleet Commanders."

"And no one is going to know where the real revolt is happening," Rick stated.

"They're either too stupid, dead, or shortly will be," the Admiral smirked.

Rick's eyes shifted to the image of Gunnar, who nodded and lifted one of his fingers, looking forward. The *Athena* and *Constellation* suddenly banked hard and streaked away from Admiral Mandell's ship.

"Where are you going in such a hurry?" the Admiral asked, taken off guard by their sudden departure. "You don't want a front row seat to history in the making?"

"Trust us," Gunnar called. "We'll have plenty of front seat."

Admiral Mandell looked back at the *Viper* as it prepared to jump to lightspeed.

"General Niker, what have you done?" he asked puzzled. Turning to Rick and Gunnar, a bright flash suddenly filled the bridge of the Admiral's Starbird. The windows instantly dimmed as another flash glared at them.

"What the...?"

"Explosions coming from the *Viper*, Sir," his science officer exclaimed.

"Where?"

"Her cargo holds."

"Helm, get us out of here," the Admiral thundered.

The Starbird flagship tilted hard as the entire side of the Valkyrie dreadnaught opened up in a brilliant eruption of molten metal. A secondary flash, brighter than the first two rippled out, engulfing most of the *Viper* as its other side cracked wide open. The explosion didn't stop, but continued its jagged journey up the dreadnaught's spine and all through its command structures. Pieces of bulkhead spun off in a furious storm of wreckage, shooting in all directions as the entire Valkyrie ship was consumed in an expanding cloud of flashing explosions.

Rick and Gunnar watched in awe at the immense power on display as it engulfed Admiral Mandell's Starbird.

"As I understand Captain Abram's report, if the Admiral's ship survives the blast, he should have a limited capacity to put up much of a fight," Rick observed quietly.

"Holy buckets, Rick," Gunnar responded astonished.

"Yeah, the simulations indicated a pretty big punch," Rick said. "But seeing it in person..."

"I told you guys," Captain Abrams called from his comm channel. "And I only fired one of them."

"Mister Mavis, do you have eyes on the Admiral?" Rick asked, still watching the fireball behind them.

"Mister Pippin has him," Gunnar announced.

Rick looked back at Toby.

"Their sensor arrays are better than ours," the science officer shrugged.

"He's still tumbling in the debris wave," Pippin announced.

"Remain on course," Rick ordered without looking at the Starbird pilot.

It took some time for the secondary explosions from the *Viper* to finally extinguish. By then, Admiral's Mandell's Starbird had been thrown clear of the debris wave and appeared to be drifting helplessly.

"Lieutenant Habba," Rick said looking up at the sensor displays over the science station. "See if you can raise the Admiral. Mister Mavis, do you think you can get me a readout on his condition before your counterpart does?"

"Already have it," Toby announced, working with his console. "His environmental, propulsion and defensive systems are all offline. Maneuvering thrusters and communications are still functional."

"Basically, in a world of hurt," Gunnar remarked, studying the readouts on his displays.

Rick carefully scanned the information flashing in front of him as Laura announced she had reestablished communications.

"Admiral, do you require assistance?" Rick asked, looking at the crackling picture of the Admiral's bridge. "We can have you teleported to our ships and tend to your wounded." There didn't appear to be anyone in the picture, only fitting displays and flashing emergency lights.

"Well played, General," the Admiral panted pulling himself up into his command chair. A deep gash over his right eye allowed blood to work its way down the side of his head. "I hope it was worth it. You have my condolences on the demise of your wives and crew."

"Sometimes, we pay a high price to do the right thing," Rick responded as the bridge door behind him snapped open. "However, this wasn't one of them." Jayda Niker and Manx Quayle stepped up behind Rick. Behind Gunnar, Audra Atlanta and Constance Fowler stepped into view.

"You planted your crew's transponders on those ships to make it look like they were there," the Admiral nodded. "Then set off their Mark Vs remotely."

"General," Toby called. "Four ships dropping out of light speed at one thirty-five forty-nine, mark three. More Starbirds."

"The rest of the fleet," the Admiral grinned painfully.

"We're being hailed," Laura announced from communications.

"Add it to this channel," Rick ordered. The display instantly changed, adding four more feeds. He recognized his fellow Starbird commanders. None of them looked very happy.

"General Niker," Commander O'Brien spoke up. "We came as soon as we heard the distress call."

"Have you been monitoring your fleet frequencies?" Rick inquired.

"Very illuminating listening," the Starbird commander remarked as the others nodded. "How can we assist in this situation?"

"Commander O'Brien," Admiral Mandell called frantically. "General Niker has disobeyed a direct order and destroyed the *Viper* and Croft

Heckla without due process. He and Colonel Conrad are to be arrested as war criminals. If they resist, open fire on them. That's a direct order."

There was a brief pause, then Rick stood up.

"Commander O'Brien," Rick began. "I'll defer to your best judgement in this matter."

The other commanders looked at each other, finally nodding.

"Admiral Mandell," Commander O'Brien spoke up, visibly agitated. "You are under arrest for high treason to the Royal Kalamarion Flight Ministry and Kalamar Central Command. You are ordered to surrender your vessel immediately."

"You can't order me to do anything," the Admiral scoffed arrogantly. "I'm the Commander of the Starbird fleet."

"If you do not surrender, we will be forced to fire on you."

"You wouldn't dare," Admiral Mandell flared.

"Easily manipulated, mindless simpletons don't know how to bluff."

Rick and Gunnar watched as the *Hyperion*, *Sulairus*, *Pandora* and *Avenger* surrounded the *Starbird*.

"General," Admiral Mandell called. "I'm ordering you to return and render aid!"

"I think you have plenty of... aid," Rick responded, motioning for Jayda to take over the weapons station. Constance gestured to the navigation station. Rick looked back and smiled, then nodded. "Captain Dayton, take us back to Shano."

*   *   *   *

The *Athena* and *Constellation* caught up to the Kendalon freighters just as they were rounding the back side of Shano only to find... nothing.

"We had *Calypso* on our scopes just before you caught up to us," a bewildered Lieutenant Gillespie reported. "Then, it just... wasn't there. I was talking to Mister Dacey and the channel just went quiet."

"Mister Pippin, since the *Constellation* has the better sensor arrays, do a full sensor scan of the entire Androlus Quasar Complex," Rick ordered.

"That could take a while," Gunnar pointed out.

"I want to know about anything out of the ordinary," Rick continued. "Mister Mavis, since we're this close, scan Shano and make sure *Calypso* didn't somehow collide with it."

"How would I even know if it did?" the science officer mumbled going to work.

"What do you figure?" Gunnar asked Rick as Lieutenant Starman turned the *Constellation* away from Shano's flaring influence.

"Before we got into it with the *Viper*, Dãsha reported that the Folding unit was still active and counting. We were a little preoccupied at the time, so I assumed it was just random telemetry."

"You don't suppose…?" Gunnar questioned.

"…The last part of the equation?"

"The part you didn't think had anything to do with what we were going to be doing?"

"…Ona included a return vector command in the equation she gave us," Rick postulated.

"I got the distinct impression she wanted *Calypso* back," Gunnar offered. "Maybe she had other plans for it."

"If that's the case, then we've got no idea where Dãsha and Dacey are or at what point in time. Even if we did know where and when, there's no way to reach them." Rick considered for several moments. Hearing the bridge door open, he turned to Toby Mavis. "Narrow your search parameters to the immediate area around Shano. Look for any temporal anomalies that might indicate a disruption in the space fabric."

Toby developed a puzzled look.

"Similar to what you might find in the vicinity of the Oneida."

The science officer nodded and went to work as Caidin shuffled toward the command chair. Rick scratched his head as he stood up.

"Caidin, I need to speak with you privately." Rick motioned to the door, but Caidin remained in place. The older gentleman gazed at the displays above the science station.

"Caidin, it's important," Rick persisted.

The older gentleman stared at the sensor scope for a moment, then looked out at Shano. The windows had dimmed so that its influence didn't overpower everyone working on the bridge. He closed his eyes, remaining perfectly still. The only sounds on the bridge were the whining of circuits. Rick watched him for several passing moments until Caidin turned around and opened his eyes.

"She's gone… isn't she?"

Rick gave the science station a glance, then looked back at Caidin, nodding slightly.

"We don't know what happened for sure. The *Stark* was tracking them from the other side of Shano and then they just… disappeared. We're checking Shano to make sure they didn't collide with it. I've also got the *Constellation* scanning their last known position for anomalies…"

"No, they didn't collide with the sun," Caidin maintained quietly.

"I just want to be sure," Rick reassured him.

Caidin turned and walked to the front of the bridge looking out one of the side windows.

"She's a millennia from here..." Caidin whispered to himself. "Doing something only she can do."

*   *   *   *

"Not finding anything at Shano, are you?" Jayda asked, leaning toward the scanning screens next to Rick. They had been searching hours for *Calypso*.

"We've scanned and rescanned it," Rick whispered, looking up at the overhead displays. "Not sure we'd know what we were looking at if we saw it. I think if *Calypso* had gotten pulled down, we would have seen something unusual in the corona flares. Gunnar's crew has looked at its last known position so many times, they're putting ruts in the space fabric."

"It's time for you to return to Kalamar," Caidin said walking into the bridge.

"I won't give up," Rick persisted.

"We have to face our current reality. All of us have important responsibilities to attend to."

"Caidin is right," Jayda agreed. "If there was something here to find, we would have found it by now. We need to get home and get out in front of this Starbird mess."

"Ona made it a point to say that there was no other that could take care of *Calypso* like Dãsha," Caidin said. "And I have absolute confidence in my wife's ability to take care of herself."

"What are you going to do about Tiana's stepmother, Nora?" Jayda asked.

"One of the responsibilities I have to get out in front of," Caidin responded quietly.

"Have you got it all figured out?" Rick asked. "Is there anything we can do to help?

"In my absence, the Mantose fortune was supposed to go to my children on their thirtieth birthday unless something were to happened to them. With Taron gone, Tiana became my sole heir. In this reality, we've only been gone a short time, so Nora will still be gunning for her. But if I were to show up..."

"That's going to require a lot of explanation," Rick remarked.

Caidin smirked, leaning against a nearby console.

"My identity will be easy to confirm and the current Board of Trustees will have no recourse but revert control of the corporation back to me, in which case, Nora will have no reason to pursue Tiana. It would be interesting to see how fast Nora comes out from under her rock when I cut off her stipend. Obviously, our sealing will be terminated."

"I should hope so. Imagine trying to account for Dãsha," Jayda smiled.

"What does it matter? Dãsha is the one I'm in love with."

"What if she never comes back?"

Caidin gazed back out at the cosmos, thinking... reaching out with his thoughts.

"I have to hold on to the hope that somehow, someday, she'll find her way back to us... to me."

Rick looked at Jayda and Caidin, then turned back to the scanning displays, gazing at them for some time.

"Mister Mavis?"

The science officer shook his head grimly. Even Caidin could see that all the scopes were barren of anything other than the *Constellation* and the freighters.

"Lieutenant Habba, get me the Colonel."

"You're on, General."

"You understand the definition of crazy is doing the same thing over and over expecting different results?" Gunnar commented.

"We've got nothing here, either," Rick replied.

"You got all the technical specifications on that Folding unit, didn't you?"

"Didn't have time. We were into operations before I had a chance."

"Seems kind of silly that we had to hurry when you think about it. Essentially, we had a time machine. You know, the things we dreamed of as kids? I think we better call this one and head home. I'm sure there are a lot of questions waiting for us back home."

Rick gave the displays another glance, still considering. He finally nodded.

"Yeah, I was just holding out hope that somehow she would find her way back here, at this time." Rick looked forward. "Helm, Navigation; take us home."

"Right beside you," Gunnar called.

Dates

Several days slipped by before the *Athena* and the *Constellation* arrived back at the orbiting space station, Newton at Kalamar. Their arrival felt quite unremarkable considering what the last two and a half years of their lives had been like. It was difficult not to tell anyone of their adventures, as explaining it would be problematic at best. It would take some time to unravel the intricate details of the Starbird debacle and Admiral Mandell's involvement with Croft Heckla and the Valkyrie. With the war effectively over, there were still plenty of political wranglings and bureaucratic posturing to go around.   Most of it created needless delays and prolonged getting to the truth. After a particularly long day of testimony, Rick, Gunnar, Jayda and Audra sat relaxing at an open booth in one of the many executive lounges of the Central Command facility.

"That was the longest couple of months I never want to go through again," Audra sighed with relief.

"Maybe you'd prefer the last two and a half years?" Gunnar suggested with a raised eyebrow.

"Funny man," Audra responded, her voice dripping with sarcasm.

"We'll know better than to stop by the Oneida Caldron ever again," Rick said.

"If we ever go anywhere near that thing again, I say we just keep right on driving," Jayda suggested good naturedly.

"I'm just glad we got through all that testimony in one piece," Audra said.

"I think the inquisitions were more focused on Admiral Mandell's insurrection," Gunnar suggested.

"I had a difficult time keeping up with everything," Audra said. "It was so convoluted you couldn't keep it straight."

"That's how government agencies and the military work," Rick admitted. "It took me two days just to outline why we remained on station in the Androlus Complex for so long."

"Seemed pretty cut and dry to me," Gunnar said. "The only reason this whole thing took so long and got so complicated was all bureaucratic rubbish heaped on top of the silly military protocols. Commander O'Brien and the others should have just blown the

Admiral to the same place we sent Croft and good riddance to the both of them."

"That would have made them no better than the Admiral," Rick countered. "They did the right thing by bringing him in and surrendering themselves to face charges for their actions. I think that alone ultimately saved their careers."

"So what's to become of them?" Jayda asked.

"They've been relieved of their commands, reduced in rank and reassigned to other places in Central Command," Rick said.

"I'm just glad the Admiral got what he deserved," Audra said.

"He'll have a long time to figure out if it was worth it," Gunnar said, giving Rick a nod and getting up.

"Hey, you girls up for another adventure?" Rick inquired, coming to his feet.

"Adventure?" Jayda stuttered, catching a gasp.

"Isn't that considered a swear word now?" Audra followed up quickly. "That's no longer a part of my dietary requirements."

"We're good. No adventure required for us girls," Jayda reassured them, wondering how serious they were.

"I'd say we got their attention," Gunnar snickered, helping Audra out of the booth.

"Definitely," Rick agreed, taking Jayda's hand.

"Please tell me you guys are kidding," Audra hedged.

"Be afraid," Jayda said with a worried look. "Be very afraid."

Gunnar closed an eye, squinting at his friend. Rick put his fingers to his forehead and looked back at Gunnar. Audra and Jayda developed an uneasy expression looking back at Gunnar and Rick.

"I totally agree," Rick said, nodding.

"Awesome minds think alike," Gunnar said, sliding his arm over his friend's shoulder.

Trying to figure out what had just happened, Jayda and Audra watched them strutting away at a brisk clip. They hurried after them, having to jog to catch up.

"What's this adventure?" Audra asked suspiciously. "I know that look. Anything from this point on is going to be scary."

"Yeah, what she said," Jayda echoed.

"Ah, I'm sensing some curiosity after all," Gunnar said without looking back. "Knew they couldn't resist a good adventure. Let's test their resolve, shall we?"

"We shall," Rick agreed, using the same jovial tone as Gunnar.

Presently, they boarded a pneumatic transport and were whisked out of the Central Command facilities and into the city. After some travel, they arrived at a station gate. As they made their way out of the terminal, they caught an open ground shuttle, taking them down a long concourse to a private entryway. Jayda and Audra kept giving

Rick and Gunnar looks and speculating between them as to their destination. Even as they were working their way through a short hall toward a bright opening at the end, they were still wondering what this so-called adventure was going to entail. As they came to the opening, Gunnar and Rick stepped through, took a step to either side and held out a hand. Jayda took Rick's and Audra took Gunnar's as they stepped out into the light of an enormous arena filled to capacity. Both women gasped, covering big smiles as they looked down into the steeply tiered seating at an empty spectator box.

"You remembered!" Jayda exclaimed, throwing her arms around her husband.

"Gunnar remembered," Rick admitted.

"I can't believe this!" Jayda yelled over the noise. "It's even the same performers! Goodness, they've got to be older than dirt. Audra, can you believe this?"

Rick and Jayda turned to the other couple, only to find them in a deep, passionate embrace.

"I guess she can," Rick snickered over the noise. He tapped his friend's shoulder and motioned to the stairs leading to the box. After getting situated, they ordered dinner; being served as they had been when they had first met. As they watched the entertainment on the main floor, they got comfortable, cuddling down in the soft chairs, letting their adventures of the past melt away until it was just the four of them, alone in their universe. They were suddenly brought back to reality by a tap on Rick's shoulder. Both couples sat up and looked at a line of officers standing on the open stairs. The lead officer spoke as quietly as the noise level would allow.

"General, Commodore Gardner would like to have a word with you and Colonel Conrad."

Rick looked at Jayda, then at Gunnar and Audra.

"It'll have to wait until after the concert. We're off duty."

"The Commodore was insistent," the officer urged.

"I said after the concert," Rick reiterated a little more forceful. "Now get your entourage back up those stairs."

The officer saluted and ushered his men back to the upper cross aisle.

"What did he want?" Gunnar asked, leaning over to hear.

"Don't care," Rick said, settling back into his seat. Jayda quickly found her place, feeling all important.

Gunnar moved back into position as well, pulling Audra back.

Even after the concert had concluded, the two couples lingered until most of the crowds had thinned. Rick gave the cross isle a glance, seeing the small group of officers waiting near the entryway.

"Well, I'm inclined to just sit here and enjoy the down time," Rick said without moving.

"Just trying to see how committed they are?" Gunnar asked, looking up at the cross isle.

"I'm sick of being told what to do and when to do it," Rick responded.

"It must be important if Commodore Gardner sent them," Jayda suggested.

"Probably a redeployment," Rick surmised. "And I'm not inclined to be in any kind of hurry for that. If they want to redeploy us, then they can wait a few minutes to do it."

Audra looked at Gunnar, then at Rick.

"Maybe we ought to see what the fuss is all about," Gunnar suggested, getting up.

Rick hesitated, not wanting to reenter the world around him. It wasn't until Jayda came to her feet and coaxed him up that he started up the stairs, being met by the lead officer. Taking a quick route back to the Central Command complex, they were brought directly to the Command offices of the Flight Ministry. Commodore Gardner grinned broadly, coming to his feet when the four officers came into the room.

"Richard, Gunnar. Thanks for coming back at such a late hour. Commander Atlanta, Commander Niker, please have a seat." He motioned to the chairs situated in front of his desk. "I know you're off-duty and no doubt you have some questions, so I'll get right to the point." He handed everyone a small viewing device and returned to his chair. "As all of you have been through the inquisitions and heard most the testimonies, given much of it yourself, I don't have to tell you the outcomes of the Starbird debacle. Needless to say, Central Command has a problem on its hands that it doesn't want to deal with."

"And what's that?" Rick asked, sounding somewhat detached.

"In Admiral Mandell's testimony, he indicated that without Croft Heckla and the Valkyrie, there is no war."

"Yeah, he gave us that sob story as well," Rick responded.

"He's correct," the Commodore stated bluntly. "With the war over, people across the galaxy want to forget about it and get on with their lives."

"You don't have to tell us that," Gunnar agreed. "We've seen enough action in the last… uhm, through this last deployment, to fill a career." Gunnar glanced at Audra, detecting a slight smile.

Commodore Gardner worked with his terminal for a moment, then turned the display around so everyone could see it.

"I've reviewed your ship's logs." He looked at Gunnar, then Rick. "All of them; including your personal logs." He paused for dramatic emphasis. "I admit I have questions; a lot of questions. But my gut tells me to just seal these up and mark them classified. Besides myself, you and your crews are the only people who know what your last mission entailed and how long you were... deployed." The Commodore let a smile drift across his lips. "I had your ship's memory banks deep scanned. I don't suppose you remember any of that math problem you mentioned?" he asked looking at Rick.

"Sorry, Sir," Rick responded.

The Commodore contemplated a moment, nodding his head.

"Very interesting stuff."

Everyone in the room smiled.

"With the sudden shift in circumstances, Central has suspended the Starbird program and stored all the functioning birds and those that were in production in a secure facility."

Rick and Gunnar looked at each other, stunned.

"Sir, they can't just mothball the program," Rick objected.

"They can, and they did," the Commodore maintained. "But they mothballed them to the Flight Ministry. Instead of being under the jurisdiction of Fleet Command, they've concluded that the Starbird design is better suited for Flight Ministry operations. I own all of them now."

The room remained silent, no one moving.

"I don't have a lot of time to devote to figuring out what to do with them, so I was wondering if you two could make some recommendations? The positions would direct any redesigns, upgrades and recommissioning, as well as finding new roles for the Assault Corsair to play in our galactic structure. A personnel training program would have to be created so they can be manned by the best in the Ministry. There would have to be someone in charge of deployment and management of all the Starbird assets. A fleet commander, if you will. Recommendations?"

Gunnar looked at Rick. He could see the wheels turning as he considered. He looked down at his viewing device then looked over at Audra. She stared straight ahead, her expression empty.

"I've taken the liberty of listing a couple of people I know personally, subject to your approval of course," the Commodore said pointing at the devices in their hands.

Rick and Jayda looked at their viewers. After a moment, Rick looked up at the Commodore's smiling face.

"You want me to head this up?" Rick asked, stunned.

"Brigadier General Niker," Commodore Gardner assured him. "And Colonel Niker. You'd have total autonomy to take the Starbird program in whatever direction you feel would be best. If I may suggest, possibly science and exploration?"

Rick nodded, speechless, then looked over at Jayda. The beaming smile told him all he needed to know. He came to his feet and reached across the desk, taking the Commodore's outstretched hand.

"Yes, of course, thank you, Sir." Rick and Jayda embraced, thrilled with the news.

"Now…" Commodore Gardner said, turning his attention to the other two. "What do you two have to say?"

Rick and Jayda turned to Audra and Gunnar.

"Well?" Rick asked, waiting to hear what they were being offered.

"General Conrad and Colonel Atlanta," the Commodore announced, smiling.

Rick and Jayda gasped at the good news. They waited for their friends to join in with the happy tidings of their promotions. Instead, Gunnar and Audra remained silent and unmoved.

"Come on, Gunnar," Rick coaxed. "You can write your own ticket. Command whatever you want, do whatever you want."

"Thank you, Commodore," Gunnar finally said. "But I'll need time to think about this."

"Gunnar, what are you doing?" Rick questioned. "This is it. It doesn't get any better than this."

"I said I'll have to think about it," Gunnar growled, getting up and setting the device on the Commodore's desk.

"Gunnar?" Audra voiced, watching him.

"I'll let you know," he said leaving the room.

Rick went to go after him, but Audra stopped him.

"No, I'll go." Audra glanced at Jayda and the Commodore, then took off after her husband.

*   *   *   *

Gunnar looked up through the mid-section windows of the Central Command building. This was the spot he and Rick had been standing in when they were about to receive their first Starbird assignments. He could see civilian traffic moving back and forth in fly-ways high above the city and above them, fighter traffic weaving around in designated maneuvering patterns. He knew the training patterns at a glance, but watched them closely anyway, looking for any deviation from the assigned patterns. His mind shifted to his training days and subsequent deployments and his progression through his career. It had been a long, but fast road. He missed being in the cockpit, but at the same time, was glad he wasn't there.

"You ok?" Audra asked quietly. She leaned back against the rail and looked at Gunnar.

"Right now, I'm feeling about as confused as I did when I woke up from being dead."

"I know the feeling," Audra smiled trying to get in his line of sight.

Gunnar smiled, shifting to her face.

"I've just been offered my ultimate career goal and I didn't say yes."

"Yeah, that sounds confusing," Audra agreed. "What are you going to do about it?"

"I promised you when we got home that this would be the end. We would retire and be together for the rest of our lives."

"You also promised me you would stay out of the cockpit. How did that go?"

"I told you," Gunnar simmered, looking away. "I did what I had…"

"Calm down… CJ told me everything," Audra admitted. "It was the right thing to do. …And I talked with Jana, or rather listened to her. I thought you told me she could hardly put two words together correctly? That girl can rattle on nonstop," Audra chuckled. "I still can't believe she actually stunned you."

"Multiple times."

"The point is, it was unfair of me to make you promise something like that."

"No it wasn't," Gunnar rebutted quietly. "You were right then and you're right now. I'm reckless. Sure it works well as a fighter pilot, but I'm not a fighter pilot anymore."

"But with this promotion, you could be again, if you wanted to."

"No, I won't get back into the cockpit under these circumstances."

"Why not? You heard Rick. You can write your own ticket. Do whatever you want, whenever you want."

"Yeah, but at what cost? Knowing the effect it has on you every time I close a canopy or fly off to who knows where."

"I was tired and emotional when I said that," Audra defended.

"You make no sense, woman," Gunnar flustered. "You give me this big speech about how scared to death you are when I'm out doing what I do best and you want me to give everything up… and now you're trying to tell me to forget all that?"

"Yeah, dying can have a profound effect on a person, besides, I'm a woman and have a license to change my mind at a moment's notice."

Gunnar scratched his head and sucked in a deep breath. Audra slid over next to him, putting her arms around him.

"Colonel, may I make a couple of suggestions that I think might address the conundrums we're facing?"

"Speak, oh wise one; the love of my life." Gunnar smiled and kissed her, then turned with her hanging on his arm and started walking as they talked.

Say Again

The blue giant, Orb, burned its hue throughout its planetary clusters, including its only inhabited planet, Commenor. Maintaining a low orbit above its southern hemisphere, a small cluster of capital and medium ships held their positions in an organized line. Further out, two Starbirds made an approach, flying directly between the formations toward the surface of the planet. As the Starbirds passed, each ship fired off a colorful salvo of rockets, some passing under the two Assault Corsairs, some overhead. As the two ships continued down into the atmosphere, several destroyers followed, forming a large Vee behind the two smaller ships. Once in the atmosphere, the Starbirds sliced through a scattered cloud layer, descending toward a clearing adjacent to the small township of Dante. Eight other Starbirds sat in a long row, four on each side of two empty landing pads.

A light crowd of dignitaries stood in a small waiting area as they watched the two Starbirds circle, then gently set down on the hard flat surface. As the dust settled, the boarding doors opened and several of the crew scurried out, assembling at the bottom of the boarding ramps. They came to attention as Gunnar Conrad and Audra Atlanta came down the *Constellation's* boarding ramp at the same time Rick and Jayda Niker came down the *Athena's* ramp. Rick motioned for his friend to lead the way as they headed for the group of people waiting for them some distance from the ships.

"I see you managed to keep yourself alive," Gunnar grinned broadly as he opened his arms to give his friend, Joanus a hug.

"It's good to see you again, Gunnar," Joanus said, returning the embrace.

"I ought to punch your lights out," Gunnar said smiling. "After what you pulled."

"Remind me what he pulled?" Rick whispered leaning close to Audra.

"The Mantose twins, Tiana and Taron," Audra whispered back.

"Got-cha," Rick nodded with a smile.

"Yeah, again, I'm so sorry about that," Joanus chuckled. "How did all that business turn out?"

"We lost Taron in a fire-fight... some time back."

"So sorry to hear that," Joanas lamented. "He was a good man."

"Got Tiana back to her father. In fact, they're both here."

"Her father? This sounds like quite the tale. I look forward to hearing that one."

"Hello, Joanus," Audra greeted, coming out from behind her husband.

"Ah, look at this! You brought the best part of you this time." Joanus took her hand and gently kissed it. "It's always a treat to have you here, Audra." He turned and shook Rick's hand. "General Niker, and this must be Colonel Niker." Joanus took Jayda's hand. "A former Kalamarion Princess. 'Tis a true honor, your Highness."

"Thank you," Jayda nodded, a little embarrassed. "But my current designation precludes you addressing me as, your Highness. It's just Colonel."

"My apologies, Colonel Niker," Joanus said, kissing her hand. He turned back to Rick. "General, before we get started, there is a little matter of business requiring your attention. Is your physician with you?"

"Doctor Mantose?" Rick responded. "Is there something wrong?"

"Nothing to worry about," Joanus replied. "If you'll follow me, please." Joanus pointed to a small building some distance from the assembly area. As the group headed for the building, Rick put his wrist communicator to his mouth.

"Lieutenant Habba, can you have Doctor Mantose join me in the Dante dispersal building."

"Aye, Sir."

"That was some fancy negotiating you pulled off, General Niker," Joanus said, touching a control on his wristband as the group approached the building. A large door opened, revealing a long corridor with a myriad of sub-hallways branching off in all directions.

"The Commenor parliament is a tough crowd," Rick admitted. "Agreeing to a Starbird training base here on Commenor was a difficult sell, and in the end, I had to use our poster boy, Gunnar, to sell it."

"When they announced they were allowing the training base here in the Dante province, I was a little worried about the idea, but the proposed location has alleviated my concerns."

"I think we've made sure it's well enough removed from any settlements."

"What about you, Gunnar?" Joanus asked. "Is your new *rustic* dwelling secluded enough from a training base?" Joanus looked back at the two Assault Corsairs parked behind them.

"I think we'll manage," Gunnar said, pulling Audra close.

"What about you, Audra? Gonna be able to live with a base that close to home?"

"The place is far enough off base in a very nice, isolated spot and he's only allowed to work on certain days of the week. The rest of the time, he's all mine."

Joanus gave Gunnar a raised eyebrow, letting a smirk go.

"Sounds an awful lot like being put out to pasture."

"Semi-retired," Gunnar smiled, looking at Audra. As they entered the building, they snaked down a series of brightly lit halls. "Don't let him take you down any narrow staircases," Gunnar warned Rick good-naturedly.

"Not going to let that go, are you?" Joanus asked without looking back.

"Gonna milk it for all it's worth," Gunnar assured him.

"I had a visitor show up in my office yesterday afternoon, said she knew you." Joanus stopped at a door.

Jayda looked at Rick bewildered.

"So why do you need our doctor?"

"Said she knew him too. Just appeared out of thin air. About gave me a heart attack. Then, she just stood there, staring at me. I thought she had teleported from a ship or something, but I could find no clearance orders."

"She?" Gunnar repeated.

Rick looked at Gunnar, mouthing Ona's name.

"Why would Ona come back here?" Gunnar asked.

"I thought she relegated herself to Hadrian?" Rick puzzled. "If she wanted to talk, she could have come directly to us."

"If she weren't so pretty, I would have had her taken to a holding cell," Joanus said.

"You're sounding a little creepy," Audra commented.

"Hey, I like looking at a good-looking woman just like the next guy," Joanus defended.

"Nice to know age hasn't dulled your senses," Gunnar snickered.

"Said she needed to stay with me until you guys arrived. I ask you, what was I supposed to do? I set her up in a guest room and brought her with me this morning." Joanus shrugged seeing Caidin shuffling up behind them. "Is this your doctor?"

"Yes," Rick smiled. "Doctor Caidin Mantose, this is Joanus Strain, Facilitator of the Dante providence of Commenor. He says he has someone here that knows us."

Caidin looked to Joanus as he turned back to the door.

"Wouldn't tell me her name," Joanus said, opening the door.

The woman stood with her back to them. She wore a dark brown buckskin style outfit with fitted soft soled moccasins. Her jet black hair looked hastily fixed up on her head. As the door swung open, she brought her hands to her hips and turned around.

"Oh," Joanus said, moving out of the way. "Didn't mention the eyes, did I? Gotta say, that was a little unnerving when I first looked up."

"Caidin," the woman said. "Richard. You're here, just as Ona said you would be."

"Dãsha..." Rick uttered, stunned.

"Dãsha, my dear?" Caidin beamed. "Is it really you?" The older gentleman dropped his cane and hurried to the taller woman, embracing her.

"It has been so long, my husband. I was starting to doubt if I would ever see you again."

Rick stared at the tall beauty. Her appearance was somewhat different now. Her jet black hair had silver streaks spiraling through it. Her complexion remained smooth and radiant, but the strains of time had etched her face. Her figure was unchanged, though without her tight fitting top, mini skirt and spiked knee-high boots, she didn't have the presence she once commanded.

"So long?" Gunnar questioned from behind. "You've been gone what, six, eight months?"

"Dãsha, what happened?" Rick asked.

"*Calypso* took Dacey and I back to the edge of Hadrian's border with the galaxies of Tolken and Argyle. Many of *Calypso's* systems shut down shortly afterwards and we began to drift, so we manually piloted the ship to the nearest inhabited planet, Gardo. We remained in orbit for many months until one Tun Osta came and took us to the surface."

"Tun?" Rick asked.

"He is a Master Thane as is Ona Tusk."

"Where do they come up with these names?" Gunnar mumbled.

"Dacey and I have been on Gardo for the past fifty-seven years."

"Fifty-seven years?" Caidin repeated.

"Through a series of unfortunate events, Dacey was made the leader of a large population of the Indigenous people warring against many others on the planet. I became the leader of their armies and his council. Many times, we saved each other from perilous situations. Through much effort and the leadership skills we developed over those years, we helped to broker peace between the populations."

"You're here, where's he?" Jayda asked.

"Dacey lived a long life, having many children to his posterity, but passed away of natural causes a month ago. I officiated at his burial proceedings on Gardo."

"So how did you get back here?" Audra asked.

"After Dacey's burial, Tun came to me and asked what I would of him. I asked to be returned to my husband. So he took me to Ona."

"What happened to *Calypso*?" Rick asked.

"Ona indicated that it had another purpose elsewhere."

Rick stepped up to Dãsha, looking up into her black eyes.

"I'm glad you've chosen to come back to us," he said, embracing her.

"My husband is here. I did not want him to suffer the loneliness that I have these many years. I have never stopped yearning for him and I am grateful to Ona for bringing me back to this time and place."

"I am so sorry," Caidin said, taking her hand.

"It has worked out for the best. In my solitude, this body you have engineered has sustained me, and I have come to terms with the Nanomech within, and the emotional construct of Danis Knox. We are one..." Dãsha cocked her head slightly. "Almost."

"Almost," Rick repeated, detecting a hint of the mechanical mannerisms that once defined her.

"Ona Tusk wanted me to convey a message to you and Gunnar," Dãsha announced.

"A message?" Rick repeated, looking at his friend.

"What's this... message?" Gunnar asked.

"She thought you'd like to know how things in Hadrian have gone in your absence."

"Is that all?" Gunnar smiled, relieved.

"You are aware of how Ona likes to communicate?" Dãsha stated.

"Yes," Rick smiled with a nod.

"While she has rarely left the Queen's side, Blinda Koss has facilitated many difficult negotiations with some of the more active, lesser known factions. An unruly Thane has proven invaluable."

Rick chuckled as Dãsha turned her gaze to Gunnar.

"Jana said to tell Gunnar that, even though Blinda does not approve, her Majesty still climbs in her spare time as she used to in Reako and on Aster."

"Silly little girl," Gunnar chuckled, giving Rick a glance.

"...and Richard," Dãsha said sharpening her gaze. "Don't get comfortable..."

"Comfortable?" Gunnar repeated, looking at his friend. "What's that supposed to mean?"

"Nor you," Dãsha said looking at Gunnar stoically.

"Wait, what?"

Dãsha turned to Jayda and Audra as they stepped forward, embracing her. Rick and Gunnar looked at each other, then at Joanus.

"I admit I had my doubts about this woman's claim," Joanus said. "But when it comes to anything to do with Gunnar Conrad or Rick Niker, it was believable enough to see if it was true. Come on, we have a banquet ready for your homecoming. Truly this will be a happy occasion."

As the group exited the building, Dãsha took Audra's arm and walked with her.

"How are you doing?"

"You're worried about me?" Audra asked.

"As I recall, you were still in some difficulty when we left Hadrian and it has not been that long since your return."

"I'm actually doing much better. I still feel a little off in the mornings, but for the most part, I'm fine."

Dãsha eyed her carefully.

"Yes, there is something different..." She cocked her head slightly. "And what of Gunnar?"

Audra looked at her husband walking ahead next to Joanus. She thought for a moment about their discussions getting to this point.

"He's fine. Death didn't affect him like it did me."

"At the time, I sensed some friction between you."

"For someone who has been in another galaxy for as long as you have, you sure remember a lot."

"Dacey and I experienced many adventures while in the Tolken. On Gardo, Tun Osta showed me that there are many different dimensions humans exist in and under the right circumstances, we can cross paths with those friends far removed from our own realities. Do not think your journeys have ended here on your home world."

"Dãsha, what are you trying to say?"

"It's not what I'm trying to say; it's what I'm trying not to say. There are other dimensions directly attached to this one and what you do in one can effect what happens in the others."

"Well, if that's not cryptic, I don't know what is," Audra said slowly, giving Dãsha a wary look. "I imagine the Nanomech inside you makes it so you remember everything."

Dãsha cocked her head the other direction.

"Like it just happened," she said smiling.

Audra returned the smile and grasped her arm a little tighter.

"We're in a good place. We're looking forward to this next phase of our lives together."

"As you should," Dãsha said. "I am truly happy for you."

"And I'm thrilled to have you back with us... It just hasn't been the same without you."

"That is understandable."

*　　*　　*　　*

First thing in the morning came a little sooner than Audra and Gunnar had hoped for. Realizing they had to hurry or fall behind schedule; they quickly finished a lite breakfast and grabbed their uniforms. As they exited their quarters, Gunnar stopped.

"You go ahead," he said, working the long row of buttons on his dark blue high collared jacket. "I'll see you in sickbay."

Audra recognized the look on his face and smiled.

"Take as much time as you need." She turned and headed down the hall, leaving him at the bridge door.

Gunnar hesitated a moment, then activated the door. The bridge was powered on with all the displays lit up and readouts flashing. Alex 7001 turned from the engineering monitors and floated over next to the command chair.

"General Conrad. Aren't you scheduled with Doctor Yamoto right now?"

"Yeah," Gunnar said with a heavy sigh. "But she's only got two hands and Colonel Atlanta is scheduled first. I just wanted to come in and see how things are here."

"I've given all the A versions a thorough inspection. Everything is operating as it should."

"I was talking about just the *Constellation*, Alex."

The droid hovered silent for a moment, then lowered his voice.

"I have taken the same care a well-seasoned fighter pilot would, to inspect and effect repairs on all of this ship's systems. This ritualistic constitutional has produced a significantly better Starbird than it was when first launched from Kalamar."

"Thanks, Alex. I'm sure the new crew will appreciate your work."

"Not likely, as human's are largely oblivious of all the technical details that go into such complex and sophisticated systems."

"I don't think you're giving these new recruits their fair due." Gunnar looked around as he stepped closer to the command chair. "Anyway, if I know this ship and its new purpose, it won't stay perfect for very long."

"The war is over, General. The *Constellation's* purpose, as is with the rest of the Starbird fleet, is for exploration and scientific discovery. Its emphasis will be exploration."

"I don't think the A model is as well suited for exploration as other ships."

"You would be correct, General. But it has certainly proven its salt in handling both roles at the same time better than anything else Kalamar has available. The new Starbird Venture series will bridge the gaps you're referring to."

Gunnar stepped up to the helm, gazing at the vast glass console, control yoke and throttles. Scanning the different stations, he finally looked back at the command chair. He walked back and sat down, fingering the controls on the armrest, then turned the chair to the science and communications console.

"Regret, Sir?" Alex asked.

"No," Gunnar responded quietly. "No regrets." He turned the chair forward, looking out the windows as the dawning sky grew brighter. "I guess I better get down to sickbay before Fuji sends a goon squad up here after me." Gunnar got up and headed for the door. "You still have the Brigadier's equation safe?"

"No one can access it without his authorization."

"I'm sure with your help, someday he'll figure out how to replicate it."

"General?" The little droid floated up and faced him. "Gunnar, it has been my honor serving with you."

"It certainly has been an adventure, hasn't it?"

"One only you could facilitate. Thank you."

"No, Alex, thank you. Who would have thought that an old fighter pilot could learn so much from a tin-plated eggplant? Seriously, you've taught me a lot."

"As have you." Alex hesitated a moment. "Goodbye, Sir."

Gunnar straightened up and saluted the droid, then turned and walked out. The ship was eerily quiet as he made the short walk down to sickbay where he found Caidin examining Audra.

"Thought you'd be done with her by now," Gunnar said, jumping up onto one of the empty examination tables.

"He's just finishing," Fuji said, stepping over to the examination table with a small device in hand. She hesitated, looking at Gunnar.

"Problem?" Gunnar asked, looking at Fuji's expression.

"It's nothing," the pretty doctor said after a short pause.

"I know how you feel about me," Gunnar whispered. "We both knew this couldn't last. But we have to be strong and let go."

"Ok," Fuji drew up. "I'm good now. Lay back and we'll do a quick scan. Wanna make sure this machine you call a body has all the gears turning in the right direction."

As he lay back the med assist arm swung out and waved over the length of his body. Blue and green lights blinked quickly across a myriad of sensors on the arm. Fuji looked up at the display over his head.

"Takes a little longer to read this due to your strange physiology."

"You don't have me memorized?"

"I'm faster at this than I was," Fuji smiled. "Thanks to you, I'm in high demand in the medical community now."

"Then why are you staying on the *Constellation*?"

"Sentimental reasons," the pretty doctor said, putting the apparatus in her hand to the back of his left ear.

Gunnar felt a slight twinge, then was looking at a tiny device in Fuji's hand.

"That's it, General... I mean Gunnar." Fuji set the personnel transponder in a protective case, next to Audra's. "Not sure how to refer to you now."

"Semi-retired means I don't need a tracker or this." Gunnar handed her his wrist communicator as he got up and turned to his wife. Caidin motioned Fuji over as Audra moved to her husband.

"This is it," Gunnar smiled, holding her hands.

"Doesn't feel real," Audra gushed.

Gunnar let a deep sigh go, looking around.

"You ok?" she asked looking into his eyes.

"Double edged sword," he answered. "Not sure how consulting for the Flight Ministry is going to work out. I'm starting to wonder if semi-retired isn't just code for, *out to pasture*."

"Yeah, I know."

"Audra, Gunnar," Fuji spoke up, looking from the device Caidin had given her. "There are some abnormalities I discovered during your final exams. Nothing serious, but I wanted to double check my findings, so I had Caidin run a separate set of tests. He came up with the same results. At this point, this is only educated speculation, but it would appear that some of the guesswork we had to do when we repaired you, has changed something in your physiology."

"Changed is a strong word," Caidin suggested. "How about shifted?"

"You guessed on us?" Audra asked.

"Is guess the right word?" Fuji asked Caidin as she looked down at the display.

"What other term would you use?" Caidin returned.

"Huh?" Audra and Gunnar blurted collectively.

"Why would you guess on Audra?" Gunnar asked, raising an eyebrow. "She's human. Nothing special about her..." He gave his wife a fast glance. "In a manner of speaking."

"Nice recovery," Audra muttered.

Fuji looked back at Caidin, who nodded.

"If I may be allowed to explain," Caidin hesitated.

"Well, out with it," Gunnar coaxed impatiently.

"Gunnar, our initial findings indicate that we may have inadvertently shifted something in your DNA sequences."

"What does that mean? I'm not going to start growing a tail or something; horns or webbed fingers and toes?"

"No, nothing like that."

"Well then, what?"

Fuji looked at Audra.

"The problem is with me?" Audra asked, pointing at herself.

"It's not exactly a problem," Fuji hesitated.

"Come on Doc, you're killing us here."

Fuji glanced back at Caidin again, with raised eyebrows.

"You said you've been feeling a little off lately; kind of sick when you wake up in the morning?"

"Yeah. Just residual stuff from the hibernation sickness, right?"

"Audra, you're pregnant."

"She's what?" Gunnar asked, stunned.

"I'm what?" Audra inquired at the same time.

Fuji grinned broadly.

"How did this happen?" Gunnar stammered.

"Did no one instruct you on how all that works?" Caidin laughed.

"Pregnant..." Audra repeated, stunned. She slowly sank into the closest chair as Fuji's communicator went off.

"Yes, General?"

"There are many similarities between Dialabron physiology and human physiology," Caidin explained. "But the genes governing the reproductive processes have just enough variations to make them incompatible. Apparently, when we made the molecular repairs, something shifted in one or both of you, aligning those variations."

"So, is there a way to know which one of us was realigned?" Gunnar asked. He noticed the dazed expression percolating across Audra's face.

"Without an extensive battery of tests that I seriously doubt either of you would submit to, there's no way of knowing for sure. I can tell you the sex if you'd like," Caidin offered.

Gunnar and Audra looked at each other, bewildered. Audra's expressions were shifting, but she remained silent. Gunnar searched her eyes for some kind of indication of her feelings on the matter.

"Maybe later," he finally said. "I think we need a little time to chew on this little gem."

"Understandable," Caidin agreed. "Do you have any questions?"

"There are probably a million of them," Gunnar responded slowly. "But right now, I can't think of a single one."

"Hey, are you ok?" Caidin asked Audra.

She looked up, still trying to process.

"Yeah," she finally said. "This is a new one for me... for both of us," she said, taking Gunnar's hand.

"I'm sure Doctor Yamoto will be following up with you in the weeks to come," Caidin reassured them.

"Do I need to make sure she..." Gunnar fumbled. "You know, sits down to rest?"

"I am sitting down," Audra mentioned quietly.

"I mean..." Gunnar sputtered. "At her age..."

"That's ancient folklore," Caidin chuckled. "The female structure has moved past age being a limiter for having children a long time ago. She'll carry and deliver just like any younger woman. The only limiter is if her age predisposes her to other medical issues that would put undue stress on mother and child. You are quite strong and perfectly capable of carrying children full term."

"That was General Niker," Fuji announced, stepping back over. "He's getting impatient. We all good here?"

Audra finally stood back up.

"Yeah, I think so."

"I'll be in contact after you get settled in your new place. We can discuss a check-up schedule; just to keep an eye on things. Caidin reassured you that there's nothing to worry about?"

"Yes, he did."

"Then, as per Regulation Thirty-B, Section Two, no unauthorized personnel shall be aboard any Flight Ministry vessel without the verbal consent of the ship's Commander. Therefore, we need to get you off this ship before Lieutenant Colonel Abrams or Lieutenant Commander Ruston finds out."

"Wouldn't want to be breaking any Ministry regulations," Gunnar said, taking Audra's hand and starting for the door. As they passed through the engineering vestibule, they stopped, looking through the Mallory doors.

"These are a great upgrade," Audra commented, letting her hands slide across the glass.

"Alex's idea," Gunnar agreed. "They are standard issue on all Starbirds now, along with the sensor arrays built into the outer skin of the ship. He also came up with several upgrades that address the challenges we faced with the systems in the A model. That is one amazing droid."

"Are you listening to yourself?" Audra smiled.

Gunnar gave her a puzzled look.

"What?"

"You used to threaten to teleport him out into space."

"Wouldn't do any good now. He can remotely operate the teleporter. Not even sure why we need to man these birds with his series and the new AIs coming out."

"Alex will be the first to admit that a human component will always be required to a Starbird's operation."

"Yeah, Rick said the same thing. So, what about us?" Gunnar developed a smile. "You really going to be ok with having a child? I mean, this is some big stuff."

Audra turned to her husband.

"We never had any kind of plan for this."

"We didn't need any until now."

"I'll be all right. Got the best doctor in the galaxy." Audra paused, thinking. "I'm sure as this all starts to sink in, you'll get to hold me while I'm crying my eyes out, and you'll have to explain why you did this to me." Audra smiled broadly.

"Hey, this is a team effort," Gunnar defended. "I'll be there for you to wipe your snot all over my shirt."

They both snickered quietly.

"I'm sure it's going to be an emotional rollercoaster for at least the first trimester."

"You'll bear with me as I try to figure out how to help you deal with all of it?"

"As long as we're together in this, I think I'll get through it just fine."

Gunnar took a deep breath, looking at the open boarding door.

"Right now, we've gotta get through the next fifteen to twenty minutes. You ready?"

"Been waiting a long time for this," Audra said, taking his arm. She stopped him before they got to the open door. "Let's just keep this preggers thing to ourselves for now, ok?"

"Why?"

"You really want all those people out there making a big fuss about it?" Audra nodded toward the door.

Gunnar considered for a moment.

"I see your point, but we're going to be in a lot of trouble when Rick and Jayda find out."

Audra leaned her head against Gunnar's shoulder.

"They'll get over it."

Pausing in the open doorway, they looked out at all the other Starbirds lined up in a row, then continued down the boarding ramp.

They were immediately met with neat line of crew members coming to attention on either side of a walkway between the *Athena* and *Constellation*. Pausing a moment, Gunnar and Audra were moved by the respectful attention. Reaching the bottom of the ramp, they turned up the aisle toward another cluster of officers standing at the end of the row. The crews pivoted in unison and saluted, then pivoted again as Gunnar and Audra started past. Noticing Frank Cooper and Greg Tanetto among the *Athena's* crew, Gunnar stepped over to them, reaching for Frank's hand.

"Find a ship of your own, Chief?"

Frank looked back at the *Athena*.

"Since Mister Moon already had dibs on the *Constellation*, I'm happy to take this one."

"Congratulations," Gunnar grinned broadly, shaking his hand and turning to Chief Billy Moon standing on the other side of the aisle.

"So what's this I hear about you two sharing engineering responsibilities on the *Constellation*?" Gunnar asked, looking at Tiana Mantose standing close to the Chief engineer.

"Have you learned nothing from watching us?" Audra joked.

"She still has a lot to learn before she can make lead engineer," Billy replied, shaking Gunnar's hand. "And after the mess I found when I got back, I wasn't about to just turn her loose on my ship."

"As I read the report," Gunnar said, giving Tiana a glance. "She pretty much saved the ship."

"Please don't feed the wildlife, Sir. She's hard enough to live with."

"And what about you, Ensign?" Gunnar smiled, moving to Tiana and shaking her hand. He looked her uniform over. "You're purposely going to share an engine room with this crabby 'ol codger?"

"Someone young needs to be in there to help him up when he falls down," Tiana chuckled. "Besides, I've kind of gotten attached to him." She held up her left hand, revealing a ring on her finger.

"Congratulations," Audra gasped, embracing her.

"Or condolences," Gunnar snickered. He turned Billy sideways. "Just checking to see if she has a gun to your back or something." Gunnar embraced the engineer, then looked at Tiana. His gaze dropped to the brilliant blue Chyropaz necklace. She smiled glancing down at the brightly shining stone, then gave him a long embrace.

"Thank you, Gunnar," Tiana whispered, kissing him on the cheek. "We're going to miss you guys." She glanced at Audra who smiled kindly.

"You won't have time," Gunnar responded, looking at Lana Nevall standing next to them. "You're all going to be way too busy exploring stuff." Gunnar looked up at the *Constellation*, then back at Billy and Tiana. "I know you'll take good care of her."

"Yes, we will," they said, looking at each other and smiling.

"Goodbye, General," Lana said, taking his hand. "Colonel Atlanta..." she said returning the embrace Audra gave her.

Gunnar looked up at the shining hull of the *Constellation* again, smiling.

"Amazing," he said, turning with Audra and continuing their walk forward. The line of officers at the front came to attention, saluted smartly, then stood at ease holding their hands clasped behind them. Gunnar came to a halt in front of the group as Rick and Jayda stepped out from the end of the line.

Jayda quickly stepped-up and hugged Audra. When they pulled apart, Jayda gave Audra an odd look.

"What's the matter?"

"It's nothing," Audra whispered, trying to shift her demeanor. "Tell you later."

"How does retirement feel?" Rick asked, patting his friend on the shoulder.

"Don't know how it feels being surrounded by all this," Gunnar smiled looking around.

"Just wanted to show you the fruits of your labors," Rick said, stepping aside. "The new Starbird Command assignments."

Gunnar smiled, stepping over to Major Lisa Dayton and shaking her hand.

"Your command assignment goes without saying, Major," Gunnar grinned. "What's your ship?"

"The *Lexington*, Sir."

"One of the best Starbird pilots ever, I have no doubt you'll command just fine. Maybe even teach these youngsters a thing or two."

"That's the plan. Thank you, Sir."

"And these two," Gunnar said, taking Major Zek Korack's hand. "You guys were at the top of the list anyway." He grabbed Major Zak Korack's hand as well. "What are your ships?"

"The *Tut*," Zek answered.

"The *Sung*," Zak followed.

Gunnar moved down the line.

"Major Dalley has taken the *Avenger*," Rick announced. "Renamed it, the *Hadrian*."

"Really?" Gunnar asked surprised. "You're gonna give up a cockpit for a command chair?"

"I thought I'd give it a try, Sir. Besides, it'll look really good on my resume," Logan said. "And..." he glanced down the line. "The Lieutenant Colonel down there is gonna need someone to fly wing with him. Keep him out of trouble."

"Piece of advice," Gunnar said looking down the line as well. "Never let each other out of your sight." Gunnar stepped to the next commander. "How in the world will the General ever be able to find anything without you on his bridge?"

"We've upgraded CORA 500 with the new AI enhancements," Captain Toby Mavis announced.

"And your ship?"

"We rechristened Admiral Mandell's ship, the *Montego*."

"Fitting," Gunnar said looking back at Jayda. He took another step, smiling. "With all the smart people moving up, who's going to know where to look for anything? Captain Pippin, what's your ship?"

"We rechristened the *Hyperion*. It's now the *Insatiable*."

"That's a great name," Gunnar said. Moving to another familiar face, he leaned in close. "Best Starbird pilot in the fleet. What's your ship assignment, Captain Starman?"

"I was assigned the *Sulairus* and rechristened it, the *Intrepid*." She looked past Gunnar at Audra who was smiling broadly.

Gunnar nodded, glancing back at Audra.

"I'm not sure there will ever be a definitive answer as to who the better pilot is," Lynette winked.

"You'll always be the best in my book." Gunnar smiled and looked at the last commander. "This is only nine," he said looking around. "Who's going to take the other ship?"

"I didn't feel good about anyone else on the list," Rick said looking behind them. "But someone has just come to mind that I believe would work really well. Tell me what you think?"

Gunnar turned around looking back at the couple behind them.

"Dãsha, would you consider assuming command of the tenth Starbird?" Rick asked.

Jayda leaned against him, trying to keep her voice low.

"Didn't you tell me she was too much of a loose cannon for command?"

"I did," Rick whispered back. "But I get the feeling she's learned how to integrate and work with people." Rick turned to Dãsha. "What do you think, Dãsha?"

The tall beauty looked at Caidin, then back at Rick.

"I will accept your invitation on three conditions," she responded, taking a step forward.

"Name them," Rick responded.

"That my service, as well as Caidin's, be conditional."

"Conditions to be determined," Rick said, waiting for the other terms.

"Whatever this ship is currently called, it will be rechristened, *Calypso*."

"Absolutely," Rick agreed immediately.

Dãsha looked down the line of new Starbird Commanders, then back at Rick.

"I wish to wear whatever style uniform I desire."

Audra and Jayda snickered knowing what she had in mind. Gunnar raised an eyebrow, wondering what the joke was as Rick took a deep breath.

"Aboard ship with no visiting dignitaries," he finally said, smiling and shaking his head.

"That was easy enough," Gunnar admitted. "Good thing she came back when she did."

"I have a feeling someone else had a lot to do with it," Rick said thinking.

Gunnar turned to Dakota.

"So, Colonel. How does the new rank fit?"

"Big shoes to fill," Dakota said, a little apprehensive. Willis straightened the silver clusters on her husband's collar.

"I can't think of anyone better to fill them," Gunnar said, pulling Audra close. "If you'll listen to some free advice. Keep your First officer close and listen to most of what she has to say."

"Most?" Audra repeated.

"They're never happy unless you give them something to complain about."

"Understood," Dakota said as Gunnar hugged him and stood back letting Audra embrace him. "Independent contracting going to be enough for you, General?"

"I think we can turn out some acceptable people."

Dakota looked beyond the line of crewmen at the new Interceptor Mark II prototype sitting next to a small shuttle.

"Going to be teaching Interceptor skills at this base?"

"Among other things," Gunnar responded glancing at Audra. "I'll have plenty of opportunities to show these new kids how to fly all kinds of things."

No he won't, Audra mouthed silently.

Gunnar turned to Rick and Jayda. He gazed at them, shaking his head slightly.

"It's hard to remember a time when I wasn't with you."

"I don't know what I'm going to do not having you around to save from something," Rick said, chuckling.

"Since it didn't happen that often, you should have plenty of time for something constructive," Gunnar snickered.

"Let's not start this up again," Audra interjected.

"Seriously," Jayda agreed, rolling her eyes.

Rick and Gunnar locked together in a tight embrace.

"Gonna miss not having you up there with us," Rick said.

"I'm available anytime you need me."

No, he won't be, Audra mouthed from behind.

The two friends exchanged embraces with Audra and Jayda, then Gunnar turned to the line of officers.

"Bitter-sweet," Audra whispered.

He and suddenly came to attention, saluting. The entire line instantly returned the salute. Gunnar and Audra then turned arm in arm and started toward the shuttle craft and Interceptor as Fuji Yamoto stepped up next to Rick and Jayda.

"It's a wonderful thing, isn't it?" the ship's doctor commented, watching them go.

"Retirement? I guess," Rick agreed in a melancholy tone.

"No, not that," Fuji said.

"Then what?" Rick asked.

"They didn't tell you?"

"Tell us what?" Jayda inquired.

"Audra's pregnant."

"Wait... What!?" Rick and Jayda exclaimed collectively, turning to the shuttle and fighter as their doors closed.

"Oops," Fuji uttered, putting her hand to her mouth. "I... probably shouldn't have said that."

"Of all the dirty, rotten..." Rick grumbled. "Everyone launch!"

The group instantly broke up, the *Athena* and *Constellation* crews scrambling on board as the other commanders rushed to their ships.

*   *   *   *

"Got away from that without a lot of fanfare," Audra sighed looking over at Gunnar's Interceptor as they throttled away.

"They would have never let us leave," Gunnar smiled as they climbed toward their cruising altitude. Leveling out, he noticed something on his tracking screens. "Rats! We've been made."

"Probably should have said something to Fuji and Caidin," Audra said looking to her right.

Gunnar looked to his left, noticing five Starbirds rise into formation next to them. Five more leveled out on the other side.

"Really?" Jayda's voice blared over their comm systems. "You were just going to leave without telling your best friends?"

"We would've gotten around to it," Audra answered.

"This just gives you a good reason to come visit... often," Gunnar said.

"Uhm, yeah," Jayda snarked. "You can be sure of that."

Gunnar and Audra watched as the Assault Corsairs on each end suddenly accelerated, crossing in front of the group, then streaking skyward followed by the next two until it was just the *Constellation* and the *Athena*. Both ships waggled several times, then pulled ahead, crisscrossing in front of the shuttle and fighter, then streaking skyward. As they banked away, one of the *Athena's* Interceptors catapulted from the ship and arched back around settling into a wing formation with Gunnar's fighter.

"Just got an alert from Joanas. There are a couple of Terrellian raider ships making a run over the southern pole of Commenor," Rick called.

"Don't even think about it," Audra warned.

"Raider ships?" Gunnar repeated.

"Gunnar, Rick, I mean it," Audra warned.

"You've got the new upgraded Interceptor prototype," Rick suggested. "It needs a proper shake down. Take only an hour or so."

"Richard Alexander Niker, you have ten Starbirds with Interceptors and umpteen destroyer class ships up there that can easily take care of a couple of Raider ships," Audra pointed out. "Gunnar is retired."

"Semi-retired," Rick reminded her.

"Come on," Gunnar protested. "It's just down to the southern pole and back. Nothing to it. Besides, there are some fancy new buttons and levers in this thing. I need to know what they do so I can teach the young'uns."

The comms hissed silence for a few moments.

"Just down and back," Audra relented sternly.

Both Interceptors suddenly bolted away, disappearing through a large bank of clouds.

"Just down and back," Audra reiterated loudly. "No engagements! Rick? Gunnar?"

The two V-winged fighters streaked into the outer atmosphere, wing to wing, steering a straight course toward Commenor's southern pole.

"I feel a little guilty not answering her," Gunnar admitted.

"Just a little?" Rick answered.

"I am retired..."

"Semi-retired. You still have a duty schedule."

"I suppose," Gunnar responded, sounding a little tentative.

"Hey, you up for this adventure?" Rick asked.

"If life isn't an adventure," Gunnar said, sounding like himself again, "then we're doing it all wrong."

Is it ever really... the end?

A Legacy

While considered one of the finest intercept fighters ever built, the T-6 Tempest had been retired from active service with the Kalamarion Flight Ministry years ago. Kalamar Command had disposed of their active Tempest inventories to outer rim echelons that contracted with Kalamar Command to effect police duties. Only a small number of the boxy style fighters remained operational within the closer systems, used mainly for training purposes. After undergoing several redesigns and upgrades, the Interceptor series was now designated, the Interceptor Mark V.

"Corona radiation indeed," Alex muttered, glancing at the information racing across his scopes. The fifteen year old adjusted the controls of his Tempest as he passed through the fading remnants of the Kidamagor comet tail. Even with energy shields, he could hear the tiny particles of frozen carbon dioxide ice crystals showering the hull of his T-6.

"You barely even got into it," a younger female voice commented over his helmet com system. "Come on," Aydra prodded. "The bet was to fly through the length of the tail, not pitter through the end of it. Don't be a wuss."

"I'm not a wuss," Alex responded casually. "I'm just testing my shields against Kida's tail."

"Dad had both of these shield systems upgraded before they were delivered from surplus," his younger sister pointed out. "You don't need to test anything. Now, you're either going to put your money where your mouth is, or you're going to wuss out and I win the bet."

"I'm not wussing out," Alex insisted, irritated. "And hey, I'm not seeing you in here doing this with me."

"If I were, I'd already be turning its orbit."

"If I run up the length of the tail, swing around Kida and come back down, will you do it?"

"If you don't quit stalling, I'm going to end up doing it anyway."

"Come on, baby sister," Alex taunted back. "Will you do it?"

"What will you give me?"

"Respect."

"No good, it's not worth anything."

"How 'bout chores for a week?"

"How 'bout I race you? Loser does the chores for a month."

"I'll land on the silly thing for that," Alex responded. "You are so on. Now, haul your sorry hind-end up here and get lined up so it's fair."

"Are your flight recorders on?" Aydra asked. "Wouldn't want there to be any wild interpretations when this is all said and done."

"Waiting on you," Alex said, maneuvering his T-6 into position.

Moments later he noticed a set of flashing Nav lights approaching at a high rate. He could barely make out the form of his little sister sitting at the controls of another T-6 as she took up her position directly beside him.

Holding steady at the tip of the corona tailings was harder than it sounded. The comet was in constant motion as it made its eternal journey around the giant blue sphere of Orb. Alex studied the readouts on the comet's composition again. Now he was glad he had stayed awake during all those Quantum Astro classes, though he had to admit it was because he had a huge crush on the teacher.

There were many different types of comets traveling through the Mila system. Most were comprised of a cocktail of various minerals mixed in frozen water or carbon dioxide gas. Traveling through deep space, most comets appeared as frozen asteroids, but as they neared a spacial body such as a sun or nova cluster, the frozen cocktail would begin to vaporize, forming violent storms of snow and ice originating from the surface. Because of the unique characteristics of Orb's energy emissions, Kidamagor's surface reacted violently by expelling large chunks of ice from its surface crust. Quickly disintegrating into tiny fragments in the comet's slipstream, the sparkling ice particles formed a great tail stretching for millions of Tanz along its flight path.

Alex detected movement to his right, noticing Aydra's wingtips wobble slightly.

"You got it?" he asked looking over at her. He saw her swing her head at him, but couldn't make out any of her features for the dark cockpit glass and the helmet she was wearing.

"The only thing you need to worry about is trying to fly straight through my wake, cause that's all you're gonna see of me."

"You know we could get in a lot of trouble just being out this far?" Alex said making several adjustments to his instruments.

"Yes, we could," Aydra agreed. "That's why we're not going to say anything to anyone except when we download the flight information to Russel so he can publish it to prove the bet."

"Up the stakes?" Alex taunted.

"Speak," Aydra answered.

"I run the tail, beat you around this hunk of dry ice and then land on it... You take all the heat for this stunt when we get caught."

Silence... Alex looked over at his little sister, sitting frozen at her controls. She finally turned her head to him, then looked back at the glistening fray of ice particles ahead of them.

"If I win, you take the heat?"

"Done," Alex responded instantly. "On my mark. Three... Two... One... Go!"

An intense blue ribbon of thruster gases streaked from the rear of both ships as they catapulted straight ahead. The fighters remained deadlocked as they raced through the ice tail of the traveling comet.

Aydra noticed her brother's ship staggering through the hail of bombarding ice. Her own ship shuddered under the storm as well. Checking her instruments, she made several adjustments.

"Better keep your exhaust dampers closed," she warned. "This ice is making things interesting in the injector chambers."

"I know how to do this," Alex grumbled tight lipped. "Make a little more room between us. The scopes can't see much of anything ahead of us."

"I don't want to get too far apart," Aydra responded with the same grit in her voice.

"We just need to make sure we're far enough away from each other that if something big comes at us, we'll have room to maneuver. I don't want you running into me.

"Ditto," Aydra grunted.

Alex glanced to his right, barely able to make out his sister as the two fighters slowly widened the gap between them. Streaking through the center of the thickening comet's tail, he glanced down at his scopes and carefully shifted his controls. He could barely make out the outline of Aydra's Tempest now as they both continued neck and neck through the center of the tail.

"How are you doing over there?" Alex asked as the buffeting increased.

"We've got to be getting pretty close..." Aydra grunted.

There came a loud *clang* at the end of her transmission giving Alex cause for concern. He could only afford a moment as he was suddenly confronted with a burst of sparkling ice shattering right in front of him. Instinctively, he ducked his head as the T-6 ploughed through. The instruments in front of him spun wildly setting off a warning indicator. Struggling with the controls, he noticed larger chunks spinning at him, most breaking into smaller sarsens and showering the fighter's energy shields. Another mass of shattered ice appeared to his left, spinning right at him. With a quick twist of the controls, he rolled right out of the way only to be struck by another wall of ice, cracks fingering along its surface as it crumbled around him. The fighter shook violently with the strike. Alex recognized the wild reaction of his instruments to the collision. He pulled the throttles back to idle and opened the exhaust dampers. Adjusting his shield energy emitters, he steered across the field of disintegrating ice. Noticing several darker spots forming ahead of him, he caught sight of stars spinning through the cascade of exploding ice.

"Aydra, are you still with me?" Silence. "Aydra?"

Correcting the spin, his instruments settled. The Tempest suddenly broke through the hail of icy grit revealing the glowing grey surface of Kidamagor. Weaving back and forth through a quickly moving field of frozen carbon dioxide, he took note of the great chunks launching from the surface of the comet.

"Aydra Janox?" he called again. "Where are you?" Silence.

Checking his sensors, he found the interference from the comet's tail was obliterating any recognizable returns. He turned for the vertical horizon of the unevenly shaped sphere. Beyond, he could see the big blue mass of Orb. Being this close, the sun's emissions were striking the comet with more intensity. Several more encounters and Kidamagor would likely disintegrate. It was hard to comprehend that a comet could shed this much material for that long and still remain intact.

Holding just inside the eruptions of ice particles propelled from the comet's surface, Alex throttled ahead. Since the comet was only the size of a small asteroid, it didn't take very long to reach the other side. Approaching the base of the tail again, he dove for the surface.

"Aydra?" he called looking at his scopes.

"There you are," his sister called back to him. "I was starting to get worried that you had ditched me. Where you at?"

"*You* were worried?" Alex responded, relieved to hear her voice. "I hear a big clang in your transmission, clearly a hull strike, then nothing, and you were worried about me ditching you?"

"Yeah, it got a little intense back there at the base. Where are you?"

"Picking a parking spot."

"Yeah, good luck with that," Aydra replied. "I can't see any surface at all. Everything is moving."

"That happens when CO2 warms up from absolute zero. "We may have miscalculated how close we are to Orb," Alex said. "I've got a nook... Touch down in ten seconds."

"Did you notice this thing is rotating?" Aydra asked.

"All comets rotate, Sis."

"Not this fast. How did we miss it?"

"Who cares?" Alex said, approaching a crumbling outcrop of frozen carbon ice. Extending his landing gear, he looked through the glass at his feet as the fighter's gear reached for a perceived solid surface. He felt the touchdown and let the T-6 settle for a moment; just long enough to look around. What surface he could see was broken and uneven, great chunks of ice launching from the comet's ever moving exterior all around him.

"Barring any other incidents," Alex said confidently. "I'll be back out in a minute and you can eat a big slice of humble pie." With that, he pulled the fighter from the surface as the nook he had landed in was starting to crumble, pieces hurling away from the surface. His headset crackled with a broken transmission.

"Alex... hit... not able... losing... pressure... sweeping..."

"Aydra? Say again, your transmission is breaking up. Are you ok?"

A streak of fear lanced through him as he throttled away from the tumbling comet and out of the CO2 flares. Moments later he was gliding along the outer layers of the tail.

"Aydra, where are you?"

He continued to scan, guiding his ship around the perimeter of the tail, calling.

"Aydra, turn on your emergency locator transponder. Launch a flare, anything I can use to find you."

Painful moments drug on as he searched, constantly adjusting his scanners.

"Come on, Sissy," he mumbled nervously. "Where are you?"

Adrenaline gushed through his insides, causing his stomach to burn. At least it felt that way. There came a fuzzed blink on the display in front of him and a crackle in his headset.

"Aydra?"

"Emergen-… won't la-… need…"

As the transmission fizzled away, Alex focused on a direction. The comet's tail was twisting as the gigantic ball of CO_2 ice rolled along its path around Orb. As he skimmed along the tail layers, he noticed a brilliant flare from inside, near the outer base of the tail. He had to concentrate hard to put the thoughts of a ship exploding out of his head as he steered in that direction. Getting closer to where the flare had occurred, his scanner display came alive again. It showed only a primary target, but at least it was something. It wasn't forming up and then disintegrating like all the ice chunks around it. Diving back into the tail, he maneuvered carefully through the thickening spray of crumbling ice chunks and pelting crystals.

"Aydra?" he called, coming up on the target. He searched the flurry of ice ahead of him but could make out only grey sheets of moving material.

"Get me out of here!"

Alex drew in a big sigh of relief.

"What happened?"

"Doesn't matter," Aydra replied, elevated. "I've got a hull breach; I've lost everything here. I'm on suit life support, and you know how long that'll last."

Alex considered quickly. These were light duty space suits; designed for civilian use, nothing prolonged. He guided his craft closer to the target, it's outline finally coming into view through the murk of blasting sheets of ice.

"Boy, are you going to get it," Alex teased, relieved to see his sister's fighter gently rolling along with the smaller chunks that continued to break apart.

"Never mind that, just get me out of here."

"Try not to take this too personal, but I'm going to give you a little nudge. We'll see how things turn out."

"A nudge?" Are you serious?"

"I can wait for a better idea if you'd like."

Alex pushed the throttle forward the moment the nose of his ship bumped against the thruster ports of Aydra's. His sister's dead fighter yawed and twisted in front of him as they moved quickly toward the edge of the tail. As the ride began to smooth out, Alex's heart settled to the point he didn't feel like he was going to wet himself. Finally, a welcome envelope of black appeared as they broke out into open space and away from Kidamagor.

"Well, that was fun. What happened?"

"You get three guesses and the first two don't count," Aydra replied, her voice shaking. "I'm surprised you got through all that in one piece. That was intense."

"Have you got anything working?"

"Everything is dead. I can't even tell if my environmental systems are working. There are at least two big holes in my glass on both sides. The only thing I can figure is my shield emitters got hit and then some bigger stuff got through to the hull."

"That'll do it," Alex remarked. "These older T-6s didn't have the integrated emitters like the newer Interceptor series does."

"Nice conversation," Aydra replied, still shaken. "Now what are we going to do? We're a long way from anywhere."

"Not to mention we're still in Orb's influence and you have zero shielding now."

"I've only got so much air here," Aydra said checking the little display on her suit's forearm.

"I guess I can keep pushing you," Alex offered. "We should be able to reach the main traffic lanes into Commenor. Someone is bound to see us."

"Yeah, let's do that," Aydra agreed. "In the meantime, what am I supposed to do about the fact that I'm running out of air?"

"That is a problem," Alex agreed, steering in the direction of Commenor. "Too bad there's no way to transfer some of mine over to you. I've got plenty."

There was silence for a moment.

"I have an idea," Aydra announced.

"Am I going to like this idea?"

"I highly doubt it."

More silence.

"Are you going to let me in on this idea?"

Alex detected movement outside. He turned on his landing lights, illuminating his sister's fighter. The swing glass to Aydra's cockpit swung open. Moments later, Aydra was working her way out and on top, moving towards the empty vertical cannon fin.

"You were right," Alex bristled. "I don't like it. Aydra, this is about as dangerous as it gets. If you lose your grip, you're going to fly off into nothing and if I lose sight of you, I can't track you."

"Just hold it steady," she grunted. "I'm pretty much peeing my pants right now and you're not much of a cheering section. Get your cockpit decompressed; I'm coming in on the copilot side."

Alex worked the environmental controls, pulling all the atmosphere inside his cockpit back into their supply tanks, then reached over to the copilot's door as his sister carefully worked her way between the thruster ports of her stricken ship. As he watched her, he checked through all his systems, stopping on a couple of menus that were flashing at him. As he kept an eye on Aydra, he studied his readouts, then set everything back to operational mode.

Aydra grabbed hold of the front of his ship and pulled herself up to the cockpit glass. This would be the scariest part, as there was nothing to hang on to. Trying to use her hands as suction cups, she meticulously planted them broadly across the glass and slowly pulled herself toward the open door. Alex noticed her hands sliding back as she carefully moved along. The inner helmet lights accentuated the petrified look on her face as she reached for the door frame. Her legs gently swung back as her hands slid across the glass away from the opening.

"Not good," Alex gulped.

Aydra pawed at the glass, reaching for the door frame as she slipped further back.

"Alex," she yelled. "Slow down!"

"I do and you'll slam into the back of your ship and likely tear that suit to ribbons."

"I'm not gonna to make it," she screamed. "Alex!"

Picking up speed as she slid back toward the top fin, she suddenly jerked to a stop. Her brother's form was sticking partially out the copilot's door, holding on to her arm.

"I've got you," he reassured her.

"But, whose got you?"

"Just relax," he said, pulling her gently back towards him. Moments later, she was grasping the door frame and pulling herself inside. It took a minute of clumsy shuffling before they were able to settle into their seats and close the door.

"Good thing you never listen to me," Alex complained.

"Yeah, good thing," Aydra responded.

As Alex worked to repressurized the cabin, he felt a sense of relief flow over him. Once the environmental systems showed green, he pulled his helmet and gloves off and looked ahead of them. In the shuffle, he must have bumped his controls. Aydra's ship was gone. He checked his scope.

There it is, he confirmed to himself.

"I say we get better suits. The military ones are bulkier, but at least they can take a little more punishment. Besides, they can hold and scrub more air." There was no response from his sister. He looked over at her; she wasn't moving. It wasn't until he detected a slight shaking in her shoulders that he realized something was wrong. He quickly removed her gloves and helmet. She raised a bare hand to her head and leaned over against her brother, sobbing.

"Hey, what's this all about?" he asked, putting a consoling arm around her.

"We nearly got ourselves killed."

"Well, to be fair, you nearly got yourself killed," Alex chuckled softly. "I just saved your sorry…"

"This is serious, Alex," Aydra shook, still crying.

"Ok, ok," he said, becoming serious. "You need to calm down. There is a lot of open space between us and home, so we're going to need to conserve as much air as we can."

"I thought you said you had plenty?" Aydra sniffled, trying to regain her composure.

"I said there's plenty over here in my ship; for me."

"What do you mean, for you?" Aydra wiped her eyes and looked down at the control display. "Just run your throttles a little higher and we'll get home in half the time. There should be plenty."

"Normally, I would say you're correct…"

"No you wouldn't. You'd deny any fact I gave you, even if the proof were sitting right in front of you."

"Well, here's the proof," Alex said, touching the screens in front of him. "Apparently, I took a little damage too." He brought up a submenu. "I must have taken something to the outer Ason manifolds. I'm losing fuel."

"Are you sure?"

"See for yourself. The tanks are bleeding dry."

"How much do you think you have left?"

"Twenty minutes at this speed… Less if I throw in the boosters."

"Get up to speed and coast longer," Aydra surmised. "How much air?"

"With both of us; about twenty hours."

"It's your call," Aydra said, sitting back and looking out her side of the cockpit.

Alex checked his navigation instruments and made a course correction, then pushed the throttles the rest of the way forward and touched the booster control. The T-6 instantly catapulted forward, accelerating. There was nothing ahead of them to see. In fact, if it weren't for the readouts in front of him telling him his velocity, it didn't appear that they were moving at all. He checked his scopes, turning on the external viewers. Behind them, Kidamagor quickly receded amid the backdrop of the blue giant, Orb. Checking his readouts, he worked to estimate how far they could get. Not liking what he was seeing, he quickly swiped away the information and turned on the emergency beacon.

"Better try to sleep as much as you can," he suggested, trying to get comfortable.

"I know, it'll stretch the air," Aydra agreed in a somber voice. She was too tried to get comfortable.

Several minutes passed before a warning indicator flashed on the display. The Ason tanks were running low. Alex reached up and shut off the alarm and sat back wishing he could see movement; something to give them hope. The only thing that gave him comfort was the display showing the shrinking sun and comet behind them. The fuel tanks finally ran empty, activating another alarm. He silenced it, then turned everything off but the environmental systems. Settling back again, he slowly drifted off to sleep.

*   *   *   *

Aydra awoke with a start, but the emotional upheaval she had been experiencing during a bad dream, lingered. She sat up and looked around. Her brother was sound asleep next to her. The control display in front of them was dark, bringing her to an instant panic.

"Alex," she nudged him while tapping on the dark display. When he didn't respond, she punched his arm. "Alex," she repeated a little more forcefully.

"What?" he grumbled, still mostly asleep.

"The ship is out of power."

"No it isn't, go back to sleep."

"Yes, it is. Nothing's working." She repeatedly jabbed her finger at the dark console.

"We're not out of power," he said, shifting away from her. "I shut everything off except environmental. Go back to sleep."

"Shut it off?" Aydra repeated. "What for?"

"Because it wasn't needed if we were asleep... Like we're supposed to be right now. Now, shut your yap and go back to sleep. You're sucking down all my oxygen."

"How long have we been asleep?"

"I don't know," Alex grumbled. "I've been asleep."

"Well, now I'm awake."

"Good for you; I'm not. Shut up."

Aydra sat quietly for several minutes until she could hear Alex's deep breathing. Fumbling with the controls in the dim light of Orb, she turned the fighter's systems back on and started cruising through the menus. Her first interest was just how much oxygen they had left. That level looked pretty good.

The Mila system wasn't very big considering the size of the sun. But that was the reason it didn't have many planets. Most had been broken up by the blue giant's gravity long ago. The few that were left were lifeless as their atmospheres had been torn away by Orb's powerful emissions. Only Commenor was left unscathed, being outside Orb's solar wind influence.

Hoping to fill her time, she continued to finger through all the control menus on the glass. Something a little concerning to her were the radiation readings. While the Tempest had operating energy shields, as Alex had pointed out earlier, the emitters could have been damaged during their blast through the comet's tail. They were still protected and looking back into the sensor history, the radiation readings were dropping as they coasted further away.

Having gone through every menu she could find in the control system, she turned on the scanning equipment. Were these surplus fighters setup for long range reconnaissance, she would be able to see any ship within the solar system, identify what they were carrying and how many people were onboard. Unfortunately, these T-6's had been stripped of most of their upgrades. Her father had reinstalled a couple of options, instructing her and her brother these were to be used only for training.

Working the scanning equipment, she hardened her gaze at the screen.

"Don't know why he doesn't like the HUD," she grumbled, giving her older brother a glance. Switching the Heads-up Display on, she looked up at the inside of the cockpit windows. A luminescent grid appeared on the inside surface with several different identified targets pocked all around the display. Additionally, a thin blue line arched through the grid lines, an indication of where the solar wind influence reached. Just outside the line was Commenor, directly ahead. Sitting back to study every detail in the HUD, she was suddenly presented with a warning indicator directly in the center of the displayed images.

Several ships were coming out of lightspeed operations close to their position. A spark of hope crept into her as information began to populate the screen in front of her.

"Uhm, Alex?" She tried to rouse her brother.

"Go back to sleep," he grumbled.

"Alex," she persisted. "We've got company."

"Uh," her brother mumbled. "What company? It's just a dream."

"No, wake up!" She punched his shoulder.

"I'm getting really sick and tired of you doing that," he retorted angrily. He sat up with a yawn, then blinked and looked at the HUD with a frown. "Why do you have that stupid thing on? I hate it."

"Here," Aydra said, pointing at the information in front of her. "Four ships. One bigger than the others; coming right at us."

"Oh, good. We're saved. Remember, you get to take the heat for all this."

"You're better at working the scanners than I am," Aydra said, gesturing to the control panel. "And please keep the HUD on so I can see what's going on."

"You know, you've got to learn how to work these things if you're ever going to have any hope of going to the academy."

"Who said I wanted to go to the academy?"

Alex shook his head disgusted and started working with the scanner controls. After a moment, the ident signals appeared.

"Uhm, this isn't good," Alex said. "Those are Terrellian raider ships. Three assaults and a container hauler."

"Are we going fast enough to outrun them?" Aydra asked.

"What do you think?" Alex replied, looking over at his sister. "They'll be on us in two minutes."

"What are we going to do?"

"I am open to suggestions," Alex said, running through the control system menus. "These ships were stripped of anything that shoots a long time ago. The only thing we have are shields."

"We've got to do something," Aydra urged.

"Out of fuel and doubled up on the environmental system with nothing to shoot back with. What would you like to do?"

Aydra sat forward.

"What's that?"

"Another ship just jumped in, right in front of us. Are we that popular?"

"It's not heading at us," Aydra noticed, poking her finger at the target. "It's meeting up with those raider ships."

They watched for a moment, unsure of their fate. Another target appeared.

"Larger one coming in, right at us," Alex announced, working the scanning equipment.

"Can't we find out who these people are?"

"Idents are coming up now," Alex said.

"Whoa, whoa, whoa," Aydra screamed as a terrific jolt rocked the helpless T-6. The blow sent them tumbling with several more violent strikes following. One of the raider ships had already caught up to them.

Alex concentrated on the display, trying to focus on the shield readings. The HUD was flashing several warning messages as well as the other ship idents.

"I think I'm going to be sick," Aydra gulped, closing her eyes against the spinning stars.

"I don't believe it," Alex said looking at the targets on the HUD, then down at the control display. "That's a Starbird..." He tapped the screen several times. "The other one is an Interceptor..." Alex watched intensely as the Starbird swung in a wide arc around them and opened fire on the Terrellian hauler, crippling it. Once neutralized, the Starbird turned back, heading after the tumbling T-6.

"I want off this ride," Aydra bellowed with her eyes still closed tight.

"Hey, calm down, that's my arm," Alex complained as his sister held a vise grip on him.

Moments later the tumble abruptly stopped as the bright white hull of the Starbird pulled alongside them.

"They've got us in their tractor hawser," Alex announced, relieved.

"Standby," a steely voice boomed over the cabin sound system. "We'll have you aboard in just a moment."

"You can open your eyes now," Alex said gazing at the sparkling hull of the Starbird. He looked up at the red phoenix on the bridge pod and the ship's name.

"This is an 'A' model... *Calypso*. Isn't that Aunt...?" Aydra asked, looking at the name.

"Dãsha," Alex finished, feeling the teleporter process begin. "You know what that means, don't you? When they figure out who we are..."

The brother and sister looked out at the Interceptor corralling the raider ships as they sparkled out. The next moment they were sitting on the teleporter pads looking up at a tall, slender woman in a white mini skirt and knee high boots. Her white hair was fixed up in a large bouffant with several black and silver locks curling down both sides. She brought her fists up, resting them on her hips as she looked down at the two teenagers. The pair scrambled to their feet and stepped off the pad, facing the woman. A small yellow droid floated forward, stopping just in front of the pair.

"They are fortunate to be alive," the droid spoke in a young, soft spoken female voice. "Orb is producing an unusual amount of gamma radiation in this zone."

"Ella 9K, are they unharmed?" Dãsha asked.

"Yes, Commander. Their vital readings are normal. Heartbeats and respirations are elevated, but I suspect that is due to their current circumstance." The droid turned and flew across the hall to sickbay.

"Welcome aboard," Commander Dãsha Mantose smiled, turning back to the somewhat dazed siblings. "You two are pretty far from home, aren't you?"

"Yes, ma'am," they responded contrite.

"You're not going to tell our parents, are you?" Aydra asked anxiously.

"No," Dãsha answered flatly. "Why would I?"

Both let out a relieved sigh and grinned.

"Thanks, Aunt Dãsha," Alex said as the three exited the teleporter room and headed up the walkway to the bridge. "Mom and Dad would pitch a fit if they knew we were this far out, unsupervised."

"Hey, would it be possible to get our other ship?" Aydra asked as they entered the bridge. "It was damaged near the Kidamagor comet."

"I am aware," Dãsha said, sitting down in the command chair. "We have been tracking you since you went into its tail."

Alex looked over at the dazzling displays at the science station as Aydra stepped up next to helm control looking out at the exchange between the Interceptor and the Terrellian raider ships. The Gen four was an amazing Interceptor to be sure, but there was something about this one that instantly stood out. The fighting skills of the pilot were amazing. The Interceptor was practically turning circles around the remaining raider ships. It didn't take long before all three were disabled.

"You've been tracking us?" Aydra asked. "Please don't tell our parents where we were."

"Oh, don't worry," a soft voice from the helm controls responded quietly. "We already know."

Alex turned to the occupant of the helm chair. Recognizing who it was, he dropped his chin to his chest and closed his eyes. Aydra looked down at the Starbird pilot turning in her chair.

"Mom!" Aydra shuddered. "You're here! Wait, you're piloting a Starbird?"

"Yes, your mother has, some skills," Audra Atlanta said smiling.

"You never told us you could pilot a Starbird," Aydra responded.

"You never asked," Audra gleamed.

"What was I supposed to ask?" Aydra retorted, still shocked to see her mother sitting at the helm of one of the most powerful ships in the Flight Ministry's inventory.

"Commander Mantose," a voice over the bridge sound system announced. "Interceptor One is requesting permission to redock."

"Thank you, CORA 2K," Dãsha responded, relaxed. She looked at Alex, then at Aydra. "Please inform... the pilot, that both his children are safely aboard..." Dãsha smiled. "...eagerly awaiting his arrival."

"That'll be Dad," Alex said looking outside as the sleek V-winged fighter swept past.

Developing a grin, Audra quietly nodded her head.

"Dãsha, if I may take my two 'very much in trouble' teenagers into your conference room..."

"It has already been prepared," Dãsha responded. "Dante Search & Rescue has been apprised of the situation. Both T-6s will be brought back to the Dante facility."

Audra smiled and got up, heading for the door.

"Thanks again for taking time out of your duty schedule to help out with this. We really appreciate how fast you responded."

"Audra?" Dãsha said, turning her chair as the bridge door opened. She noticed a familiar form passing through the open Mallory doors at the far end of the hall; he didn't look very happy. "It was a pleasure to have you back at the helm, if only for a little while."

Audra looked back at the interior of the bridge.

"Thank you, it's great to see you again," Audra said, herding her children into the conference room before there was an exchange in the hallway.

"Are you guys all right?" Gunnar asked anxiously as the conference room door closed behind him. He threw his arms around Aydra, holding her as a wave of relief flooded through him.

"Dad..." Aydra gasped, barely able to breath.

"We're fine," Alex said as his mother wrapped her arms around him.

"Dad…" Aydra gasped for air. "Dad… need air… must breath." She tapped his shoulders with both hands.

"Sorry," Gunnar said, releasing her. "We were just so worried. Do you see this?" he said pointing to their mother. His voice had suddenly shifted.

"Here it comes," Alex muttered.

"Hold it out, Dear," Gunnar said pointing at Audra's head.

Audra fingered her hair, separating several streaks of white.

"This one is called Alex…" she said holding out the locks. "This one has Aydra written all over it."

"Your mother was a beautiful woman, with pretty dark brown hair, but you two are going to be the death of her."

"Hey, I'm still beautiful," Audra demanded, looking at her white strands.

"Of course you are, Sweetheart," Gunnar reassured her, turning back to his kids. "What were you thinking?"

"Dad, I can explain," Alex said, straightening up. "This was all…"

"My fault," Aydra cut in. "I dared Alex to do it."

"Daring your brother to do something stupid doesn't absolve him of doing it. What happened?" Audra asked, sitting down.

"I did a fly through at the end of the tail to see what the conditions were like," Alex started. "Then I dared Aydra to a race through the middle of the tail…"

"The middle?" Gunnar burst. Audra quieted him with a look.

"…and around the orbit to land," Alex continued, "then back out."

"You landed on it?" Gunnar erupted. "Lee Alexander, have you learned nothing in your Astro classes? It's a comet! Mostly frozen carbon dioxide, water, dust and a crap ton of other stuff that can turn volatile in an instant. They're great when they stay frozen, but…"

"Yes, Dad," Alex rebutted. "I know what they're made of and I know they become unstable when they're exposed to too much heat. Believe me, we both know."

"So what happened to the other Tempest?" Audra asked, trying to avert an argument.

"I tried to beat Alex to touchdown," Aydra said. "But experienced a hull breach and total power failure. Alex came in and pushed me out of the tail."

"Pushed you out?" Gunnar repeated intensely.

"What else was I supposed to do?" Alex asked

"Not gone in there in the first place," Gunnar retorted.

"How did you get her out?" Audra cut in, trying to hold down the energy in the room.

"Aydra transferred over to my ship when it was determined she was running out of air," Alex said.

Visibly upset, Gunnar clinched his fists and muttered something indistinguishable. Audra smiled, coming back to her feet. She stepped over to her husband and took his arm. He instantly relaxed, his fists melting into open hands as she splined her fingers into his.

"Very risky operation, considering those are civilian suits," she said, squeezing her husband's hand.

"Very risky," Gunnar echoed. Audra squeezed his hand a little harder. "One snag in the wrong place and it could have torn."

"I don't know what else we could have done under the circumstances," Alex admitted, contrite.

"You could have never gone in there in the first place," Gunnar smoldered. Audra jerked on his hand slightly. He gave her a glance, then looked back at Alex and Aydra. "But, your responses to the situations you put yourself in…" He felt another jerk. "…were spot on. It's exactly what I would have done."

Alex and Aydra smiled broadly, looking at each other with relieved expressions.

"You really landed on it?" Gunnar smiled, throwing an arm around his son's neck. "What was it like?"

"Careful, Dad," Alex grimaced with the near choke hold.

"Sorry."

"Yes, it was pretty intense. If those T-6s didn't have energy shields, I would have never made it."

"I would hope not and I would hope that you would have better sense than trying to take something that doesn't have shields into those conditions."

"Uncle Rick said he was in something a lot worse without energy shields," Aydra offered.

"What ship was that?" Gunnar asked, trying to think back. Audra leaned in close.

"Hadrian… Jayda rescue… Acuity… Drake mining facility… funnel maneuver?"

"Oh, yeah," Gunnar mumbled back.

"So," Audra spoke up. "I hope this little adventure was worth the price we're going to pile on top of whatever the bet was you two made with each other."

"Don't worry about that," Aydra said smiling. "I think we've learned our lesson."

Audra and Gunnar looked at their children for a moment, then smiled and embraced them again.

"All right," Audra said, pointing at one, then the other. "You two work out your bet and your father and I will work out what comes in addition to it."

"Now, back to the bridge and watch how incredibly cool your mother is at the helm of a Starbird," Gunnar said as they started for the door.

"Yeah, about that," Aydra said, giving them a suspicious eye. "Mom was a traffic controller for the Flight Ministry. How is it she knows how to pilot a Starbird?"

"It's classified," Gunnar grinned broadly, giving Audra a wink.

"Classified?" Aydra quibbled.

"Classified," Audra repeated.

"Dad works at a Ministry training facility," Alex pointed out. "You don't think Mom might have picked up a thing or two?"

"Yeah, but a Starbird is a little more than 'a thing or two'," Aydra agreed, standing firm with folded arms.

"Yes, it is," Audra said, ushering her two teenagers out into the hall. "And someday, if you get Ministry clearance, we'll consider telling you about some of the things your father and I used to do when we were young and dumb. Now, off ya go; I'll be right there." Audra held Gunnar back as Aydra and Alex headed straight into the bridge.

"How 'bout I go back to flying off halfcocked and you yelling at me for doing it?" Gunnar complained, enveloping Audra in his arms.

"I'm thinking wormhole operations are less stressful than raising teenagers," Audra agreed.

Gunnar shivered, looking into her big brown eyes.

"Maybe a fight to the death with Blinda Koss?" he suggested sarcastically.

"All that's still classified," Audra hushed him, grinning broadly.

"Too bad," Gunnar agreed. "Can you imagine trying to explain to our kids that their parents breached the problems of time and space?"

"... and that Mom and Dad were both killed doing it?" Audra snickered.

"Yeah, but the beauty is, the kids would never believe any of it. Just like they don't believe their mother can pilot a Starbird."

"I'm afraid both our children are just like us," Audra said, kissing her husband.

"I just hope when they head out on their own, they'll have their mother's good sense tucked away somewhere under that thick skin of theirs."

"They got that thick skin from their father."

"And their good sense from their mother."

"Oh, General, you always know just what to say." Audra grinned broadly as they kissed again.

"I love you," Gunnar breathed back. "Now, go on. Show your kids how piloting a Starbird is done."

"Meet you back over Dante?"

"It'll be an adventure."

"Always has been," Audra agreed, turning for the bridge as Gunnar headed back to the fighter access in Engineering.

*　　*　　*　　*

"You won the bet," Aydra said, following Alex around the bridge. "I was supposed to take the heat for it."

"Yeah, but I'm your big brother and should have had more sense than to let you goad me into something so dangerous. I should have just done it myself and you stayed outside to make sure everything went ok."

"Result would have been the same," Aydra said. "I would have come in after you anyway."

"I think that's the *wisdom* thing Mom was talking about. No wonder Dad named you after Aunt Jayda. You kind of like doing your own thing."

"Ever notice how many aunts and uncles we have that aren't actually related to us?" Aydra suggested as she and Alex stopped at the science station for a look.

*　　*　　*　　*

"Permission to take the helm?" Audra asked, stepping back into the bridge and stopping next to Dãsha.

"I was under the impression that disciplining teenagers required an enormous amount of energy and a long, boring tongue lashing," Dãsha said cocking her head slightly.

"Oh, don't worry," Alex said with a grin. "We just got the opening credits."

"Yeah," Aydra agreed quickly. "The long boring stuff comes when we get home."

"And forever more after that," Alex finished.

"You'd think you two would have figured all this out before you went off and did something stupid," Audra snickered.

Dãsha looked at the siblings, then at their mother.

"I suspect the adventures for these two are just getting started." Dãsha gestured to the empty control station up front. "The helm is yours."

Aydra and Alex watched with amazed admiration as their mother sat down at the helm and the rear console slid out into position behind her.

"I really like the new firmware updates you've done here," Audra smiled, tapping several points on the vast glass in front of her.

"Alex 7001 loaded and configured them himself," Dãsha responded.

Alex gave Dãsha a glance, then looked forward to his mother at helm with a suspicious eye. "Please tell me you didn't name me after a droid..."

Audra looked over her shoulder.

"Well... you're not, but if you knew the droid, you wouldn't be asking that." Audra said, taking hold of the manual controls and looking ahead as her husband's Interceptor swept past the windows in front of them. "Ready for light speed, Commander."

"I'm certain you and Gunnar know your way home," Dãsha smiled, cocking her head slightly.

"We certainly do," Audra grinned broadly, passing a wink to her children and pushing the throttles forward. Several recovery ships came into view as the Starbird bolted into light speed toward Commenor.

Is it ever really, the End?

The Prolog
A Casual Conversation

The four main characters of the Starbird trilogy, Jayda and Richard Niker, Audra Atlanta and Gunnar Conrad, sit down for a conversation with the author.

*   *   *   *

"Jayda Thron Niker, you're a Princess of Kalamar. What does that entail?"

"You're aware that I abdicated my royal title a long time ago?"

"Yes, in order to seal with Rick. We'll discuss that in a moment."

"Ok, duties of a Kalamarion Princess. It's been so long since I've had to explain it. It wasn't until I was about twelve years old that I had to attend any formal social engagements. They were exceptionally boring. We had to wear these huge, lavish dresses. I didn't feel any different than going to a friend's birthday party except for the clothes we were wearing."

"Did you have to go to a lot of big parties?"

"Are you kidding? The older I got the more parties I had to go to and the bigger they got."

"So when did you start having to do the kind of things you would associate with royalty?"

"Right after my sixteenth birthday. My parents had me doing liaison work for charities and catastrophic relief work."

"So is that all your royal duties entailed; service work?"

"Heavens no. Most royals do a lot of emissary and ambassador work. I was no different. Even as a teenager I was sent to broker relations with several different cultures, from the Oganons in the Mewar system, to the Castle Peak culture on Tilauris Fifteen."

"But you didn't stay with it, why not?"

"It felt like I was at odds with who I was and what I wanted to be."

"I don't get it. Isn't being a Princess every girl's dream?"

"I don't think there's a culture out there that doesn't have a skewed sense of a 'happily ever after.' Certainly there are parts of royal duties that live up to all the hype, but for the most part, you give up a little bit of yourself the further along you go. For me, I'm glad I was able to see it early on and make a course correction."

"You shifted to military service."

"Shifted to what a royal was allowed to do in the military."

"They wouldn't let you serve in combat?"

"Are you kidding? I had to pull some strings and tell a couple of little lies just to get into Traffic Control. By the time my parents caught wind of what I was doing, I had already been working there for several weeks. When they saw how much I loved it, they agreed to see how things went."

"Traffic Control was where you met Audra?"

"Yes, she trained me and remained my immediate supervisor through my whole tour of duty."

"How was she to work with?"

Jayda looked at Audra and grinned broadly.

"About the meanest supervisor ever!"

"Mean?" Audra responded with a laugh. "Maybe firm, but never mean."

"No, seriously," Jayda giggled. "She was very patient with everyone she worked with and always seemed to have everything together. I remember early on we were in a training exercise when all of a sudden, Long Range started tracking several squadrons of Valkyrie ships inbound. Everyone in the room was freaking out, but Audra knew exactly what to do. What really impressed me was that she didn't just shove everyone aside and try to do it all herself. She could have easily done so, but she calmly worked everyone in the room, making assignments and helping them accomplish those assignments. Yeah, there were a couple of instances where she took control just so she could make a couple of the green pilots feel at ease. She really had a way of talking on the communications channels."

"You think pretty highly of her."

Jayda looked over at Audra again and smiled.

"Yeah, we're best friends."

"Speaking of, let's talk about Rick. You gave up your royal status, to seal to him. Tell me about that."

Jayda snickered, looking over at her husband.

"You've met him. What's to tell? I was too young to seal when I first met him, but it didn't take me very long to fall head over heels for him. When I became of age, I didn't have to think twice about giving up being a royal. I will tell you my parents weren't very happy about it, but after a while, they came around."

"Any regrets?"

"About what? Abdicating royalty or sealing to Rick?"

"Yeah."

"No, but there are some things we've experienced during our lives together that give us pause. However, most of it is classified now."

"You can't talk about it?"

"Do you have Flight Ministry clearance?"

"Uhm... I'm the author...

"So..."

"Shouldn't I have all access?"

"Let me ask you a question."

"Shoot."

"You created my character. Who do I embody? Where did I come from? I think everyone here would love to know about their creation."

"Rick's character needed a companion. She was originally conceived as a Princess of Kalamar. I felt she needed to be set apart from any of the other characters, but still relevant. All of you were taken from real life. Jayda came from a girl my little brother had a huge crush on as a teenager. She was perfect for what I had in mind. Does that answer your question?"

"Yes."

"Ok, Rick, your turn. You're somewhat of an enigma."

"How do you mean?"

"You didn't waste any time getting after life."

"No, I don't suppose I did."

"Let's see... Brought up on a rainforest planet. The son of a vegetation farmer."

"Somebody has to grow it."

"An accomplished Star pilot at age eight. Went on your first adventure shortly after getting your Intergalactic Certification, mastered laser sword and particle ring combat, entered Kalamarion academy before the legal age and made flight leader shortly afterwards."

"You're telling me stuff I already know. Is there a question in there somewhere?"

"Sorry, I'm just trying to get to something a little more personal that people might not have gotten from what's on record. Let's see... Oh, here's a good one. I think this is mentioned only once in passing, so it probably isn't remembered. CORA 500..."

"Yes, an amazing piece of tech."

"But you didn't create her, did you?"

"No, but nobody needs to know that. I have plans to see that she is developed further along with the Ambuquate 7000 series. I have no doubt those two AIs will be an integral part of the Starbird program for some time to come."

"Yes, they are amazing, aren't they? But what the readers really want to know is where did she come from? If you didn't create her, who did?"

"I don't really know exactly where she came from. When I first learned to fly, my dad picked up an old, wrecked alien fighter from salvage he intended to turn into a Gilchrist racer. He didn't have the time or credit to make the needed modifications, so it sat behind one of our outbuildings for years. I pulled its mainframe and got her working. Her memory banks weren't intact enough for me to figure out where she came from. Only that she was of alien origin. Didn't matter to me. She was easily adaptable to the environments I was working in. CORA is an acronym for Computer Operations Range Assistant. I was going to name it something really flashy, but she wouldn't have it. Imagine, an AI with attitude. Her primary design was for extended long range recon operations. I've had her in all my ships, including every fighter cockpit I've ever flown. She's sorta been my guardian ever since.

"Your best friend is Gunnar."

"Yes, he is."

"Of course I am," Gunnar agreed quickly. "When has that ever been in doubt?"

"You two are nothing alike other than you're both amazing pilots."

"I wouldn't lump me into that category. Don't get me wrong, I'm good, but Gunnar is the amazing one." Rick gave Gunnar a good-natured nudge to the shoulder. "This guy can fly anything and make it look like child's play."

"It's true," Gunnar chuckled confidently.

"But you've been flying longer than Gunnar."

"What does that have to do with anything?" Rick sat up straight and pointed at his friend. "This guy has the gift."

"So what is it about Gunnar that makes him who he is to you?"

"You know better than anyone."

"The readers would like to hear it from you."

"Yeah," Gunnar joked, looking over at his friend. "Explain it to the readers."

"Gunnar is everything I'm not. I need to have a clear path to everything I do, whether it be behind the controls of a fighter or ship or working out a scientific problem. I guess you could say I'm kinda anal about it. He's always fond of telling me that trying to do something the same way and having it fail every time is a sign of a psychotic mind. This guy doesn't rely much on preplanning or a set way of attacking a problem. His propensity to rush into everything without thinking much about what might happen is both his strength and his weakness. I wished I could just throw caution to the wind like he does. But it's a formula that works for him and is what has made him such an extraordinary pilot and individual. So, my turn to ask you something. How do you see me and Gunnar?"

"Alter egos of each other. It would have been too perfect to have all your personality traits rolled into one character and quite frankly, quite unbelievable. I created you first because at the time of your conception as a character, I didn't dare hope or dream of putting myself into such an adventure. That would have seemed too transparent. It made more sense to create someone who I aspired to be like. As things progressed, I felt a little more at ease casting myself as a secondary character. By the time this storyline had matured into what it has become, I was fully confident this adventure could be carried to its conclusion without you getting in the way of each other. I think you complement each other rather well. You're best friends, but even more than that, a believable pair that work well together. Having said all that, you all need to know I didn't create all of you by myself. Every character helped in their own metamorphosis. They all had a hand in their own destinies and I'm grateful for that."

"I wouldn't have thought an author could or would do that," Audra commented.

"You'd be surprised how much discussion, even arguments there are with different characters about how they should be and what they should say and do."

"I don't recall having any arguments," Audra said. "Discussions, yes, but no arguments."

"That's because your persona fit the character so well."

"Who was my persona?"

"I'm a little nervous about answering that..."

"Why be nervous?"

"It's difficult to explain. You were conceived in the image of a young lady I dated in High School. I took a lot of the feelings I recall having for her and used them for you and Gunnar. Everything about you is what I remember of her."

"Does she know this?"

"You don't think I would have told her beforehand?"

"You still talk to her after all this time then..."

"Sure, why not? We're just really good friends now."

"So, who was Gunnar patterned after?" Jayda asked.

"Isn't it obvious?" Audra interjected.

"Myself."

"Then all this makes perfect sense," Jayda responded.

"Ok, so what was it about this young lady that invoked such a strong connection that you modeled me after her?" Audra asked, smiling.

"All the character traits I remember about her. Funny, pretty, strong willed but compassionate and smart. She is so smart; you have no idea. Well... actually, you probably do."

"Given the explained parameters, it's only logical."

"I guess that's one of the things that really attracted me to her at the time. As a High School teenager way back when, I knew she was far smarter than me and I really appreciated it. She never held it over me or flaunted it. She just was what she was and I was proud to be seen with her on my arm. At the time, I never wanted anything more than what we were to each other. She had more maturity than I did and could see things I couldn't. She was always very kind to me while we were together."

"Yes, I can see how you feel. It's plain to see why Gunnar and I have the relationship we have. By the way, I am... really smart. All this science stuff you have me doing, takes a really smart person to figure out and do."

"You can pilot a Starbird and you know your way around its engine room."

"I'd just as soon not if I can help it. I am still human ya know..."

"Yes, aren't you though..."

"Aren't you supposed to be asking me some questions?" Audra reminded with a grin.

"Yes, sorry, we sort of got drawn away. Your middle name? The story never does say what it is. Only that you never used it."

"My parents certainly used it; all the time. My siblings and I were always getting into trouble."

"Ah, the whole, bold full name announcement."

"Lane..."

"Audra Lane Atlanta. I like it."

"I'm partial to it. Anything else?"

"Yes. Would you explain why you kept your maiden name and didn't take Gunnar's, like Jayda did with Rick?"

"Certainly..." Audra raised her eyebrows and smiled broadly. "Such personal questions..." She gave her husband a glance. "It was Gunnar's idea initially. We talked about it quite a bit. Rick and Jayda had their own notoriety. Jayda was a Princess of Kalamar and Rick was a talented warrior and scientist. Gunnar was famous for his exploits as a fighter pilot. But he wanted to make sure I had every opportunity to excel in my professional career on my own merits and not by any associations with him."

"That makes sense. Thank you for clearing that up. Now we come to the last one."

"Saving the best for last, eh?" Gunnar smiled.

"Keep telling yourself that. Maybe when this is over, we'll both believe it."

"Oh, don't worry," Audra snickered. "He believed it long before this interview."

"There are a couple of things I'd like to ask that never got explained or covered."

"What more is there to know about me? What you see is what you get. I don't believe in being something I'm not."

"Fair enough, but as your alter ego, I wanted to express how I feel about you."

"That sounds kind of self-serving, doesn't it?"

"On the surface it probably does, but I think when you hear what I have to say and ask, it won't sound that way at all."

"Shoot, Turbo."

"Do you remember when I created you?"

"Yes, it was before the No Name mission from the Old Republic chronicles."

"No further back than that?"

"I recall I was about ten years older than you were," Gunnar responded. "You had an enormous amount of creative energy going on and could scarcely get it out."

"It was nearly impossible to channel. Initially, it was just supposed to be Rick and Jayda. She was a bit different back then."

"Wait, what?" Jayda cut in, surprised. "I don't remember being different."

"No, you wouldn't, would you?"

"How was I different?"

"Princess Katherine Gardner; Kate for short. A fiery redhead with a temper. I decided a calm, blonde Jayda Thron would work better."

"It was a wise choice," Jayda agreed, nodding her head.

"You and Rick were my only *in* to this world I created in my imagination."

"Transforming that world into a trilogy is an amazing feat. But I get the feeling you wanted to ask something else."

"Why did you hate droids so much? You gave Alex 7001 so much grief at first, but then by the end, you were good friends?"

"I was pretty hard on him at first, wasn't I?"

"Hard on him?" Audra burst. "You nearly shot him out into space a couple of times."

"Yeah, I wasn't very patient with him?"

"Uhm," Rick hedged. "No, you weren't."

"So, what was your deal?" Jayda asked.

"I'm sure Rick remembers," Gunnar started.

"I remember," Rick affirmed matter-of-factly.

"It stems more from the AI concept than any one droid," Gunnar began. "The ministry wanted to see how well autonomous AIs performed in combat and had a Gen II AI installed in my T-6. Rick tried to duplicate CORA's AI programing design, but there were some glaring problems. I flew with it for ten sorties. As far as maneuvers, they were as precise as they could be. The Gen II's ethical... slash, moral reasoning quickly became an issue."

"Classified?"

"Not really," Rick interrupted. "Just some tweaks that needed to be made."

"Tweaks?" Gunnar responded solemnly. "I had to take control multiple times when it tried to fire on transport traffic. There were a bunch of other issues that soured me on the whole idea of autonomous AI."

"I think it was one of the few times you were ever angry with me," Rick commented. "But I knew if you could just give the 7000 series a chance, you'd change your mind."

"Alex turned out to be top notch," Gunnar agreed.

"Ok, let's move on. Tiana and Taron Mantose were wearing Chyropaz orbs when you picked them up on Commenor."

"Yes," Gunnar agreed. "Annoying rocks."

"Why annoying?"

"They were like mood rocks or something. Always glowing different colors and stuff."

"In the right hands, a Chyropaz is a sort of seer stone," Rick informed everyone. "The images it can project aren't necessarily what's going to happen and quite often are misunderstood as actual events."

"After you got the twins to the *Constellation*, you were talking with them. Tiana was sort of coming on to you..."

"Wait," Audra interrupted. "She did what?"

"Come on," Gunnar defended. "I already told you all about her."

"Oh, yeah, sorry."

"At the end of your conversation, you were looking into her Chyropaz and you both saw a couple kissing at a sealing. It was you and Tiana."

"Wait, you were doing what? I don't remember you saying anything about that," Audra chided suspiciously.

Gunnar shifted uncomfortably.

"Must have slipped my mind. It was her sealing to Billy. A kiss for the bride. Nothing wrong with that."

"Was it on the lips?" Audra asked with a hint of derision.

"You were there! It was on the cheek as I hugged her."

"Let me ask something a little more serious."

"Please," Gunnar beckoned.

"Have you and Audra reconciled yourselves to what is ultimately going to happen to both of you? As she is human and you are not, you'll likely outlive her by a considerable amount. After everything you've been through, how will either of you continue on without the other?"

"Considering out past adventures, it's a tough question that's taken a long time to discuss." Gunnar and Audra looked at each other and

smiled. "We considered Dãsha and Caidin Mantose. Dãsha will certainly outlive her husband because of her physical makeup and his age. As you indicated, I will likely outlive Audra. But that's only if our human courses remain in a state of equilibrium. Life will take us to wherever life will take us and if one of us should depart this existence before the other one, we've determined to live the best we can without the other knowing that we'll be with each other in the next plane of existence. I imagine if we can survive what we've gone through in Hadrian…"

"Classified," Audra nudged.

"Don't care," Gunnar responded looking back. "Pretty sure Robert has clearance. If we can get through that and then make it through parenting, whatever comes after that will be a walk in the park. Does that answer your question?"

"Yes, I think so…"

"So, what would you like to tell us?"

"Feeling kind of silly now…"

"Why? We're all friends."

"That's true, but this is kind of emotional stuff."

"Give it your best shot. We promise not to make fun of you."

"I guess it's that this is the end of our adventures together. I'm gonna miss all of you more than I care to admit."

"Go ahead and admit to it," Rick said. "You'll feel better."

"I can't imagine my life without you. I've known all of you most of my life, and now it's time to say goodbye. I've laughed and cried with you; lived and died with you. I've seen the wonders of the universe with you and been there for all your failures and triumphs. I'll miss the love you guys express to each other, feeling that love myself. I'll miss Jayda being Rick's rock; his anchor. I'll miss Audra being Gunnar's life energy, his calming influence. I'll miss Rick being everyone's pillar. I'll miss him being the first one that conveyed me into this world I wanted to be a part of. I'll certainly miss Gunnar. How he can do hard things just because they are the right things to do. I'll miss his reckless abandonment to any given situation he is faced with. I'll miss his sense of humor in the face of overwhelming odds. I shall miss the adventure. I don't know if I can express any better in words how I feel about you guys. Sometimes, I wish I could fly off with you, wherever you go and whatever you do. I wish I could always be with you. I guess that's what I wanted to say. Thank you, Jayda, Rick, Audra and Gunnar. You've certainly helped make my life… an adventure."

"If that's the case, then you must be doing it right."

Acknowledgements

As this project comes to a close, I would like to thank some of the people who helped make it possible. There are so many that have contributed through the years. From its first conception back in the mid-1970s to the time I pulled the tattered, handwritten, spiral bound pages out of its dusty book box and started working on it. First, to my children. I would tell them the story and they would enjoy listening, even reading some of it. I've already mentioned my youngest son's love of science fiction and all the feedback he would give me on many of the subjects throughout this entire series. Even my middle son contributed to some of the ideas I used to create this adventure. My two daughters have been instrumental by pushing me on beyond the finish of the trilogy by showing me that audio books would be a great adventure.

To all the many friends that encouraged me through this adventure. To my Beta readers. Their invaluable input to the original manuscripts of all three books was so important. Without it, this story couldn't have been as exciting. To my cover artist. She is an amazing artist, somehow able to transfer the visions in my head onto a cover in a way that compels readers to open the book just by looking at it. To all these people, I convey my eternal thanks.

Finally, the most important of all, my editor and wife. Her job was essential to the success of this trilogy. As an editor, she could look at this story from a different perspective, seeing things I couldn't. As she reads many different genres, she has developed great insights on what a novel should look like from a technical aspect. She can also see how a plotline should move and flow, understanding how to get it from point A to point B in a logical fashion. As my wife, she understands how to communicate to me in a way that I can understand and help me make the adjustments needed to polish the story into the adventure it needs to be. That's not to say it has been easy. Are you kidding? There has been a lot of give and take; growing to understand each other in ways we never dreamed of. Thank you Lola for your insights, your expertise and patience. It's been a great adventure! I love you.

Robert James Schultz

Robert is a graduate of Ricks College, now called BYU-Idaho. He currently works for BYU-Idaho AV Productions as the Head of Engineering. Robert holds a private pilot's certificate and works on general and experimental aircraft avionics at the Rexburg Airport in his free time. Water sports, RV camping and riding motorcycles are some of his favorite hobbies. He is a 2013 and 2015 Ironman Coeur d'Alene Triathlon finisher. Robert has loved writing since his early teens, most dealing with the science fiction genre. Married to Lola Hazel VanLeishout in 1983, they are the parents of five children and the Grandparents of six. Robert and Lola reside in Sugar City, Idaho.

If you liked this story, please take a moment and write a short review and post it on Amazon and Barnes and Noble. Just input the title into the website search bar.